CPM
THE NEW CRITICAL PATH METHOD

The State-of-the-Art in Project Modeling and Time Reserve Management

DENNIS H. BUSCH

PROBUS PUBLISHING COMPANY
Chicago, Illinois

Library of Congress Cataloging-in-Publication Data Available

ISBN 1-55738-117-8

Printed in the United States of America
BC
1 2 3 4 5 6 7 8 9 0

Table of Contents

CHAPTER 1.	Project Modeling—Introduction	1
CHAPTER 2.	The Project Model	23
CHAPTER 3.	Date Math and Calendars	41
CHAPTER 4.	The Forward Pass	51
CHAPTER 5.	The Backward Pass	79
CHAPTER 6.	Project Time Reserve	105
CHAPTER 7.	Calendar Reserve and Free Reserve	131
CHAPTER 8.	Modeling Variants: The Good, The Bad, and The Ridiculous	147
CHAPTER 9.	Change and Status in the Modeling Process	167
CHAPTER 10.	The Role of the Computer	195
CHAPTER 11.	Additional Roles of the Computer	215
CHAPTER 12	Modeling Techniques in a Practical Project Environment	243
CHAPTER 13.	Summarization and Vertical Integration	279
CHAPTER 14.	Horizontal Integration	311
CHAPTER 15.	Output Product	331
CHAPTER 16.	Resource Modeling	363
	Epilogue	393
APPENDIX A.	Arrow (AOA) versus Precedence (AON) Modeling Techniques	399

APPENDIX B.	**PERT versus CPM**	**403**
APPENDIX C.	**Basic Requirements of a CPM Computer Processor**	**407**
APPENDIX D.	**Qualitative Evaluation of Project Models**	**419**
APPENDIX E.	**Hammock Structures as Independent Project Models**	**425**
APPENDIX F.	**Glossary of Acronyms and Abbreviations**	**429**
APPENDIX G.	**Modelfile Composition**	**441**
APPENDIX H.	**Conventions**	**447**
	Index	**453**

CHAPTER 1

Project Modeling—Introduction

PROJECT PLANNING AND MANAGEMENT

The term *planning* covers a broad spectrum of activities. People plan their days, their meals, their budgets, their careers, their vacations; some even make plans for interment. Businesspeople plan for capital investments, sales and profits, return on investment, work schedules, taxes, facilities and tooling. Virtually nothing goes on that does not involve some sort of planning, but the purposes and the processes of planning differ widely. A capital plan and a work plan look vastly different, are produced by radically different methods, and serve quite different purposes.

For all these differences, many still believe planning to be a simple activity. *If everyone does it without any training, then it cannot be very hard to do.* This may be true of much day-to-day planning, but it should not be accepted as a generalization. Many forms of business planning require scientific methods performed by highly skilled practitioners.

In particular, project planning employs such a science. The process focuses the definition of the word *planning* considerably, limiting both the object of the planning and, to a great extent, its nature and methods.

To begin our study of project planning, we must define the term *project*:

> A project is any deliberate effort that achieves finite objectives through a series of tasks that are interdependent due to the interrelationships of the work and/or shared resources.

In the past, only massive undertakings of large corporations were generally considered to be projects in the full sense, but projects are common enterprises of virtually every business. Many people work on projects without thinking of their work in those terms. By the above definition, virtually everyone works on projects, so all should understand the science of project planning.

Each business must decide how much and what kind of effort to expend to plan and manage its projects. To answer this question, we hold to an inviolable principle of project management that *all of the functions undertaken in connection with a project must be essential to meeting its contractual objectives and/or provide a payback to the project.* That is, the results of the labor and any other expenses (for computers, software, etc.) must add equal or greater value to the project. Since the planning process does not contribute directly to the output of a project, it must pay the project back through cost savings, and these paybacks must exceed the expenses of the planning process. Returning value added to the project is the prime objective of the planning effort.

Effective project planning is a series of disciplined steps necessary to define a project adequately for proper management.

1. The planner determines the objectives of the project. To define the work of a project, one must first understand and define all of its expected results or products, and determine when each must be produced.
2. The planner determines the work required to achieve the project objectives. This work must be defined from the start condition and run through the completion of all objectives. It must then be broken down or decomposed into finite tasks, each with specific start and finish points. In total, this *is* the project.
3. The planner determines the resources (labor, material, equipment, facilities, etc.) necessary to accomplish the work, and when each must be available.
4. The planner coordinates the work to be performed as efficiently as possible in terms of resource utilization, sequencing the tasks such that prerequisites become available before they are needed, while still meeting all project objectives. In this step, it is necessary to define how each finite task will be performed relative to every other. This combines steps 1, 2, and 3 into a viable scheme of the sequence and flow of work. The results of this step can be called the *Work plan* or the *Project plan.*
5. The planner overlays project structures [hierarchies such as the responsible organization, Work Breakdown Structure (WBS), etc.] to assign

or apportion the work to appropriate groups of workers and other related management structures.

6. The planner routinely publishes information about the plan, including any changes to the plan and its status or measured performance, to all project personnel.
7. The planner tracks delays in the work along with any changes in the program scope. Each instance must be analyzed to assess its impact and then new plans must be developed to avoid or mitigate any damage to the balance of the project.

These steps seem simple and often taken for granted. Project managers frequently give the project planning process too little attention. They perform these fundamental steps only superficially, leaving the finite tasks and their interrelationships only vaguely defined. The resulting nebulous plan starts the project without a clear course, limiting the ability to measure and control the work.

Such a sloppy planning process provides no payback, no value added (NVA), even for its limited cost. In the next budget allocation, the planning effort is slighted because of its poor return, so the situation is self-perpetuating. This will happen until management recognizes that their ability to manage a project is directly related to the quality of the planning that goes into the project.

Planning is a key discipline that can provide the best opportunities for cost savings on a project. Only by thorough and proper planning, however, will it realize these savings.

REACTIVE VERSUS PROACTIVE MANAGEMENT

Project management is the control of the critical technical, schedule, and cost parameters of a project in order to affect its outcome positively. It is important to understand that one cannot control each of these three parameters in isolation; they all depend on one another. It is almost possible to say that they are the same thing measured in different units. At any rate, they are not independent variables and cannot be managed as separate entities.

Some claim that uncertainty in projects makes attempts at rigorous planning methods futile. This is absurd. Risks and uncertainties increase the need to plan. Otherwise, we concede the outcome of the project to fate, and allow circumstances to control it instead of its managers.

When management allows a project just to happen and reacts to circumstances, this is called *reactive management.* This is the easiest scheme of management; it is also the most costly. The fundamental problem with reactive management is that delaying the response to a problem until it becomes imminent eliminates many possible solutions. These lost options are generally the cheapest options.

We can do better than reactive management. *Proactive management* seeks to manage an endeavor with an eye on its future, using the past only to define present conditions. Proactive management parameters are based on the remaining project. All managers would undoubtedly choose the proactive option and most believe that they mange their projects in this way. It is not, however, a simple matter of choosing a method. A supposed proactive manager who fails to develop an adequate project plan, or to assess impacts of current status and changes to the plan relative to the uncompleted portion of the project and makes business decisions accordingly, quickly begins practicing reactive management.

Managers must back up their decisions to manage proactively with proper actions, which are rooted in good planning. They must initiate the project with adequate planning and continue this process throughout the entire course of the project to achieve proactive management. They must manage the residual effort of the project, not the residue.

It would be nice to believe that a project could be perfectly planned, allowing completely proactive management. Reality is not that perfect, though. Uncertainty and risk always complicate efforts. This prevents totally proactive management, but it should not prevent us from trying. We can manage as much of a project as possible proactively, leaving much more time and resources to react to unforeseeable complications. Project management must wage a constant battle to retain control of a project. This leads to a single, perhaps startling, principle:

> The project modeling process is the only proactive tool ever developed for project management and control. Therefore, if a project is not using this methodology, it is managed reactively.

This is a bold statement, but our exploration of the planning process will prove its validity.

THE PROJECT MODEL

Project management and planning have perplexed intelligent people for centuries. Engineers attempted to solve this problem in the same fashion that they solved other engineering problems. They developed a method that embodies all of the key steps of project planning (steps 1-7 above) in a mathematical modeling process which we shall call *project modeling*. Through certain processes we will treat these mathematical properties as dependent variables and in so doing produce impacts in one caused by another.

Project models are structures composed of the essential planning parameters of a project, many of them represented mathematically. This model and its related mathematical processes, help us to accomplish the seven steps of project planning. In fact, it is difficult to separate this project modeling process from project planning as described above. The objectives and functions are virtually the same.

Fundamentally, building a project model requires decomposing the project into finite tasks (step 2) necessary to accomplish its objectives (step 1). The relationships between the tasks and the objectives must be defined for the model (step 4), and the modeler can add the resource requirements for each task (step 3), though this may be optional. Steps 1 through 4 of our planning process match the steps for building a project model.

Later we will see that, through the balance of the modeling process, we can also accomplish the last three planning steps. We will find that project modeling provides us a means of accomplishing the seven planning steps and, more importantly, it does so through a very disciplined and structured process. It also provides proactive management information.

Building a good model of a project is by far the most difficult undertaking in the entire process. Think about it: The planner must define all of the tasks necessary to accomplish objectives 2 years, 5 years, even 10 years into the future. The planner must then define the tasks' durations, their relationships to one another, and necessary resources. This is a monumental task and can only be successfully accomplished to certain degrees. Predicting the future to perfection is an impossibility.

Uncertainty makes these tough chores even more difficult. Many a downstream task depends on the outcome of tasks performed in the near term. In many ways, the task of building a project model is like trying to capture a cloud. This does not diminish our efforts, it only tells us that our process cannot be static; it must change as the project evolves.

Project planning is an iterative, unending process, but at the same time, our plans cannot be open ended. Initial (baseline) decisions establish a total plan, from all starts to all finishes, however high the probability that the plan will change, possibly many times. This gives the project firm direction, which is essential to its effective management. A ship without a rudder and a project without a plan are perfect analogies. A firm plan is vital to the change process as well, providing a stable condition relative to which to control and measure change. Deviations, additions, and deletions of work to accommodate changes can be determined and coordinated with the original model. A revised project plan that encompasses both the old residual and the new requirements results.

The interrelationships of the work also require initial choices among several possible ways to complete a series of tasks. The relationships between two tasks may not be constant; they can be rigid, soft, variable, or all three at the same time. An initial configuration of interfaces must be developed that reflects the compromises required to generate the initial plan. These interface compromises compare schedule objectives against each task's prerequisites, all in the most efficient manner in terms of cost and risk. Change affects the interfaces just as it does the tasks. As the project evolves, so must the interfaces to keep the plan or model consistent with the project's current direction.

The degrees of detail of a plan frequently vary. Tasks and interfaces in the near term (over the next year, for example) are defined much more specifically than those in the far term (perhaps three years away). The model reflects this mixed level of detail in a *rolling wave task decomposition.* In general, this means that spans of many near-term tasks range from 20 to 60 days. The relatively fewer far-term tasks usually span longer periods. Interfaces follow the same pattern. As time passes, and the far term becomes the near term, former long-term tasks become better defined, in their turn being further decomposed along with associated interfaces.

If all this could be done accurately with little trouble, then planning and controlling a project would be simple, but uncertainty and complexity complicate this chore. Add to this the dynamics of project changes and lapses in schedule performance, and the process becomes monumental, laborious, and exacting. This is the reality with which the modeler must cope.

In another light, project modeling is a method by which to simulate a project through mathematical properties and relationships. We reduce the variables that affect a project to as few as possible, then simulate the essential remaining parameters as dependent mathematical variables. Data from this standard, systematic process can be analyzed to aid management decisions. *The payback to the project from the modeling effort is the savings resulting from these decisions.*

Understanding these decisions and the analysis undertaken to arrive at them is a prime objective of our study. Only through comprehension of all the modeling processes can we achieve insight into an understanding of these decisions. Then we can begin to exploit the project management process to the fullest.

PROJECT TIME RESERVE

Before we can begin to understand the intricacies of the project modeling process, we must first understand its basic purpose. The results of the mathematical modeling process is the derivation of a value called *float* (sometimes called *slack* or *margin*). The importance of this value requires detailed examination.

A very simple five-activity project will help illustrate the principles (see Figure 1-1). Boxes represent the tasks of the project, Task1, Task2, Task3, Task4, and Task5. Each box gives a time estimate for the accomplishment of the task: Task1 will take 20 days, Task2 will take 40 days, etc., measured in workdays (excluding weekends and holidays). These estimates are the *durations* or *spans* of the tasks.

The tasks are performed in sequence, with one activity finishing before the next can start. This relationship is represented in the diagram as lines connecting the finish end of one box to the start end of the next.

Figure 1-1

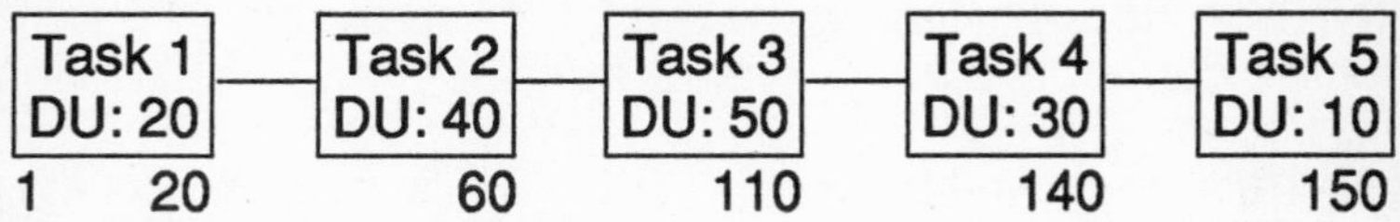

This series of tasks constitutes a *path of activities,* defined simply as a sequence of tasks that flows in some fashion from a start (e.g., Task1) to a finish (e.g., Task5). In our entirely sequential example, we can add up all the spans along this path to determine its total span. This summed value is called the *cumulative sequence of the path* (CSP). In our example the CSP is 150 days (20 + 40 + 50 + 30 + 10).

This is useful information, but it does not help us make any decisions until we relate it to some other value. We are in the position of a gold miner who can produce gold for $350 an ounce. The miner must relate this cost to the going price of gold to estimate the mine's profitability. If gold is selling for $400 an ounce, then the miner is in business; if the price of gold is $300 an ounce, the miner has nothing but a tax shelter or an expensive hobby. In both cases the work costs the same amount, but its cost compared to its market price determines profit or deficit.

In planning our project, we compare the time required to accomplish the necessary work to the time available for its accomplishment. Our example project takes 150 days. If we have 170 days available, we enjoy an excess of time compared to the work, but if the project must be accomplished in only 130 days, we face a time deficit. In the first case the project is starting off in a fairly healthy condition; in the second case the same project is in serious trouble. It is by comparing a path's cumulative span to the time available for its accomplishment that we can assess whether we have a profit or a loss.

We will call the difference between time requirements and availability *time reserve*. The equation that derives time reserve is key to the understanding of the project modeling process. It is:

$$PTR = TA - CSP$$

where

PTR	=	Path Time Reserve
TA	=	Work Time Available to the Path
CSP	=	Cumulative Sequence of the Path

Path Time Reserve equals the work time available to a path less the cumulative sequence of the path.

This value (PTR) is the resultant data of the project modeling process.

Let us use this equation to determine the time reserves of our examples. In the first case, with 150 days of work to be accomplished in 170 days, the equation yields:

$$PTR = TA - CSP = 170 - 150 = +20$$

We have a positive reserve of 20 days. That is, we have 20 days more time available than the work requires.

In the second case, we must accomplish the same 150 days of work in only 130 days, giving the equation:

$$PTR = TA - CSP = 130 - 150 = -20$$

We have a negative reserve of 20 days. That is, the work requires 20 days more than we have available, which indicates a problem. In our first example, a positive time reserve value indicates a healthy project; in the second, the negative value tells us to take steps to alter the project to finish it in the allotted time.

A negative time reserve means, not that we have irrevocably missed a schedule objective, but that we will miss one if the project continues according to the current plan. Once we define and understand the situation, we should be able to find a course of action to resolve it, provided we don't ignore the warning signs and wait. Problems, as a rule, do not resolve themselves; this is the business of management. One of the key principles of proactive management is to resolve problems as soon as possible:

The more time a project has to react to a problem, the greater the choices of available solutions.

Early action allows us to weigh possible solutions and select the cheapest or the safest or a combination of both. But the longer we wait to resolve a problem, the fewer our choices, and many times we lose the best, cheapest solutions. This is the real price of reactive management.

Historically, time reserve has been called *float*. Apparitions, loans, and battleships float—and the correlation of this term to a project seems nebulous. This term may have come from boating, suggesting that a project with positive float is high in the water, while one with negative float is sinking. Whatever its source, the borrowed term *float* seems inadequate to describe the difference between two time values. Time reserve is a much more accurate and descriptive term for the concept.

This term also gives a good parallel to *cost reserve*, which is a percentage of the contract value that project management usually holds back to buy resources to solve unexpected problems. In the same way, time reserve is uncommitted time in the project span that can accommodate solutions to project problems.

The project modeling process then, yields a new project parameter (time reserve) that can be used to proactively measure true schedule health. Once we find it, however, we must manage and control it carefully, if it is to be of any use.

THE SEVENTH PLANNING STEP

We have seen that the effort of carefully building and then maintaining the project model accomplishes four of the seven basic planning steps. Publishing the data resulting from the model would achieve the sixth, and arranging (ordering) the data according to the project structures would accomplish the fifth. Virtually all of the objectives of planning can be met by the construction, maintenance, and publication of the project model.

For all this effort, however, this data is only the input, the initiation of the modeling process. As yet, we have not directly discussed how project modeling accommodates the seventh step of planning:

7. The planner tracks delays in the work along with any changes in the program scope. Each instance must be analyzed to assess its impact and then new plans must be developed to avoid or mitigate any damage to the balance of the project.

Ideally, we could build a project model at the onset of our endeavor, translate this to a schedule, and then carry out the project. Planning would be complete before work began. This requires two conditions, however:

1. There are no changes to the plan.
2. There are no deviations from planned performance.

Those with any experience will chuckle at such wishful thinking. The hard reality is that projects, by their very nature:

1. Undergo continual change to suit both internal and external requirements. Project length and more complex technical requirements exacerbate this condition.
2. Require constant adjustment due to the impact of failures to meet schedules of detail tasks.

Managers struggle constantly to mitigate the adverse effects of these two forces, and good planning provides their best chance for success. Here the modeling process pays its biggest dividends. As the project's requirements change and schedules slip, the model can be altered to accurately reflect the progress (or lack thereof) and evolution of the project, not so much to keep score of the progress, but rather to evaluate the impacts of changes on the remainder of the effort (the seventh planning step) and to maintain a viable plan to meet project objectives. The planning process must continue throughout the life of a project.

To understand the analysis for the seventh planning step in more detail, we first break it down into three distinct parts:

1. Alter the model to reflect current status and change.
2. Determine the impact of each change or slip on specific downstream activities and assess the impacts on time reserves.
3. Test workaround solutions to ensure that they solve problems as intended without creating new ones.

We maintain the model by altering the characteristics of existing activities and interfaces, adding entirely new activities and interfaces, and/or deleting activities and interfaces. The objective is to constantly, accurately simulate the project's current condition. As the project changes, the model must keep pace.

After we have altered the model to simulate the current condition, then we analyze the impact of changes through several means. First we compare the schedule resulting from the altered model to the previous schedule. Any differences reflect potential schedule impact.

We can also compare the current and previous time reserves to measure the true impact of a change. If the time reserves remain positive, we can still meet all schedule requirements with the current planned task sequence and durations. We still have more time available than the cumulative sequence of each path. But, if one or more paths have deteriorated into a negative time reserve, then one or more scheduled objectives are in jeopardy.

For example, if a slip or change reduces time reserve of a path from 30 days to 10 days, the impact is severe. If the time reserve of another path declines from 10 days to a negative 10 days, this is even more severe. The same 20-day erosion took the path from a condition of more time than work to one of more work than time, jeopardizing at least one schedule objective on this path. By such logic, we can rate the effects of several slips and/or changes on the project's reserves from the most severe to the mildest.

We can model workaround solutions just like any change. These could reflect added tasks, deleted tasks, changes to the scope of existing tasks, changes to the interrelationships of tasks, and/or other effects. The difference is that an external force does not induce the change, but rather an internal effort initiated to resolve a problem. We must test this solution to ensure that it achieves the desired results without any adverse impact in any other area. To do this, we alter the parameters of our project model by the same process by which we assessed the impact of external changes. We then compare the modeled schedule and time reserve results to our original plan.

As part of this process, we need to study the impact on time reserve of failure to perform tasks as expected. As an example, let us examine the relationship between the delay of a task on a path and time reserve on the path using our original problem. This is graphically illustrated in Figure 1-2.

After completion of Task3, 140 days after the project's start, we determine that Task1 actually took 30 days rather than 20 days as planned, Task2 actually took 50 days rather than the planned 40, and Task3 actually took 60 rather than the planned 50. The original plan called for completion of the first three tasks after 110 days, but it actually took 140 days.

We still have 40 days of work to do by the original plan, which would mean that we can now expect to finish the project on day 180 (140 + 40 = 180), a 30-day delay. Since the project began with 170 days of time available, it is now in obvious trouble. We began with a positive 20 days of time reserve, but we now face a negative time reserve of 10 days (170 – 180 = –10).

The path increased from 150 days long to 180 days long. The 10-day slips on each of the first three tasks combined to push the revised cumulative sequence of the path beyond the time available, which did not change. The original time reserve (+20 days) was consumed by the time Task2 was completed.

Figure 1-2

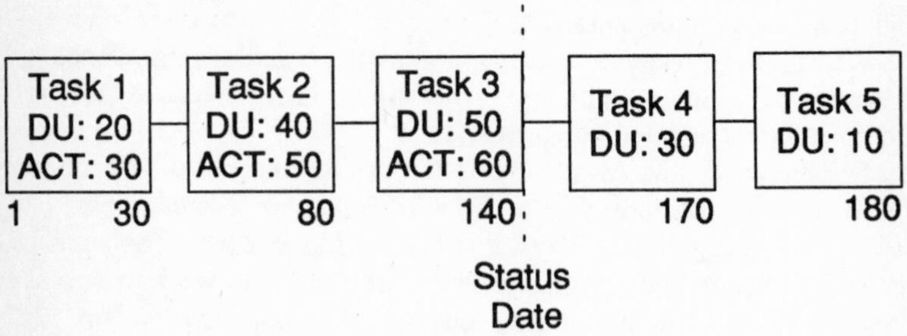

This analysis combined history (30 + 50 + 60 days) with our projected plan (30 + 10 days) to arrive at the cumulative length of the path. Let us reexamine this problem from a purely proactive standpoint. We have 40 days of work remaining (30 + 10 = 40), and we have used 140 days of our original 170 which leaves us only 30 days (170 – 140 = 30) in which to complete the project. This also leaves us a shortfall of 10 days.

Our analysis arrived at the same answer either way we examined it because in the total path length analysis, we added a constant (time elapsed = 140 days) to both values [(140 + 40 = 180 days CSP) and (140 + 30 = 170 days TA)]. This leads to the important conclusion that:

Time reserve is always measured relative to the remaining work effort and the remaining time available.

It is always a forward-looking, and therefore proactive, parameter. The time reserve value helps us to manage and control the remainder of the project, because it is measured relative to the remainder of the project. This is the most important feature of the time reserve parameter, and the key to its thorough exploitation. This is where the payback resides.

Analysis of our example showed that as of the current position (day 140), the project is in serious trouble. We have only 30 days left to complete 40 days of work and only drastic measures will save our end date. Monitoring progress all along would have made this trouble visible much earlier. Although Task1 slipped 10 days, this amount of time was in itself not a true indicator of project health, as it provided no insight into the remainder of the project. To assess overall project health requires relating our slip to a project or path parameter like time reserve.

Because our example featured a single, linear path, the slip on Task1 eroded the time reserve of the project (path) by the same 10 days. Since we had only 20 days to begin with, this slip represents a 50 percent reduction. To measure the severity of our problem, we must compare this loss to the amount of the project completed. Comparing the time reserve erosion (50 percent) to the amount of completion (only 20 percent) reveals a severe problem because the rate of erosion significantly exceeds the rate of progress. If the reverse were true, then we would not have a problem. We not only need to examine the erosion of the reserves, but we also need to compare the percentage of erosion to the percentage of path completion. This is the *effective rate of time reserve erosion.*

In the complete analysis of time reserve, we must:

1. Compare the times reserves of the model to the previous condition.

   Example 1: If a project experienced the same 10-day delay, but with 100 days of time reserve, we would not judge this a serious impact. The same 10-day time reserve erosion in this case represents only a 10 percent reduction.

   Example 2: A 30-day erosion of time reserve on a path that originally had 100 days of time reserve has less project impact (30/100 = 30 percent) than a 10-day erosion on a path which originally had only 20 days of time reserve (10/20 = 50 percent).

2. Compare the amount of time reserve erosion in the model to the rate of project completion.

   Example 3: A project on which 50 percent of time reserve is eroded in the first 20 percent of the path faces a more severe problem than the same 50 percent erosion occurring after 60 percent of the path is complete.

These examples show that we must analyze impacts to time reserves in several ways. We will find that comparing the amount of time reserve erosion and the rate of erosion relative to the percentage of project completion reveals the most useful information about project impact. This can be summarized in the fundamental principle of time reserve analysis:

> The amount and rate of erosion of time reserves are true indicators of project impacts resulting from project perfor-

mance or change. These parameters measure impact relative to the remaining portion of the project, and are therefore proactive indicators.

Time reserve is more than a number—it is management data. It is useful only if it informs managers' decisions. Managers need to understand how much time reserve a project has, where in the project this reserve exists, and how slips and changes affect these conditions. They must also understand the decisions they can make based on this information and how these decisions address their problems.

In summary, to evaluate the effects of changes, we analyze:

1. The potential impact to the schedule.
2. The amount of time reserve erosion relative to original reserves.
3. The rate of time reserve erosion versus progression.
4. The adequacy of remaining time reserves to meet project objectives.
5. The possible effects of any workaround solution.

TIME RESERVE DECISIONS

A manager has a problem any time a path has a very low, zero, or negative time reserve value, or when the time reserve is eroding at an unacceptable rate. To solve this problem, the manager must increase the numeric value of the right-hand side of the time reserve equation (PTR = TA – CSP) by either:

Increasing the time available to the path, or
Decreasing the cumulative sequence of the path

We can increase the time available to a path in three ways:

1. Finish the path later.
2. Start the path earlier.
3. Accomplish some of the path work during nonwork periods (overtime—holidays, weekends, etc.).

Since finishing the path late means missing a schedule objective, we try to avoid the first option. Unfortunately, if we find no other solution to the problem, or if we ignore our problems, we will probably be left with this option.

We can exercise the second option only before starting the path and if we can overcome all the obstacles to an early start. Careful, thorough planning at the onset of a project allows the exercise of this option, and it is in fact used quite often, as when, for example, a firm prereleases funds to initiate procurement of resources with long lead times.

Overtime

The most common option for increasing time availability, though, is the third one. Working overtime (off-shift, Saturdays, Sundays, or holidays) generates more work time between a start date and a finish date, increasing the time available to accomplish a sequence of work.

Let us examine how much time overtime is available. An average month includes 21 work days that can be applied to the accomplishment of a project. The rest of the month can be considered overtime. This includes four weekends, and usually a holiday. We could also add up to 30 second shifts and 30 more third shifts. Summing available hours, an average month includes only 168 hours of straight time and 552 hours of overtime, a 3-to-1 ratio. Practically, though, only a portion of this overtime is actually usable, but this portion is still a large and valuable resource.

We could consider overtime as a project time reserve in itself. It is finite and calculable, and project management can use it to solve schedule problems. This is discussed in Chapter 7. Excessive reliance on this technique is one of the tell-tale signs of reactive management, though, especially if it is used excessively as the project's completion approaches. Further, overtime almost always adds cost to the project. Reactive managers often salvage schedule objectives in the eleventh hour through excessive use of overtime, creating a cost overrun. The unenlightened may see this as a cost problem rather than a planning problem.

Modeling techniques help us to determine how much overtime is necessary and where to selectively apply it. They also suggest its use early on. This allows judicious, optimal use of overtime. Modeling techniques also reveal alternative solutions to schedule problems besides overtime.

In the search for alternatives to overtime, we must remember that the time available parameter is but one half of the time reserve equation; the equation also includes the cumulative sequence of the path. To increase time reserve through this parameter, we need to decrease its value. We can do this in three ways, as well:

1. Decrease the span of a task or tasks by reducing work scope.
2. Decrease the span of a task or tasks by effectively adding resources to work on these tasks.

3. Rearrange the sequence of the work to perform it in parallel rather than series.

Though managers can and sometimes do solve schedule problems by the first option, they only avoid the issue. Only through thick rose-colored glasses can we see accomplishing a job on time by eliminating part of the job as anything but a partial failure. Much as we would like to avoid this option, however, we may be forced to use it in the absence of any other solution.

In the second option, to decrease the spans of tasks along the path by effectively applying additional resources to them, the key word is *effectively.* This option again frequently characterizes reactive management who also salvage schedules through massive injections of labor, material, etc. It is easy to throw bodies at a problem, but this guarantees only that the accomplishment of the tasks will cost more. It may even increase the task's span.

As a general rule, however, applying additional resources to a task or series of tasks should allow us to complete them in less time. True management skill is needed to apply enough additional resources to complete a task or tasks efficiently in a shorter span of time.

The application of additional resources will always cost the project additional money, so this option really gains schedule reserve at the expense of cost reserve, ultimately reducing profit. We may have to sacrifice some profit to achieve schedule objectives. When we must use it, modeling processes can show specifically where to apply the additional resources to meet needs and solve time reserve problems. At least in this way we are as efficient as possible and minimize adverse impacts to profit.

The third option, altering the sequence of work, is frequently the best way to increase time reserve, but it is the most complicated. We can probably alter a series of tasks that we had originally planned to accomplish in a sequence to accomplish at least some of them in parallel. This requires a very thorough analysis of the tasks, and it yields a quantitative change to task relationships.

For a simplistic example, assume we planned to accomplish three tasks, each 30 days long, in series. This path would take 90 days to complete. Obviously, if we performed these tasks in parallel or at least partially in parallel, we would complete the total span in less than 90 days. The amount of compression depends on the quantitative change of the relationships (see Figure 1-3).

Suppose that information obtained from the workers tells us that the second task could begin with minimal risk 20 days into the first task, overlapping the last 10 days of Task A and the first 10 days of Task B. Suppose that we also determine that Task C could begin with minimal risk 20 days after Task B begins, overlapping the last 10 days of Task B and the first 10 days of Task C. The total sequence from the start of A to the finish of C would be only 70 days long (the first 20 days of A, the first 20 days of B, and all 30 days of C). We added 20 days of time reserve to this path of tasks (10 days from the overlap of A and B plus 10 days from the overlap of B and C). By rearranging the sequence of work, we can accomplish 90 days of work in 70 days without the use of smoke or mirrors.

One significant consideration affects our ability to overlap tasks. During the period of overlap, we need additional resources, including adequate personnel, material, facilities, etc., to perform both tasks simultaneously. The total resources of the project have not increased, but the levels of resources required during specific periods of the project have changed.

Figure 1-3

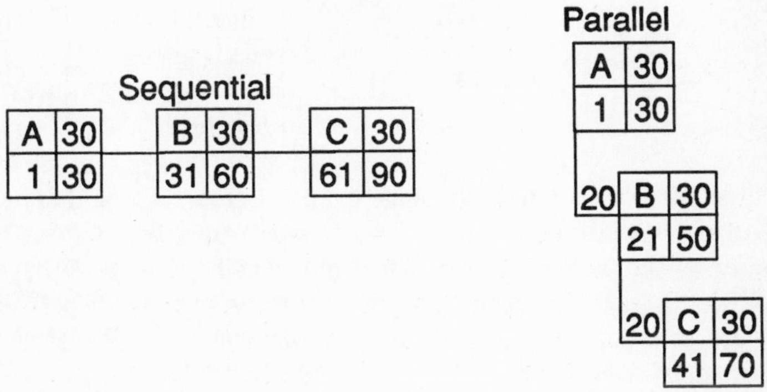

The planning decisions for a project that we base on modeling processes are essentially the same decisions that we must make without modeling: overtime, level of resources, work sequences, etc. Without a rigorous modeling process, however, we most likely make decisions in reaction to lack of schedule performance and/or by intuition. Neither of these are based on substantive data about the balance of the project; they are instead reactive mechanisms. The project modeling process on the other hand provides a discipline and rationale for schedule decisions based on the impact to the remaining project. It gives us a proactive tool: the time reserve mechanism.

THE CPM PROCESS

Determining time reserve by comparing cumulative path sequence to path time available is a valid, simple means to understand the concept of time reserves, but it does not provide a practical way to calculate time reserve in anything other than a very simple model. A project model is no longer considered simple when two or more paths merge or intersect. This occurs any time a task has more than one input and/or any time a task is itself an input to more than one task. Since most projects entail multiple relationships among tasks, complexity is the normal condition.

To understand why the current process cannot handle path intersections, let us merge our original simple example with a second path.

Figure 1-4

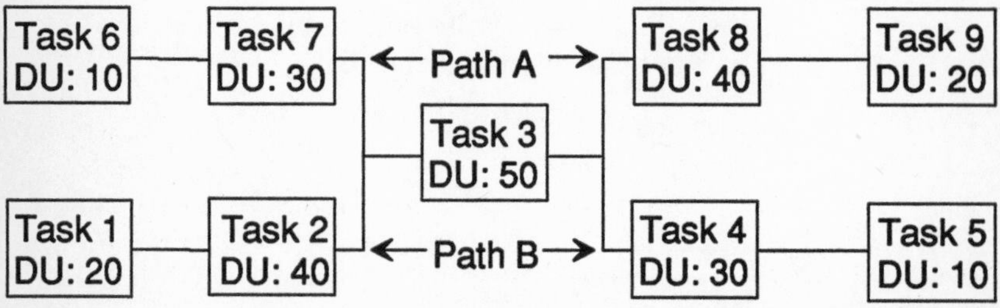

Here Task3 is common to both paths. On the surface, it would appear that path B drives the start of Task3 as 60 days of work must be completed along this path prior to the start of Task3, while along path A only 40 days must precede it. This assumes that both paths start on the same day. If path A were to start 30 days after path B, then it would drive the start of Task3. If the finish of Task1 were delayed by 20 days, path B would again drive the start of Task3. The point is that any time paths converge, any path can drive the common point, depending on the lengths of the paths, their sequences, when each started, and how each actually progresses.

Each new wrinkle changed the lengths of the paths. When both paths started simultaneously, path B was 150 days long (20 + 40 + 50 + 30 + 10) while path A was 170 days long (10 + 30 + 20 days waiting for path B to catch up + 50 + 40 + 20) even though the path required only 150 days of actual work. The difference is 20 days tagged onto the length of path A waiting for path B to catch up before Task3 could start.

When path A started 30 days after path B, path A took 150 days to complete (10 + 30 + 50 + 40 + 20) while path B took 160 days (20 + 40 + 10 days waiting for path A to finish + 50 + 30 + 10). The starting point for each path makes the difference here. Paths' starting points relative to one another can affect their cumulative spans provided they merge at some point.

When path A started 30 days after path B and the finish of Task1 was delayed by 20 days, path A became 160 days long (10 + 30 + 10 days waiting for path B to catch up + 50 + 40 + 20). Path B was 170 days long (20 + 20 days for the delay of Task1 + 40 + 50 + 30 + 10). This illustrates how lack of schedule performance can affect the lengths of other paths at their points of intersection.

All this manipulation of path length soon becomes very cumbersome, and the bigger and more merged paths in the model, the more difficult it becomes to determine a path's cumulative length. A model of a few hundred activities can have many times that number of paths. This makes our path length process impractical, so we must develop some alternative method of calculating a path's time reserve.

This alternative process is called the *Critical Path Method (CPM)*, although the name can be deceptive. This term leads many people to the mistaken assumption that the process devotes itself to finding something called a "critical path." Despite the ominous and important sound of the word *critical*, identifying this path is actually a secondary purpose of the practical CPM process.

Subsequent chapters will explore this process in detail, but for now it is enough to understand that CPM helps us derive a project's time reserves—this is its primary purpose. As we thoroughly examine this process, it is important always to keep in mind the definition of time reserve. Though we will change the method of its calculation, the value still represents the relationship between the length of time a *path* of work requires and the time available to accomplish it. Time reserve is still the proactive parameter that allows us to assess the schedule health of a sequence or path of activities.

SUMMARY

Project Planning

Project planning is a disciplined approach to formulating viable project plans and then managing the successful accomplishment of the project. A project is any deliberate effort that achieves finite objectives through a series of tasks that are interdependent due to interrelationships of the work and/or shared resources.

Project planning is accomplished in seven principal steps:

1. Determine the objectives of the project.
2. Determine the tasks required to achieve the project objectives.
3. Determine the resources necessary to accomplish the work.
4. Coordinate and arrange the work as efficiently as possible, sequenced such that each task's prerequisites are accomplished before they are needed, while still meeting all project objectives.
5. Overlay project structures to assign and describe the work to the appropriate groups of workers.

6. Publish information about the plan, including any changes to the plan and its status or measured performance to all project personnel.
7. Assess progress and the impact of changes, and adjust the plan as necessary.

Project management is the control of the critical technical, time, and cost parameters of a project in order to affect its outcome positively. Reactive management is responding to project circumstances as they arise. Proactive management is control of the project with an eye on the unfinished portion of the project. The project modeling process is the only proactive tool ever developed for project management and control.

The Project Model

Project modeling is a mathematical, procedural means of accomplishing the seven steps of project planning. The modeling process yields Project Time Reserve (PTR), the key parameter to exploiting the process' benefits. Project time reserve is the difference between the time available (TA) to accomplish a sequence of work and the cumulative length of time necessary to accomplish the sequence of work (CSP):

$$PTR = TA - CSP$$

The Seventh Planning Step

Project planning must be continuous because:

1. Both internal and external requirements cause continual change.
2. Projects sometimes fail to meet detailed schedule objectives.

The seventh planning step attempts to avoid or mitigate these problems by:

1. Altering the model to reflect current status and changes.
2. Determining the impact of changes and schedule slips on downstream activities and on time reserve.
3. Testing solutions to ensure that they solve problems without creating new ones.

We evaluate the impacts of schedule change and performance by analyzing:

1. Potential effects on the schedule.
2. The amount of time reserve erosion relative to original reserves.
3. The rate of time reserve erosion versus progression.
4. The adequacy of remaining time reserve to meet project objectives.
5. Any workaround solution to ensure that it solves intended problems without creating new ones.

Time Reserve Decisions

We can resolve time reserve problems by one of two methods:

1. Increasing the time available (TA) to a path.
2. Decreasing the cumulative sequence of the work (CSP).

We can increase the time available to a path by:

1. Finishing the path later than planned.
2. Starting the path earlier than planned.
3. Accomplishing some work during nonwork periods (overtime).

We can decrease the cumulative sequence of a path by:

1. Decreasing the span of a task or tasks by reducing work scope.
2. Decreasing the span of a task or tasks by increasing the resources working on these tasks.
3. Decreasing the cumulative span of work by rearranging the sequence of the work.

The Critical Path Method (CPM)

The critical path method is a means of calculating time reserves that allows for the convergence of project paths.

CHAPTER 2

The Project Model

AN APPROACH TO THE MODELING PROCESS

Now that the basic purpose of the modeling process is defined, we can work toward an understanding of the process. To some, it may seem unnecessary to wade through all of this detail, but we can never use the process properly without it. Only by understanding the principles of modeling in detail and ensuring that our models accord with them can we obtain valid data about our project.

Therefore, we must laboriously examine each technique of project modeling to determine its proper structure and uses, and then combine them to understand their interaction. Beyond that, we must learn how to apply all of this to complex practical applications. At this point, we will state the characteristics and requirements of each modeling technique as rules. We will prove the validity of these rules in due course.

CHOICE OF MODELING TECHNIQUES AND A DATE CONVENTION

Before we can begin to examine the project modeling process, we must make some choices. Over time, various modeling techniques have been developed:

- PERT (Program Evaluation Review Technique)
- CPM
- Arrow
- Precedence

Any discussion about advantages and disadvantages of any of these techniques eventually boils down to personal or application preference. They are all acceptable modeling techniques. Though differences might make one preferable to another for a specific job, anything that can be done in one technique can be done in the others. In fact, they all employ basically the same process.

To draw an analogy, this book could have been written in French or German. Both are complete languages and both adequately communicate ideas, but they accomplish the task of communication in very different ways. An individual or group may prefer one language over the other, and legitimate reasons may dictate the selection of one over the other, but either could, in fact, do the job. The mistake would be to write the book in both French and German, as this would make it exceedingly lengthy, redundant, and quite confusing.

The various modeling techniques, while they provide the same information, differ enough that trying to explain something in one technique and then repeating it in another would quickly confuse the reader. It is enough to understand that to choose one set of techniques is not to condemn the others. We can choose our method from four combinations:

1. PERT in Arrow
2. PERT in Precedence
3. CPM in Arrow
4. CPM in Precedence

We have selected the last combination: CPM in Precedence. We offer brief explanations of the PERT and Arrow techniques in Appendixes A and B to demonstrate the differences between PERT and CPM, and between Arrow and Precedence, to establish the basis for interpolation. Someone who wished to model in one of the other combinations could transpose the principles stated here to the alternate technique.

During each stage of our examination we encounter terminology and conventions that require explanation. One of these is a date perception. Since much of the input to and output from the modeling process takes the form of dates, we must state a convention for dealing with the point in time of a start date versus the point in time of a finish date.

When we say we will start a task on June 1, we mean we will start at the *beginning* of the work period on June 1. If we say we will finish a task on June 1, we mean we will finish at the *end* of the work period on June 1. A start date of June 1 and a finish date of June 1 are one work unit, or day, apart.

We will maintain this *start-finish convention* in our modeling analysis. This may seem a moot point, but without awareness, it can lead to confusion in our mathematical processes. It will become very important once we begin date math.

A CPM model describes a project through three elements: activities, events, and constraints. The model must represent all of the pertinent properties and characteristics of the project through one of these elements.

ACTIVITIES

Modeling breaks the effort or work of a project down into finite, discrete tasks called *activities*. The sum total of all activities is—must be—the total project. Activities consume time and/or resources; this is their primary distinction. Each activity must be decomposed to the point that:

1. Its start and finish ends are measurable and/or they define the points of interface into or out of the task.
2. Its relationship to the project structures can be represented as a single value in these structures (e.g., one organization, one WBS, one location, etc.) This is not a modeling requirement, but a control requirement and will be further explained in Chapter 11.

These two requirements must guide the definition of the detail tasks of a project. This greatly facilitates building a model, and also greatly enhances the control of the project.

The modeling methodology provides a mechanism called a node to define all three model elements. In Precedence convention, we define every activity as a *node*. We will later define the other two elements relative to nodes.

Precedence model diagrams represent nodes as rectangular boxes compartmentalized into smaller rectangular boxes. These compartments hold specific information about each activity called *attributes,* which are separated into five categories:

1. Model attributes
2. Schedule attributes
3. Structural attributes
4. Resource attributes
5. Event attributes

We will eventually explore all of these categories, but for now we will limit the discussion to the first *(model attributes)* and fifth *(event attributes)*. The many attributes in these two categories will all be introduced and defined over the next several chapters. Some of the model attributes of an activity are: the textual description (abbreviated AD) of the task; the estimated span of time to accomplish the task, called the *duration* (abbreviated DU); and start and finish dates (which take many forms).

Each node (and the activity it represents) is assigned a unique series of alphanumeric characters that identify it. This *node indentification code* (abbreviated NI) is yet another activity model attribute. Every node has at least two attributes: the node identifier (NI) and the activity duration (DU). These are the *base node attributes,* and are the essential parameters of all activities.

Figure 2-1

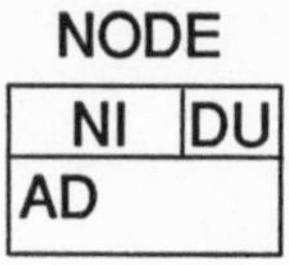

EVENTS

The second project model element, the events, define the points of accomplishment of the project. They have no durations; they are points in time describing states of being (e.g., Task A complete or Task A start) and consume neither time nor resources. An *event* is the completion or initiation of an activity or a series of activities. Precedence defines events as *node ends.*

Events are actually subrecords of nodes: one subrecord describes the point of initiation and one describes the point of completion. The attributes of events are thus subattributes of the activity or node. Some of these attributes include: the node identifier (the same as its activity identifier), flags, a title, and dates. The flags define the graphic shape of each event's marker on Gantt barcharts and model plots and the highest level at which to report its schedule (see Chapter 13). For the start-end event, the flag is called the *Start Event Flag (SEF)* and the flag for the finish-end event is called the *Finish Event Flag (FEF).* A short, textual description (SED for the start-end event, FED for the finish-end event) provides a title for each event. Each event takes its dates from the appropriate activity end. Figure 2-2 illustrates an activity node with its ends designated as start and finish events.

Figure 2-2

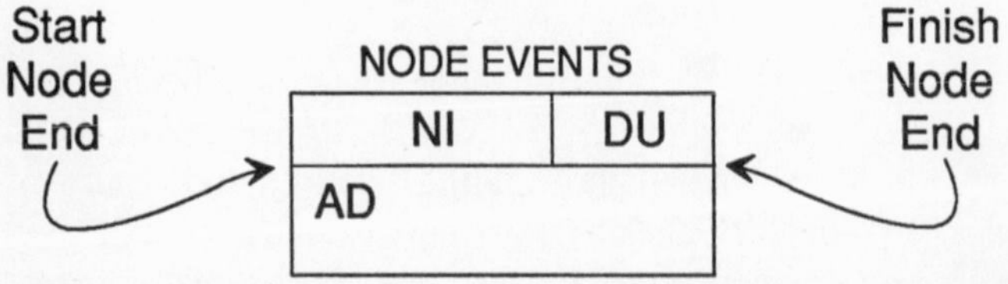

Since the event is a subrecord of the node, its data also appears in the node box. Node compartments contain some of the data, while other information may appear above the appropriate node end in additional graphic shapes (e.g., triangle, arrow, or diamond) defined by the SEF or FEF parameter. Next to each event symbol, some of its attributes such as the event description (SED or FED) appear as illustrated in Figure 2-3.

Figure 2-3

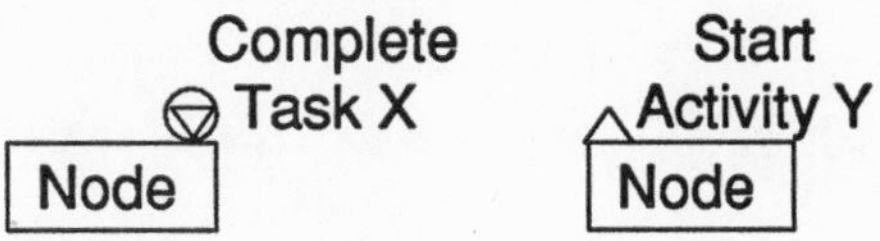

We must emphasis two key concepts.

1. All activities (and therefore nodes) have two inherent events: the start of the activity at one end and the finish of the activity at the other.
2. Some attributes of these two events are common to the activity and some attributes are unique to each event.

Some events mark the start or completion of two or more independent paths. No one activity (node end) will suffice for this event, which results from several independent nodes. In this case, it is best to set up a separate node with no span (duration = 0) and interface all appropriate activities into or out of it. We call this an *events node*. When we establish such a special node, we must remember that it still has two ends and, therefore, two events: one measuring the preceding nodes' finishes and the other measuring the succeeding nodes' starts. This is illustrated in Figure 2-4.

In the example, event AU could measure the completion of three independent paths: CU-PB, ZN-FE, and MO-NI. The sample data show that this completion date would be July 29, when *all* three paths finish. This is the latest finish date. Note that it appears in the finish end of events node AU.

By our date convention, this finish date means the end of the working day on July 29. The event that marks when the next tasks (AL, AG, or CO) could start could not occur on the 29th, but rather the morning of the next working day, August 1. (July 30 is a Saturday and the 31st is a Sunday.) Note that the start end of events node AU reflects this date. This event measures when the follow-on tasks *can* start, not when they actually start.

Figure 2-4

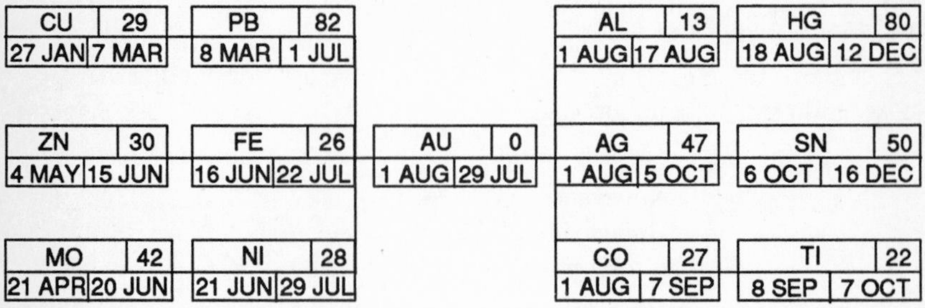

The diagram may seem confusing because the dates of the events node are not in the correct chronological order reading from left to right. We must remember that the finish end of the events node records the finish of its predecessors, while the start end records the start of its successors; the later start date appears at the left or start end of the events node, while the earlier date appears at the right or finish end. An events node (DU = 0) carries both dates and the user then must decide which one is pertinent.

Events nodes differ from activity nodes only in that they have no activity attributes. They represent no activities, but still feature two events, just as every other node does. Usually, we focus attention on only one end of an events node, either the start end or the finish end. Once we have determined which, we simply ignore the information in the other half.

We must understand the definition of events to use them properly. This has caused one of the principal misunderstandings of Precedence modeling. We have defined events here to prevent the problems that often arise from this question.

Another term, *milestone,* is often thought to be synonymous with the term *event*. Although all milestones are events, the reverse is not true. A project has twice as many events as nodes (one start and one finish for each node). This much detail frequently overwhelms tracking and control, especially for higher levels of management. Therefore, we select important events called milestones from the total for reporting. The significance of milestones is indicated in the schedule level value in the SEF or FEF parameters. Higher schedule levels (indicated by numerically smaller values) represent more important milestones.

We have introduced and basically defined the first two model elements, activities and events, that describe the physical project. Activities represent all work of the project; events represent the initiation or completion of that work, enabling us to judge our schedule performance. Now we turn to a discussion of how all of these activities (nodes) and events (node ends) relate to one another.

CONSTRAINTS: AN OVERVIEW

The third element of the model, called *constraints* (also precedences, ties, interfaces, inputs, outputs, or connections), reflects the interdependency of activities and events. Constraints are key to a project model, for three reasons:

1. They are unique to project modeling methods. No other scheduling methodology uses finite, direct interfaces.
2. They tie all of the events and all of the activities together, transforming a collection of independent activities and events into a cohesive plan. They define the total planned relationship of the work.
3. They allow us to relate and measure the progress and completion of each activity with respect to the rest of the project.

Precedence diagrams represent constraints as node-to-node relationships, constraining one node in a sequence to another. One constraint can interface only two nodes. An activity with several interfaces will require one constraint for each relationship.

The attributes of the constraint define the finite, exact relationship between two nodes in terms of four key values. The first and second key attributes, the node identification (NI) codes, specify which two activities (nodes) the constraint joins. The unique value of this code for every node provides exact definitions of the two nodes to be interlaced.

Besides identifying the two nodes in the constraint, we must also identify their sequence, noting which must occur before the other can proceed. This first or preceding node is called the *predecessor* of the constraint (PN). The second, succeeding node is called the *successor* of the constraint (SN). Constraint relationships flow from predecessor to successor.

Project models follow a fundamental graphic convention:

Time always flows from left to right of a diagram.

The left-to-right convention applies to both graphic and tabular representations. The predecessor node should always lie to the left of the successor node, in the same order as the activities will flow. Consequently, when a constraint is listed as a pair (e.g., AWUNA-ATUAA), the first node identifier always indicates the predecessor (AWUNA), and the second node identifier always indicates the successor (ATUAA). Figure 2-5 illustrates a constraint and some of its attributes.

Constraints reflect all activity and event relationships in a project by connecting one end of the predecessor node to one end of the successor node. Since

Figure 2-5

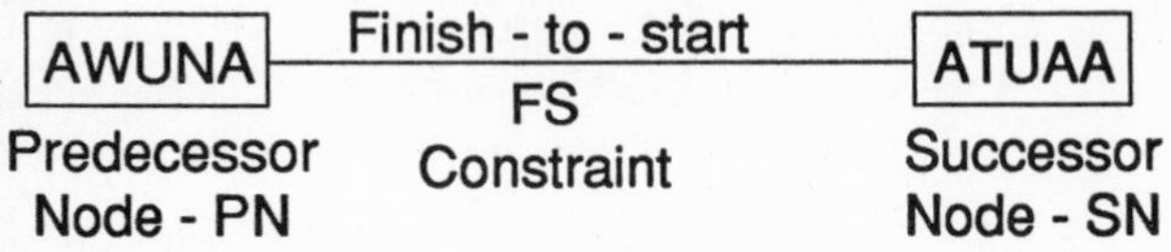

each node has both a start end and a finish end, any two nodes can have four combinations of interfaces. These are called the *constraint types,* and each represents a specific kind of relationship between two nodes.

The workers who perform a task best know the requirements to start and finish it. In modeling terminology, the workers can best determine a task's predecessors. The receiver of this interface which in this case is the asking task is the successor of the constraint.

Every predecessor falls into one of two basic categories: inputs necessary to begin a task and inputs necessary to complete a task. These are called respectively *predecessors of an activity's start* and *predecessors of an activity's finish.* Once we understand that an input is necessary, we must determine whether it is needed to begin the task or at some point during work on the task. This ultimately defines the point of interface with the successor.

Combining the source of the predecessor to its interface with the successor identifies the *ends* of the nodes that are constrained. This defines the third attribute of a constraint, the *constraint type*. Each constraint type identifies a very specific relationship between two activities or events, as we will thoroughly explain later.

The fourth attribute of a constraint is a numerical entity coupled with the constraint type to further define node relationships. This integer value, called the *lag of the constraint*, measures the amount of delay of the successor node relative to the predecessor. This provides a finite quantification of the task interrelationship. The term *delay* has a specific meaning for constraints. The base rule for all lag values states that:

> A lag value on a constraint must represent the accomplishment of work.

Lags resemble durations of nodes in that they represent the accomplishment of a task. The source or purpose of lags varies among constraint types.

Constraint Type: Finish-to-Start (FS)

The first and simplest type of constraint, the *finish-to-start constraint,* binds the finish of the predecessor to the start of the successor. This means that the predecessor task must finish before the successor task can start. This very specific, finite, sequential relationship prevents the successor task from going on simultaneously with the predecessor.

This constraint type is abbreviated FS and appears in diagrams as a line running left to right from the finish end of the predecessor node to the start end of the successor node. Figure 2-6 illustrates two examples of correctly drawn finish-to-start constraints and one that violates the left-right convention.

Figure 2-6

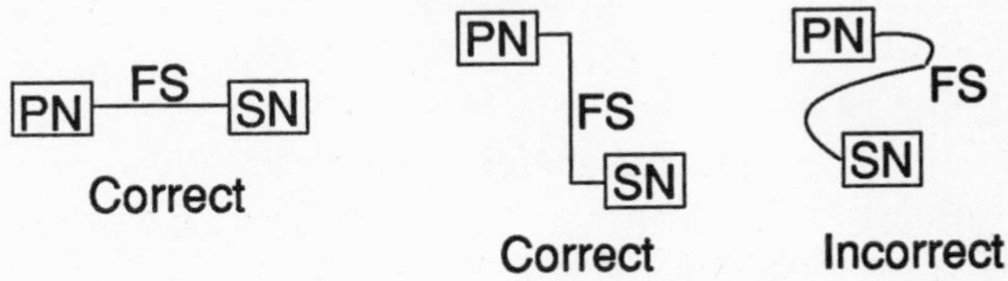

The lag value on a finish-to-start (FS) constraint should be zero unless some activity falls between the two tasks. This lag activity would fall into one of the following categories:

1. A delay caused by a physical or chemical process such as:
   a. Concrete curing
   b. Microbe growth
   c. Imposed delays (e.g., a contract provision requiring delivery 30 days prior to an event)
2. Work that is not deemed significant enough to warrant a node identifier, but that still affects time consumption, for example:
   a. Packing and shipping
   b. Transportation time
   c. Queueing

In either case, an activity (node) with a duration equal to the constraint lag could represent the lag activity just as well. We simply choose to model the work on the constraints instead. This is illustrated in Figure 2-7.

We model delays that do not represent the accomplishment of effort using other element attributes, which we will introduce later.

Figure 2-7

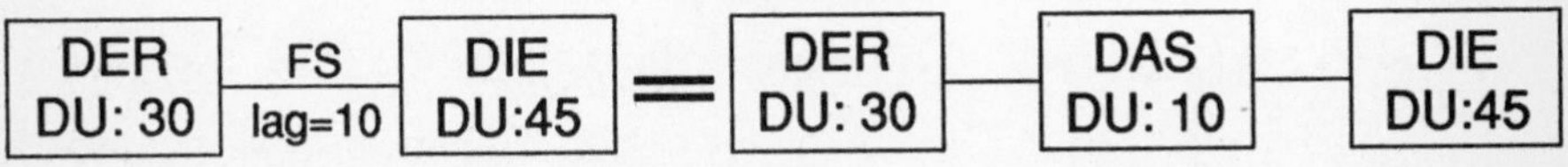

Constraint Type: Start-to-Start (SS)

All inputs are not generated at the completion of a task. Some may be generated after a portion of a task is accomplished. If the completion of a portion of one task is an input to the start of another, we have the second constraint type, the start-to-start (SS) constraint. A start-to-start constraint allows a successor activity to start once its predecessor starts and the time lag passes. A start-to-start constraint might more appropriately be called a middle-to-start or partial-to-start constraint. The amount of the predecessor (measured in the same units as its duration) that must be accomplished before the successor can start is the lag of the constraint.

Figure 2-8 illustrates an example of this type of constraint. Activity A (the predecessor) takes 30 days to accomplish and Activity B (the successor) takes 45 days to accomplish. The first ten days of work on A must be completed before B can begin. This is the lag of the constraint. The successor task, B, cannot start until 10 days after task A starts. If the start of A slips, then the start of B will also slip, provided it has no other, later predecessor. In standard notation, we would describe this constraint as follows: Constraint: A-B SS 10.

Figure 2-8

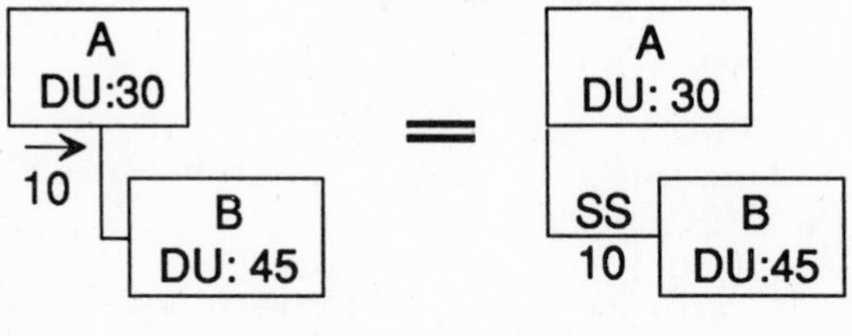

All too often the chronological sequence of a schedule is used to create constraints. The desire to start activity B 10 days after the start of A does not in itself create a start-to-start constraint from A to B with a 10-day lag. A constraint exists only if the start of B depends on input from A. If this input is not the end result of A, then it must be generated somewhere between the start and end of A (the lag portion). In the absence of these two conditions, the relationship between tasks A and B is probably not necessary and no constraint should bind them. This type of rigor is required in developing constraint relationships to maintain valid impact analysis and yield accurate time reserve values.

We must note two subprinciples here:

1. Every start-to-start constraint requires a lag value.
2. This value represents a portion (less than the total) of the span of the predecessor.

A start-to-start constraint with no lag means that the successor can start the instant the predecessor starts, or that the predecessor contributes nothing essential to the start of the successor, as no time is specified in which to generate it. More than likely both tasks in such a relationship share a common predecessor, and separate constraints should bind it to both tasks.

The start-to-start constraint's lag cannot exceed the span of the predecessor, since, by definition, it equals some portion of the predecessor. If, in the previous example, the lag were 40 days this would mean that the first 40 days of a 30-day task must be accomplished before the successor can start. This absurdity indicates an invalid constraint.

The acid test of the validity of a start-to-start constraint is whether the relationship can be replaced by its modeling equivalent. To do this, break the predecessor into two tasks, the first with a span equal to the lag of the constraint and the second with a span equal to the remaining duration (predecessor DU – lag). The successor activity remains unchanged. Two finish-to-start constraints then finish the model: one joins the first partial task to the second and the other joins the first partial task to the successor, as depicted in Figure 2-9.

Figure 2-9

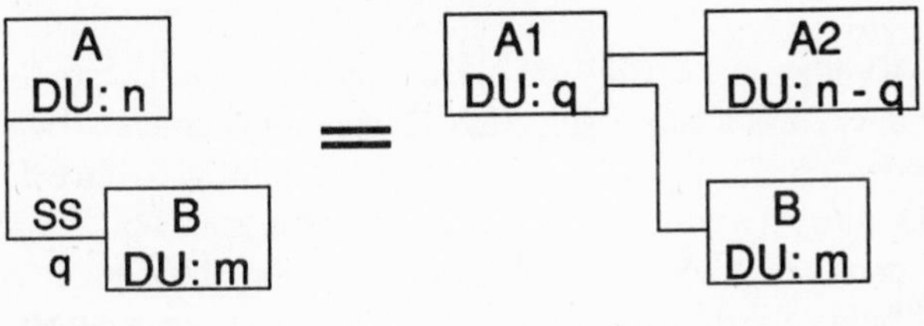

The only difference between these two methods of notation is that the second shows the exact point of interface. To achieve this, we must define one more activity and one more constraint. The modeler must find a compromise between detailed depiction of the interfaces and the number of nodes and constraints. Following this second, more detailed procedure throughout a complex project model would add significantly to the size of the model, and therefore the work of data administration.

Constraint Type: Finish-to-Finish (FF)

In effect, a start-to-start constraint allows related activities to overlap, whereas the finish-to-start constraint binds them strictly in sequence. In both of these types of constraints (FS and SS), predecessors initiate their successors. Sometimes, however, the input of a predecessor is needed not at the start of a task but rather at some point during the progress of the task in order to complete it. To model this type of interface requires a finish-to-finish (FF) constraint.

This constraint type prevents the successor from finishing until its predecessor finishes and the lag time passes. Practically, this relationship could be described better as a finish-to-middle or finish-to-partial constraint. Like the start-to-start constraint, the lag of the finish-to-finish constraint specifies a portion of one of the tasks. The SS constraint's lag specifies part of the predecessor, but the FF constraint's lag specifies part of the successor—the amount of time needed to complete the effort of the successor upon completion of the predecessor.

Figure 2-10

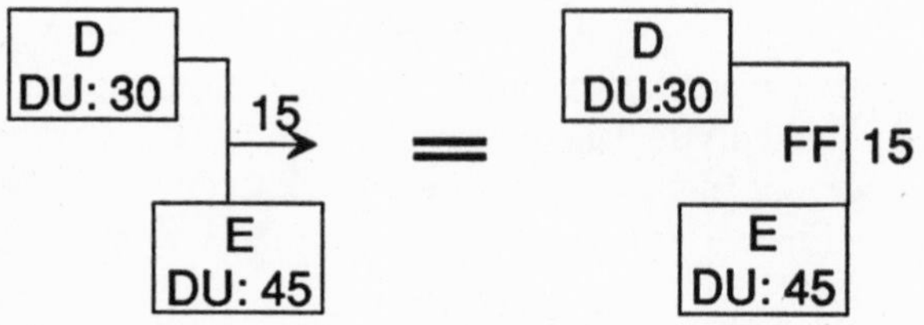

Figure 2-10 illustrates a finish-to-finish constraint with a predecessor, Task D (span of 30 days) and a successor, Task E (span of 45 days). Task E can start independently of Task D, but 15 days of work remain in E once D finishes. If the finish of D slips, then the finish (though not the start) of E could slip, depending upon its other predecessors. The math of the CPM process will verify the logic of defining the finish-to-finish relationship in this manner.

This definition gives rise to two subprinciples:

1. All FF constraints require a lag value.
2. This value cannot exceed the duration of the successor, as it is a portion of that task.

An FF constraint with no lag would allow the successor to complete the instant the predecessor finished, requiring no more time to finish the successor than the predecessor. This means that the successor does not depend upon input from the predecessor and no essential relationship binds them. Instead, two such tasks would probably share a common successor.

The lag of a finish-to-finish constraint cannot exceed the successor's span, since it specifies a portion of that span.

We could replace the FF constraint with comparable but more extensive modeling, as illustrated in Figure 2-11. First, we would decompose the successor into two tasks, the first with a span equal to the portion of the successor that can precede the predecessor's input (successor DU – lag), and the second with a duration equal to the lag of the FF constraint. The predecessor undergoes no change. An FS constraint from the original predecessor to the second task and another FS constraint from the first partial task to the second partial task complete the model.

These techniques differ only in that the second shows the exact point of interface. To achieve this, we must define one more constraint and one more activity (node). The modeler must make another compromise between detailed depiction of the interfaces and the number of nodes and constraints in the model.

Figure 2-11

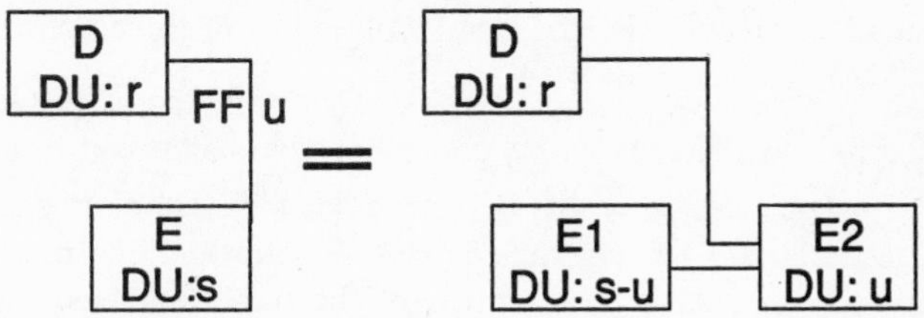

Constraint Type: Start-to-Finish (SF)

Thus far we have discussed finish-to-start, start-to-start, and finish-to-finish constraints; this leaves only one possible connection of node end to node end: the start of the predecessor to the finish of the successor (abbreviated SF). This type of constraint prevents the successor from finishing until the predecessor starts and the lag time passes. Practically, this means that a portion of the predecessor must be completed before the remaining portion of the successor can finish. This might be better described as a middle-to-middle or partial-to-partial constraint.

The SS and FF constraints also dealt with partial tasks, but this constraint type deals with both a portion of the predecessor and a portion of the successor. This makes the lag a complex number, combining part of the predecessor span and part of the successor span. We do not deal with an end of either node and the constraint lag is totally buried in the two activities, so nothing distinguishes the examples illustrated in Figure 2-12.

This ambiguity makes SF constraints impractical, but we can simply decompose either the predecessor or the successor into two activities and, with combinations of the other three more specific constraint types, model either example. We can avoid the SF constraint and use only the more practical constraint types, FS, SS, and FF, to model the relationship between any two activities.

Figure 2-12

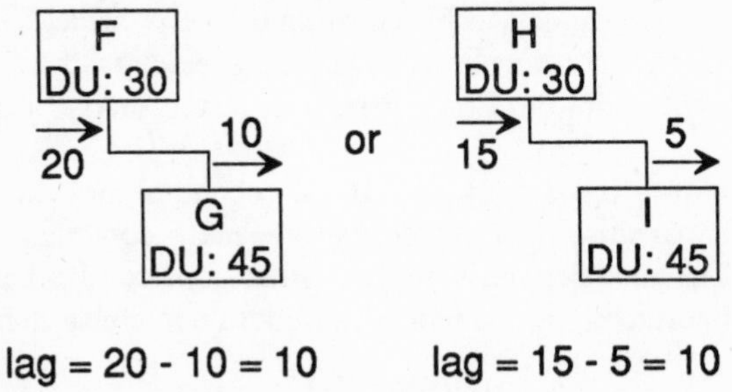

To summarize, practical Precedence involves use of three constraint types:

1. Finish-to-start (FS), in which the successor task can start once a predecessor finishes. Normally this constraint type has a lag of zero.
2. Start-to-start (SS), in which the successor task can start once a predecessor starts and the lag time passes. Practically, this means that the successor can start once a portion of the predecessor equal to the lag time finishes.
3. Finish-to-finish (FF), in which the successor task can finish once the predecessor finishes and the lag time passes. Practically, this means that a portion of the successor equal to the lag time remains to be completed after the predecessor is finished.

Multiple Constraints between Two Tasks

Proper modeling requires that the modeler fully understand the three practical constraint types, and also how to use these constraints in combination. A start-to-start constraint between two nodes frequently accompanies a finish-to-finish constraint between the same two nodes. A start-to-start constraint alone indicates that the ultimate finish of the predecessor has no impact on the finish of the successor. Indefinite delay in the predecessor's finish would not delay the successor in any way. We encounter this situation only very rarely. It is quite common that a successor task can start only after completion of a portion of a predecessor, the conditions for an SS constraint, while the finishes of the same two tasks are also bound together, which indicates an FF constraint. It is much more common for two tasks to be bound only by an FF constraint than only by an SS constraint.

This is the rationale for handling multiple relations between two activities. As a project progresses, these multiple relationships become more and more common as we must accomplish work in parallel in order to meet schedule requirements.

Let us examine a simple example, shown in Figure 2-13. Two activities (nodes J and K) are initially sequentially (FS) constrained. In the preferred, least-risk course, the predecessor (J) would finish before the successor (K) started. During the course of the project, however, the predecessors of these two tasks begin to slip, eventually eroding the Path Time Reserve until it is no longer positive. We must act in order to recover the program.

We determine that tasks J and K can be accomplished in parallel rather than in series. This will reduce the total time to accomplish the two tasks by the amount of time they can be overlapped. As we have seen, however, overlap can occur at both ends of the node.

We must first determine what necessary inputs the predecessor (J) provides to the successor (K), that is, how much of J must be completed before K can start. This relationship yields a start-to-start (SS) constraint with a lag value that represents the work to be accomplished in the predecessor (J) before the successor (K) can begin.

But this does not completely define the relationship, as it gives no information about the ability of the successor (K) to finish based on the predecessor's (J) finish. With only a start-to-start constraint, the predecessor's (J) finish could delay indefinitely without affecting the successor's (K) finish in any way. The model must account for the relationship between the predecessor's (J) finish and that of the successor (K). This yields a finish-to-finish constraint with a lag value representing the work remaining in the successor (K) once the predecessor (J) finishes. Two constraints, one start-to-start and one finish-to-finish, are necessary to completely model the relationship between these two tasks.

Figure 2-13

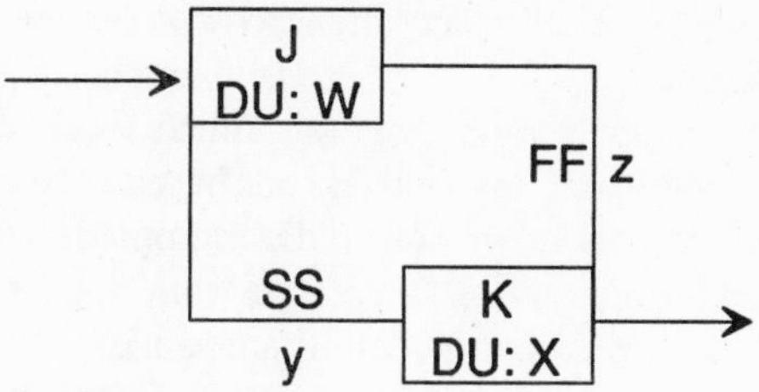

MODEL PARAMETER SUMMARY

It is interesting to note that it took almost twice as much text to define constraints as to define both activities and events. This says something about the complexity of constraints versus the other two model elements. Activities and events are much more visible and concrete. If we could truly visualize a project, however, we would see that the interfaces, the interrelationships between activities and events, create its complexity. Further, the lack of attention to the interfaces is typically the source of a project's troubles.

This major mind set change comes with the project modeling philosophy. We cease trying to manage a project as a simple conglomeration of activities and events, and instead place considerable emphasis on managing the relationships and interfaces, the constraints between the activities and the events in the project.

This leads to one of the basic principles of the project modeling philosophy, the conclusion that the key to managing a project is managing the interfaces. Stated formally:

> The key to managing a project lies not so much in the control of activities and events, as in the control of the project's interfaces (constraints).

Unfortunately, developing good interfaces or constraints is the most difficult task in project modeling. Sometimes they seem nebulous. Sometimes they lie deeply buried, especially when workers do not understand the constraints of their tasks. Nevertheless, success depends on discovering the interfaces among a project's activities, modeling them, and then controlling them as if they were absolutes. In the beginning of a project, it is especially important that the model reflect the best case, or in other words, the least-risk interfaces. Later, when necessary to recover time reserve or schedule, we can alter the interfaces to accommodate workaround solutions, probably increasing risk. Constraints, like tasks, can and will change. They must be as dynamic as the project. This is the proper attitude for modeling.

Yet, however thorough a model or how much work we devote to it, we must realize that the model and any analysis are based on estimates. There is no magic, no crystal ball, no perfect way of predicting the future. Our model simply represents our best understanding of the project.

Some conclude from this that modeling is too inexact and too much work. This classic excuse is tantamount to admitting that planning is futile, that every project is better off at the mercy of fate. Others believe that scientific methods, though imperfect, offer the better course. They believe that if the effort to control the variables and to guide the future of the project solves only a few problems in advance, the effort is justified.

It is this steadfast refusal to allow fate to control a project that separates the planner from the scheduler. True planners concede circumstance and uncertainty no control over the outcome of the project. They struggle to ensure that project managers retain as much control as they can. To achieve this, they must learn to model properly and then they must model and analyze and model and analyze and . . . .

SUMMARY

Through project modeling, we isolate Path Time Reserves. The model must be structured and maintained properly to yield valid values of time reserve. We have chosen the *Critical Path Method (CPM)* in *Precedence* as our modeling technique.

Activities are discrete units of work that consume time and resources. We must break the work down into activities so that:

1. The start and finish of each activity are clearly defined as the points of interface into or out of the tasks.
2. Each activity can be represented in the project structures as a single value.

Nodes represent all three model elements in the modeling processes. In Precedence, activities are nodes. Each activity in the model must have a node identifier (NI) and a duration (DU). An activity can have many other attributes.

In Precedence, events are node ends. Every node has two inherent events; one at its start and one at its finish. Milestones are events that have been assigned significance by the project.

Interfaces between activities and events appear in the model as *constraints*. In Precedence, constraints define node-to-node relationships, where a single activity interfaces with another. The first or left node in a constraint is called the *predecessor*. The second or right node in a constraint is called the *successor*.

The constraint type defines which predecessor node end (event) interfaces with which successor node end (event). Of the four types of constraints, only the first three have practical use:

- Finish-to-start (FS)
- Start-to-start (SS)
- Finish-to-finish (FF)
- Start-to-finish (SF)

Time delays associated with constraints are called *lag values*. Where these values appear, they must represent the accomplishment of work. Generally, finish-to-start (FS) constraints should not have lag values, unless we choose to represent the accomplishment of a simple task on the constraint.

Lags on start-to-start (SS) constraints represent overlaps among tasks, that is, the portion of the predecessor that must be accomplished before the successor can begin. All start-to-start constraints should have lag values.

Lags on finish-to-finish (FF) constraints also represent overlaps among tasks, that is, the portion of the successor that remains to be accomplished once the predecessor finishes. All finish-to-finish constraints should have lag values.

Start-to-finish constraints both end and begin in the middle of tasks. Their lag value is a complex number derived in part from the predecessor and in part from the successor. This makes this type of constraint too convoluted and cumbersome for practical use.

Whenever a start-to-start constraint binds two tasks, we should investigate the possibility of a finish-to-finish constraint between the same two tasks and vice-versa.

Constraints are the key element of the project modeling method. This distinguishes it from all other forms of planning and scheduling. The key to managing a project lies not so much in the control of activities and events, as in the control of the project's interfaces (constraints).

CHAPTER 3

Date Math and Calendars

INTRODUCTION

Date math arises in all aspects of the modeling process. Problems constantly require that we add two dates together, subtract one date from another, or add durations to dates. In all of these operations, we can only consider work days. As commonly portrayed, dates are not readily usable (6 Aug., 89 or 8/6/89) for mathematical processes. Further, there is no simple way to determine the number of workdays between any two distant dates. Somehow, we must translate dates into numerical values that we can manipulate with simple mathematical equations. This is the rationale for date math.

DATE MATH EQUATIONS

The primary date math equation gives the relationship between the start date of a task, its span, and its finish date:

Finish = Start + Duration
FIN = ST + DU

Simple algebraic rearrangement allows us to calculate any value in the array based on the other two:

DU = FIN − ST
ST = FIN − DU

We must further modify this equation to accommodate the common date convention that start dates reflect the beginning of the work period and finish dates reflect the end of the work period on the day specified. Simply stated, we start work on the morning of the start date and finish in the evening of the finish date. To accommodate this convention we alter each of the equations slightly as both the start date and the finish date are workdays so the duration includes both. These altered equations appear:

$$\begin{aligned} FIN &= ST + DU - 1 \\ ST &= FIN - DU + 1 \\ DU &= FIN - ST + 1 \end{aligned}$$

The 1 in all of these equations is measured in the same units as the duration, whether that is days, shifts, hours, fortnights, or millennia (in the case of geology models). The math will accommodate any work unit, but the simplest, most understandable unit to deal with is workdays. For our purposes, the *unit* will always be workdays, but remember that date math principles are identical for all units of time.

Examining the fundamental date math equation points out a principle for choosing between addition or subtraction of a unit to avoid violating the start-finish date convention correctly:

> When calculating a finish date from a start date, subtract one time unit. When calculating a start date from a finish date, add one time unit.

This rule will come into play repeatedly as we derive the modeling equations.

Let us work an example with the fundamental equation. A task starts on the 35th day of a project (at the beginning of the day) and takes 10 days to accomplish. The base equation then tells us when this task can finish:

$$\begin{aligned} FIN &= ST + DU - 1 \\ &= 35 + 10 - 1 \\ &= 44 \end{aligned}$$

The 10-day task starting on the 35th day would not finish on the 45th day, as we might have expected. The start-finish date convention reminds us that both the start date and the finish date are workdays, so the 10-day task actually finishes on the 44th day. We can check this easily by counting it out: 35-day 1, 36-day 2, 37-day 3, 38-day 4, 39-day 5, 40-day 6, 41-day 7, 42-day 8, 43-day 9, and 44-day 10.

In a second example using one of the equation variants, a task that will take 35 days must finish on the 139th day of a project (at the end of the day). When should the task start?

$$\begin{aligned} ST &= FIN - DU + 1 \\ &= 139 - 35 + 1 \\ &= 105 \end{aligned}$$

Again, the start-finish date convention reminds us that both the start date and the finish date are workdays, so we must add one workday to find a start date of the 105th day (at the beginning of that day).

In a third example, a task that finishes on the 3,695th day of a project began on the 3,636th day. What is its duration?

$$\begin{aligned} DU &= FIN - ST + 1 \\ &= 3{,}695 - 3{,}636 + 1 \\ &= 60 \end{aligned}$$

The task has a span of 60 days.

As we get into the modeling processes, we will begin to work with a series of equations, all of which stem from these date math principles.

COUNTING CALENDARS

The start and finish dates of tasks are generally not measured in numerical time units, however, but as standard calendar dates (specified by a day, a month, and a year). We will need to translate calendar format data into numeric format data in order to perform the date math. Then, if the answer is a start or finish date, we must translate the numeric value back into standard date format.

We can accomplish this through *calendar counting*. We first define a calendar period that encompasses the entire span of the project. It should begin several months prior to the project's initiation date and run several months past the scheduled project completion date. This ensures that the period includes any date that we could possibly need.

With this period defined, we begin by assigning the first calendar work unit (in this case workdays) the number 1. In Figure 3-1, this falls on the 11th of the first month. We assign the second work unit (the 12th) the number 2, the third work unit (the 13th) the number 3, and so on, through the entire calendar. We skip all weekends and holidays not intended to be workdays. In Figure 3-1, the 16th, 17th, 23rd, 24th, 30th, 1st, and 7th are weekends, and the 22nd is a holiday. When we go from the last workday of one month (the 29th) to the first workday of the next month (the 2nd), we advance the calendar count by one unit (from 14 to 15). This counting process may cover several years, amassing work unit totals of four, and even five figures.

Figure 3-1

Month #1

S	M	T	W	R	F	S
					1	2
3	4	5	6	7	8	9
10	1 11	2 12	3 13	4 14	5 15	16
17	•6• 18	7 19	8 20	9 21	22	23
24	10 25	11 26	12 27	13 28	14 29	30

Month #2

S	M	T	W	R	F	S
1	•15• 2	16 3	17 4	18 5	19 6	7
8	20 9	21 10	22 11	23 12	24 13	14
15	25 16	26 17	27 18	28 19	29 20	21
22	30 23	31 24	32 25	33 26	34 27	28
29	35 30	36 31				

To translate a calendar date into a numeric value, we simply look up the proper date on the calendar and determine its equivalent numeric value. We can then enter this value into the appropriate algebraic equations. Likewise, we can translate the results of date math back into date values by reversing the process.

For example, let us calculate the finish date of a task that starts on the 18th of the first month and takes 10 days to accomplish. Referring to Figure 3-1, the 18th of the month translates to the 6th workday. Adding this value to the task's duration in the appropriate equation, gives a finish date of:

$$\begin{aligned} FIN &= ST + DU - 1 \\ &= 6 + 10 - 1 \\ &= 15 \end{aligned}$$

The 15th workday translates back to the 2nd of the next month, the finish date of the task. Similarly, all date problems follow this process:

1. Translate all dates to the equivalent numeric value.
2. Enter the numeric values into the appropriate equation, and solve the problem mathematically.
3. Translate all numeric values representing dates back into calendar dates.

Generally, more than one calendar applies to any project. For example, a subcontractor may not consider the 22nd of the month a holiday, intending to work on that day. Calculations involving any activity performed by this subcontractor must use a calendar that counts the 22nd as a workday. Starting from the same base date, this subcontractor's calendar would count the 22nd as the 10th workday, the 25th as the 11th workday, and so on. Figure 3-2 shows the completed calendar including the 22nd of the first month as a workday.

Figure 3-2

Month #1

S	M	T	W	R	F	S
					1	2
3	4	5	6	7	8	9
10	1 11	2 12	3 13	4 14	5 15	16
17	•6• 18	7 19	8 20	9 21	10 22	23
24	11 25	12 26	13 27	14 28	•15• •29•	30

Month # 2

S	M	T	W	R	F	S
1	16 2	17 3	18 4	19 5	20 6	7
8	21 9	22 10	23 11	24 12	25 13	14
15	26 16	27 17	28 18	29 19	30 20	21
22	31 23	32 24	33 25	34 26	35 27	28
29	36 30	37 31				

If this subcontractor were to accomplish the same task from the previous example, the solution would yield slightly different results. The start date (the 18th) would still equate to the 6th workday and the equation would still yield the 15th workday as the task's finish. When we translate this value back to a date using the second calendar, however, the task finishes on the 29th of the month. This is because the subcontractor worked the 22nd of the first month. Our counted calendar takes this into account such that we determine the correct answer.

Normally we must use a six-day week calendar that includes Saturdays as workdays to account for Saturday overtime on selected activities and paths. Typically, this type of calendar ignores Saturdays that immediately follow holidays; these are still considered nonworkdays. If Friday the 22nd is a holiday, then Saturday the 23rd is also a nonworkday. We accommodate this by simply skipping any day we do not wish to include as a workday in the count. Figure 3-3 shows the counted six-day week calendar. In this case, we also included the 22nd and 23rd of the first month as workdays.

Figure 3-3

Month #1

S	M	T	W	R	F	S
					1	2
3	4	5	6	7	8	9
10	1 11	2 12	3 13	4 14	5 15	6 16
17	•7• 18	8 19	9 20	10 21	11 22	12 23
24	13 25	14 26	15 27	•16• 28	17 29	18 30

Month #2

S	M	T	W	R	F	S
1	19 2	20 3	21 4	22 5	23 6	24 7
8	25 9	26 10	27 11	28 12	29 13	30 14
15	31 16	32 17	33 18	34 19	35 20	36 21
22	37 23	38 24	39 25	40 26	41 27	42 28
29	43 30	44 31				

This calendar, would give yet another answer for our simple example. The task begins on the the 18th of the month, the 7th workday in this calendar. Using the principle equation, the result is:

$$
\begin{aligned}
FIN &= ST + DU - 1 \\
&= 7 + 10 - 1 \\
&= 16
\end{aligned}
$$

Translating day 16 back to a calendar date, we find the task will finish on Thursday the 28th of the month, 10 working days from the 18th of the month, including that day.

The answers in these examples varied because of the differences in the workday patterns of each of the calendars. As long as spans are measured in estimated work units and we use the appropriately counted calendar, the date math process will account for all variations in work patterns. This valuable process provides the mathematical basis for the modeling processes. Our basic understanding of counted calendars and date math are integral to the project modeling process.

Figures 3-4 and 3-5 are two calendars, both defined over the span of a year. However, each defines different workday patterns. The calendar in Figure 3-4, has a five-day week pattern (skipping Saturdays and Sundays) with eight additional holidays. The calendar in Figure 3-5 has a six-day week pattern, but it shows the same eight holidays. In addition, it excludes the Saturdays immediately following the holidays as workdays (January 2, April 2, November 26, and December 24). These two calendars will be used in the solution of all subsequent problems.

Figure 3-4. Calendar: 5-Day Week, with 8 Holidays

MO	S	M	T	W	TH	F	S
JANUARY						1	2
	3	1 4	2 5	3 6	4 7	5 8	9
	10	6 11	7 12	8 13	9 14	10 15	16
	17	11 18	12 19	13 20	14 21	15 22	23
	24	16 25	17 26	18 27	19 28	20 29	30
	31						

MO	S	M	T	W	TH	F	S
FEBRUARY		21 1	22 2	23 3	24 4	25 5	6
	7	26 8	27 9	28 10	29 11	30 12	13
	14	31 15	32 16	33 17	34 18	35 19	20
	21	36 22	37 23	38 24	39 25	40 26	27
	28	41 29					

MO	S	M	T	W	TH	F	S
MARCH			42 1	43 2	44 3	45 4	5
	6	46 7	47 8	48 9	49 10	50 11	12
	13	51 14	52 15	53 16	54 17	55 18	19
	20	56 21	57 22	58 23	59 24	60 25	26
	27	61 28	62 29	63 30	64 31		

MO	S	M	T	W	TH	F	S
APRIL						1	2
	3	65 4	66 5	67 6	68 7	69 8	9
	10	70 11	71 12	72 13	73 14	74 15	16
	17	75 18	76 19	77 20	78 21	79 22	23
	24	80 25	81 26	82 27	83 28	84 29	30

MO	S	M	T	W	TH	F	S
MAY	1	85 2	86 3	87 4	88 5	89 6	7
	8	90 9	91 10	92 11	93 12	94 13	14
	15	95 16	96 17	97 18	98 19	99 20	21
	22	100 23	101 24	102 25	103 26	104 27	28
	29	30	105 31				

MO	S	M	T	W	TH	F	S
JUNE				106 1	107 2	108 3	4
	5	109 6	110 7	111 8	112 9	113 10	11
	12	114 13	115 14	116 15	117 16	118 17	18
	19	119 20	120 21	121 22	122 23	123 24	25
	26	124 27	125 28	126 29	127 30		

MO	S	M	T	W	TH	F	S
JULY						128 1	2
	3	4	129 5	130 6	131 7	132 8	9
	10	133 11	134 12	135 13	136 14	137 15	16
	17	138 18	139 19	140 20	141 21	142 22	23
	24	143 25	144 26	145 27	146 28	147 29	30
	31						

MO	S	M	T	W	TH	F	S
AUGUST		148 1	149 2	150 3	151 4	152 5	6
	7	153 8	154 9	155 10	156 11	157 12	13
	14	158 15	159 16	160 17	161 18	162 19	20
	21	163 22	164 23	165 24	166 25	167 26	27
	28	168 29	169 30	170 31			

MO	S	M	T	W	TH	F	S
SEPTEMBER					171 1	172 2	3
	4	5	173 6	174 7	175 8	176 9	10
	11	177 12	178 13	179 14	180 15	181 16	17
	18	182 19	183 20	184 21	185 22	186 23	24
	25	187 26	188 27	189 28	190 29	191 30	

MO	S	M	T	W	TH	F	S
OCTOBER							1
	2	192 3	193 4	194 5	195 6	196 7	8
	9	197 10	198 11	199 12	200 13	201 14	15
	16	202 17	203 18	204 19	205 20	206 21	22
	23	207 24	208 25	209 26	210 27	211 28	29
	30	212 31					

MO	S	M	T	W	TH	F	S
NOVEMBER			213 1	214 2	215 3	216 4	5
	6	217 7	218 8	219 9	220 10	221 11	12
	13	222 14	223 15	224 16	225 17	226 18	19
	20	227 21	228 22	229 23	24	25	26
	27	230 28	231 29	232 30			

MO	S	M	T	W	TH	F	S
DECEMBER					233 1	234 2	3
	4	235 5	236 6	237 7	238 8	239 9	10
	11	240 12	241 13	242 14	243 15	244 16	17
	18	245 19	246 20	247 21	248 22	249 23	24
	25	26	250 27	251 28	252 29	253 30	31

Figure 3-5. Calendar: 6-Day Week, with Holidays

MO	S	M	T	W	TH	F	S
JANUARY						1	2
	3	1 4	2 5	3 6	4 7	5 8	6 9
	10	7 11	8 12	9 13	10 14	11 15	12 16
	17	13 18	14 19	15 20	16 21	17 22	18 23
	24	19 25	20 26	21 27	22 28	23 29	24 30
	31						

MO	S	M	T	W	TH	F	S
FEBRUARY		25 1	26 2	27 3	28 4	29 5	30 6
	7	31 8	32 9	33 10	34 11	35 12	36 13
	14	37 15	38 16	39 17	40 18	41 19	42 20
	21	43 22	44 23	45 24	46 25	47 26	48 27
	28	49 29					

MO	S	M	T	W	TH	F	S
MARCH			50 1	51 2	52 3	53 4	54 5
	6	55 7	56 8	57 9	58 10	59 11	60 12
	13	61 14	62 15	63 16	64 17	65 18	66 19
	20	67 21	68 22	69 23	70 24	71 25	72 26
	27	73 28	74 29	75 30	76 31		

MO	S	M	T	W	TH	F	S
APRIL						1	2
	3	77 4	78 5	79 6	80 7	81 8	82 9
	10	83 11	84 12	85 13	86 14	87 15	88 16
	17	89 18	90 19	91 20	92 21	93 22	94 23
	24	95 25	96 26	97 27	98 28	99 29	100 30

MO	S	M	T	W	TH	F	S
MAY	1	101 2	102 3	103 4	104 5	105 6	106 7
	8	107 9	108 10	109 11	110 12	111 13	112 14
	15	113 16	114 17	115 18	116 19	117 20	118 21
	22	119 23	120 24	121 25	122 26	123 27	124 28
	29	30	125 31				

MO	S	M	T	W	TH	F	S
JUNE				126 1	127 2	128 3	129 4
	5	130 6	131 7	132 8	133 9	134 10	135 11
	12	136 13	137 14	138 15	139 16	140 17	141 18
	19	142 20	143 21	144 22	145 23	146 24	147 25
	26	148 27	149 28	150 29	151 30		

MO	S	M	T	W	TH	F	S
JULY						152 1	153 2
	3	4	154 5	155 6	156 7	157 8	158 9
	10	159 11	160 12	161 13	162 14	163 15	164 16
	17	165 18	166 19	167 20	168 21	169 22	170 23
	24	171 25	172 26	173 27	174 28	175 29	176 30
	31						

MO	S	M	T	W	TH	F	S
AUGUST		177 1	178 2	179 3	180 4	181 5	182 6
	7	183 8	184 9	185 10	186 11	187 12	188 13
	14	189 15	190 16	191 17	192 18	193 19	194 20
	21	195 22	196 23	197 24	198 25	199 26	200 27
	28	201 29	202 30	203 31			

MO	S	M	T	W	TH	F	S
SEPTEMBER					204 1	205 2	206 3
	4	5	207 6	208 7	209 8	210 9	211 10
	11	212 12	213 13	214 14	215 15	216 16	217 17
	18	218 19	219 20	220 21	221 22	222 23	223 24
	25	224 26	225 27	226 28	227 29	228 30	

MO	S	M	T	W	TH	F	S
OCTOBER							229 1
	2	230 3	231 4	232 5	233 6	234 7	235 8
	9	236 10	237 11	238 12	239 13	240 14	241 15
	16	242 17	243 18	244 19	245 20	246 21	247 22
	23	248 24	249 25	250 26	251 27	252 28	253 29
	30	254 31					

MO	S	M	T	W	TH	F	S
NOVEMBER			255 1	256 2	257 3	258 4	259 5
	6	260 7	261 8	262 9	263 10	264 11	265 12
	13	266 14	267 15	268 16	269 17	270 18	271 19
	20	272 21	273 22	274 23	24	25	26
	27	275 28	276 29	277 30			

MO	S	M	T	W	TH	F	S
DECEMBER					278 1	279 2	280 3
	4	281 5	282 6	283 7	284 8	285 9	286 10
	11	287 12	288 13	289 14	290 15	291 16	292 17
	18	293 19	294 20	295 21	296 22	297 23	24
	25	26	298 27	299 28	300 29	301 30	302 31

Many other combinations could have been defined: seven-day week with no holidays, seven-day week with holidays, five-day week with no holidays, six-day week with no holidays, five-day week with Thursday and Friday as the weekend, etc.

Working with these two calendars offers the experience necessary to handle any number or combination of other calendars. Some examples complete our study of date math and calendars.

EXAMPLE 1: How many workdays fall between May 2 and November 4 using a five-day week?

$$
\begin{aligned}
DU &= FIN - ST + 1 \\
&= 216 - 85 + 1 \\
&= 132
\end{aligned}
$$

EXAMPLE 2: How many workdays fall between May 2 and November 4 using a six-day week?

$$
\begin{aligned}
DU &= FIN - ST + 1 \\
&= 258 - 101 + 1 \\
&= 158
\end{aligned}
$$

Note that the difference in workdays between Examples 1 and 2 equals the number of Saturdays between the two dates. Nonworkdays in one are workdays in the other.

EXAMPLE 3: When does a task finish that starts on June 8 and has a duration of 120 days using a five-day week?

$$
\begin{aligned}
FIN &= ST + DU - 1 \\
&= 111 + 120 - 1 \\
&= 230 \text{ or November 28}
\end{aligned}
$$

EXAMPLE 4: When should a task start that ends on September 15 and takes 90 days assuming a six-day week?

$$
\begin{aligned}
ST &= FIN - DU + 1 \\
&= 215 - 90 + 1 \\
&= 126 \text{ or June 1}
\end{aligned}
$$

SUMMARY

Date math is an essential element of project modeling. Its fundamental equation states:

$$FIN = ST + DU - 1$$

The start-finish date convention requires that we subtract one time unit where duration includes both dates.

We must translate dates to numerical values in order to properly perform date math. We do this by counting calendars. A counted calendar in a calendar that:

1. Covers the entire period of the project.
2. Assigns a numerical value to each work unit.
3. Determines these values by counting from the first work unit to the last.

In a typical date math problem, we:

1. Translate all dates into numeric equivalents using the appropriate calendar.
2. Solve the problem mathematically using the proper equation.
3. Translate any resultant integer values that represent dates back into the appropriate calendar dates.

CHAPTER 4

The Forward Pass

The actual modeling process begins with a subprocess called the *forward pass*. The forward pass is a process: part rules, part mathematics, all applied with considerable discipline. It is also a subprocess within the overall project modeling process. Alone it can yield useful management information about a project. It provides its most significant benefits, however, as a part of the greater project modeling process.

Unlike many scientific modeling processes, the forward pass involves only simple math, including a great deal of addition and a little subtraction. The process itself cannot be considered simple, though. The combinations and variations of its many facets create considerable complexity. It is complicated even further by the sheer number of activities, events, and constraints to which it is applied in even the simplest of projects.

As the word *forward* implies, this process considers the progression of time from an earlier point to a later point. Conventionally, we portray time graphically as flowing from left to right. The forward pass follows this convention, flowing from left to right with respect to time. This gives us the first rule of the forward pass:

> The forward pass process begins at the starts of a project and proceeds chronologically from left to right through a project model.

The term *starts of the project* will be discussed later in this chapter. For now, let us say they are basically tasks with no predecessors.

The forward pass process provides the schedule for each node in the model as a function of its predecessor or precursor relationships. More specifically, we determine when a task can start based on the predecessors of its start and we determine when a task can finish based on the predecessors of its finish.

The predecessors determine the earliest dates by which tasks can happen. These dates could appropriately be called the *can dates* of each node, as they specify when a task can start and can finish. They are more commonly called the *early dates*: the Early Start (ES) and Early Finish (EF) dates. The forward pass process calculates these two dates for every node in a model.

FORWARD PASS EQUATIONS

Duration and Finish-to-Start

The second rule of the forward pass takes the form of a set of five equations to provide the specific dates generated by particular types of predecessors. Sample problems best illustrate the applications of these equations. Figure 4-1 illustrates the first problem, a simple sequential series of four tasks: ALPHA, DELTA, GAMMA, and OMEGA. Between these four activities fall three finish-to-start constraints: ALPHA-DELTA, DELTA-GAMMA, and GAMMA-OMEGA. The nodes specify the duration of each task. The model assumes five-day weeks with eight standard holidays. The project should start on February 1.

Figure 4-1

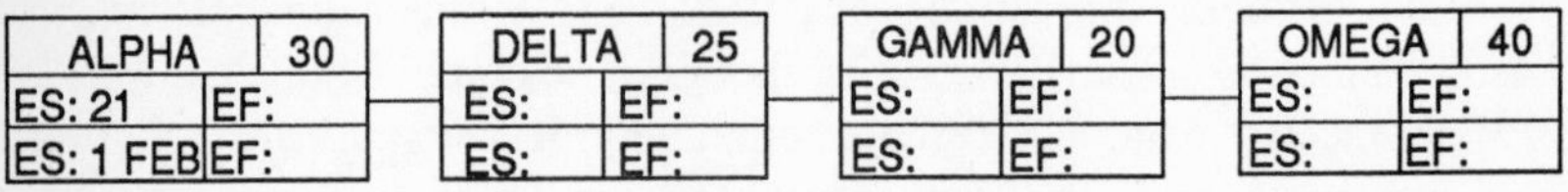

We must begin at a start, a task with no predecessors, at the chronological left of the model. In this case, we start with task ALPHA, and then proceed to the right through the entire model. To accommodate date math, we translate the start date (February 1) into its numeric value on the appropriate calendar. February 1 is the 21st workday, so the Early Start of ALPHA is the 21st day. With ALPHA's duration of 30 days, it will finish on:

$$\text{FIN} = \text{ST} + \text{DU} - 1 = 21 + 30 - 1 = \text{workday } 50 \text{ (March 11)}$$

This leads to the first specific equation of the forward pass process, which calculates the Early Finish of an activity based on its Early Start and duration. This equation is a direct derivative of our base date math equation:

$$\text{DU Equation: EF of Node} = \text{ES of Node} + \text{DU} - 1$$

A corollary of the first rule arises from the left-to-right flow of the forward pass. This is one of the more commonly forgotten rules.

> Always calculate the early start of a task prior to its early finish, or the start of a task is always a predecessor of its finish.

To continue with our problem, the calendar shows that the 50th workday occurs on March 11. Task ALPHA starts on February 1 and takes 30 days to complete. We can expect to finish it on March 11. Stated another way, the Early Start (ES) of ALPHA is February 1 and its Early Finish (EF) is March 11.

The finish of ALPHA enables its successor, task DELTA, to start. ALPHA finishes on the 50th workday, at the *end* of that day. If DELTA starts immediately after, it starts the next morning, on the 51st day. The calendar tells us that ALPHA finishes on a Friday, so DELTA cannot start on the next calendar day (Saturday, March 12), but on the next workday, Monday, March 14. A counted calendar keeps this straight.

If for some reason, the finish-to-start constraint between ALPHA and DELTA had required a lag of one day, then the Early Start of DELTA would be delayed by one workday to March 15. A two-day lag would result in a March 16 Early Start for DELTA. The start-finish date convention, the Early Finish date of the predecessor, and the constraint lag value all affect the Early Start of a successor in a finish-to-start constraint. Combining the effects of each provides the forward pass equation for finish-to-start constraints:

$$\text{FS Constraint: (ES of SN)} = \text{(EF of PN)} + \text{lag} + 1$$

This equation states that, with a finish-to-start constraint, the Early Start (ES) of the successor (SN) equals the Early Finish (EF) of the predecessor (PN) plus the lag value (if any) plus 1. Recall that the lag measures any delay in the constraint, and we add 1 to accommodate the start-finish convention. In our example, this equation would yield:

$$\begin{aligned}\text{(ES of DELTA)} &= \text{(EF of ALPHA)} + \text{lag} + 1\\ &= 50 + 0 + 1 = 51\end{aligned}$$

The calendar gives a date for the 51st workday of March 14, which matches our previous logic. This is the Early Start of DELTA. Note that the combination of the counted calendar and the equation takes all necessary factors (e.g., holidays, weekends, durations, lags, constraint types, start-finish convention, etc.) into account to calculate the correct date.

To further explore this equation, let us add to our example a lag of five days on the ALPHA-DELTA finish-to-start constraint, the equation would then yield:

$$\text{ES of DELTA} = \text{EF of ALPHA} + \text{lag} + 1$$
$$= 50 + 5 + 1 = \text{workday } 56$$

The 56th workday, March 21, would be the Early Start of DELTA. This math represents completing task ALPHA, waiting five workdays, and then starting task DELTA.

Let us continue with the original problem, where DELTA starts on the 51st day (March 14). Since this task has a duration of 25 days, the EF duration equation leads us to expect it to finish on:

$$\text{EF of DELTA} = \text{ES of DELTA} + \text{DU} - 1$$
$$= 51 + 25 - 1 = \text{workday } 75$$

The 75th workday, April 18, becomes the Early Finish (EF) of DELTA, after which GAMMA can start. Using the finish-to-start equation, we find:

$$\text{ES of GAMMA} = \text{EF of DELTA} + \text{lag} + 1$$
$$= 75 + 0 + 1 = \text{workday } 76$$

The 76th workday, April 19, is then the Early Start (ES) of GAMMA. This task takes 20 days to complete, so it will finish on:

$$\text{EF of GAMMA} = \text{ES of GAMMA} + \text{DU} - 1$$
$$= 76 + 20 - 1 = \text{workday } 95$$

The 95th workday is May 16. This is the Early Finish (EF) of GAMMA, after which OMEGA can start, per the finish-to-start relationship:

$$\text{ES of OMEGA} = \text{EF of GAMMA} + \text{lag} + 1$$
$$= 95 + 0 + 1 = \text{workday } 96$$

The 96th workday, May 17, is the Early Start (ES) of OMEGA. OMEGA takes 40 days to finish, so it will finish on:

$$\text{EF of OMEGA} = \text{ES of OMEGA} + \text{DU} - 1$$
$$= 96 + 40 - 1 = \text{workday } 135$$

The 135th workday, July 13, is the Early Finish (EF) of OMEGA, as well as the expected finish of the project. We have now calculated when each activity can start (the Early Start of each) and finish (the Early Finish of each) based on the given constraints, activity spans, and a known project start date. Figure 4-2 shows the completed forward pass of this model as a graphic plot.

Figure 4-2

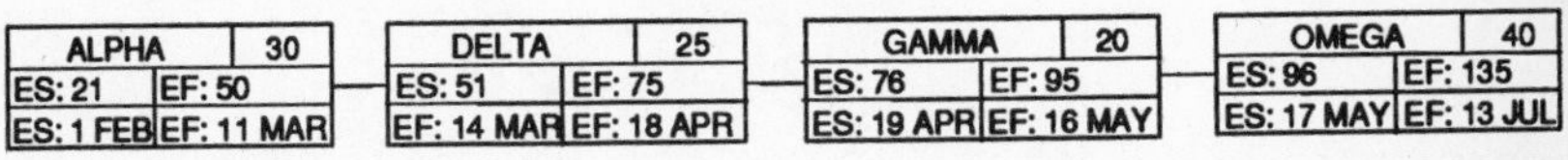

The total project, in this case a single path, takes 115 days to complete (the sum of the durations: 30 + 25 + 20 + 40 = 115). If we started on February 1, we could expect to finish on:

$$\begin{aligned} \text{EF of path} &= \text{ES of path} + \text{DU} - 1 \\ &= 21 + 115 - 1 = \text{workday } 135 \end{aligned}$$

Day 135 translates on the calendar to July 13, the same date we found with the forward pass. This basically verifies that the whole equals the sum of its parts. If for some reason the start of the project were delayed to February 15 (day 31), then we could expect the project to finish on:

$$\begin{aligned} \text{EF of path} &= \text{ES of path} + \text{DU} - 1 \\ &= 31 + 115 - 1 = \text{workday } 145 \end{aligned}$$

The project end date (its Early Finish) would slip to July 27 (workday 145). The Early Start and Early Finish for all other tasks in the project would slip accordingly, as illustrated in Figure 4-3.

Figure 4-3

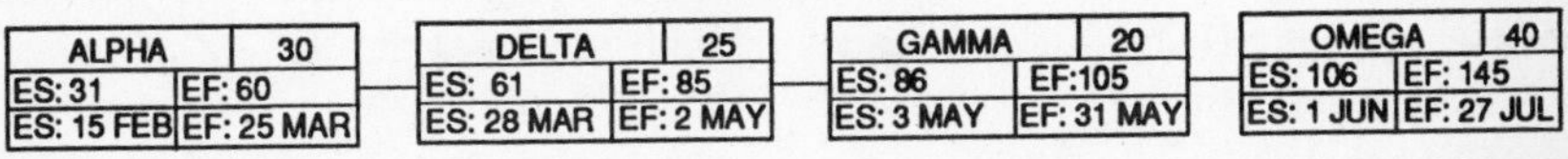

We now examine equations for the three other constraint types: start-to-start (SS), finish-to-finish (FF), and start-to-finish (SF).

Figure 4-4

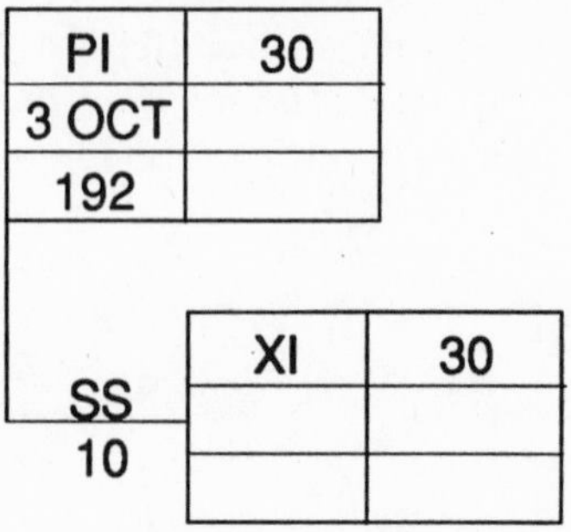

Start-to-Start Forward Pass Equation

Figure 4-4 illustrates a start-to-start constraint where task XI can start 10 days after task PI starts. The forward pass equation for the start-to-start (SS) constraint is:

SS Constraint: ES of SN = ES of PN + lag

The equation mathematically expresses the idea that the Early Start (ES) of the successor in a start-to-start constraint depends on the Early Start (ES) of the predecessor and the lag of the constraint. Since both dates in the equation are start dates, we need not add or subtract a unit for the start-finish convention.

To calculate the early dates in the example, we must first determine the Early Start of PI, then calculate the early dates of XI. The example assumes the Early Start of PI as October 3, the 192nd day on the five-day week calendar. The above equation gives an Early Start for XI of:

ES of XI = ES of PI + lag = 192 + 10 = day 202

We can expect activity XI to start on October 17 (day 202), 10 workdays after the start of PI. The Early Finishes of both activities depend on their Early Starts and durations:

EF of PI = ES of PI + DU − 1
= 192 + 30 − 1 = day 221 (November 11)

EF of XI = ES of XI + DU − 1
= 202 + 30 − 1 = day 231 (November 29)

Figure 4-5

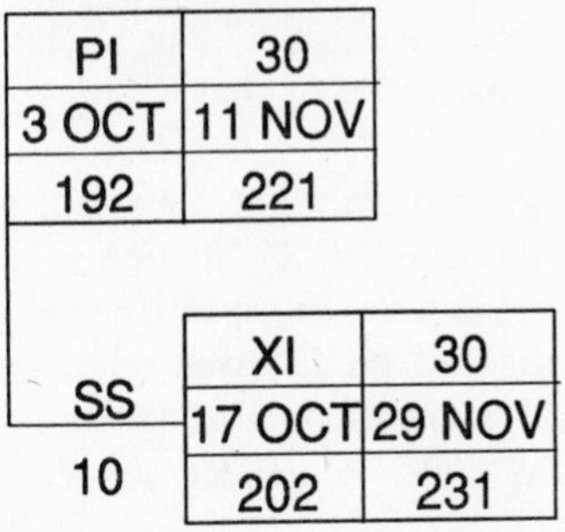

The completed forward pass of this simple model looks like Figure 4-5. We find two paths in this problem: the first goes completely through activity PI, and the second goes through the first 10 days of PI and the entire span of XI. Considering just the activities shown, the first path is 30 days long, and the second is 40 days long (10 + 30).

Finish-to-Finish Forward Pass Equation

A similar equation covers finish-to-finish (FF) relationships in the forward pass process. In this type of constraint, we need to find the effect on a successor's Early Finish of its predecessor's Early Finish and the lag on the constraint. Since it deals with two finish dates, this equation does not require a start-finish convention adjustment.

FF Constraint: EF of SN = EF of PN + lag

In the following example of a finish-to-finish constraint, task TAU cannot finish until 15 days after task RHO finishes. This relationship is illustrated in Figure 4-6.

Figure 4-6

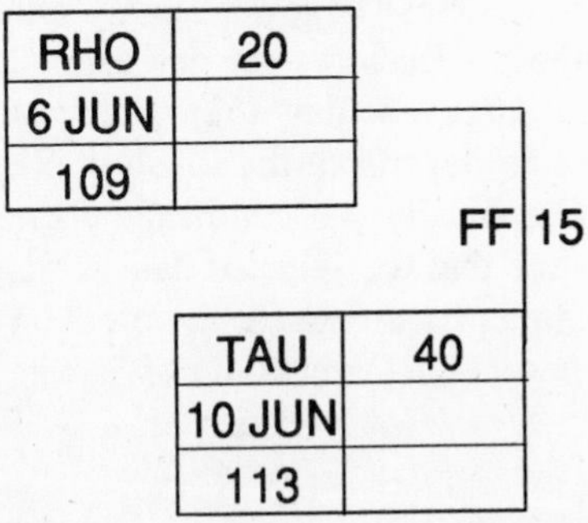

The Early Finish of RHO is not, however, the only predecessor affecting the Early Finish of TAU. This task cannot finish until 40 days after it starts as well. Two predecessors drive the Early Finish of TAU, and since a task can have only one Early Finish date, one of the predecessors will have to take priority.

When two or more predecessors directly affect an early date, we have a *multiple predecessor relationship* (MPR). In our example, the finish of TAU depends on both its own start and the finish of RHO. Therefore, we must choose the earliest date that satisfies all predecessor relationships. This cannot occur until the latest predecessor is satisfied. This leads us to the third rule of the forward pass, the rule of the multiple predecessor relationship:

> If a node end has more than one predecessor, calculate each relationship independently, and then select the latest of the early dates.

Two examples will help demonstrate this rule. First, let RHO start on June 6 (workday 109) and TAU on June 10 (workday 113). The Early Finish dates of RHO and TAU, respectively, would be:

EF of RHO = ES of RHO + DU − 1 = 109 + 20 − 1 = day 128

This is equivalent to July 1. The Early Finish of TAU will then be a choice of:

EF of TAU = ES of TAU + DU − 1 = 113 + 40 − 1 = day 152

or

EF of TAU = EF of RHO + lag = 128 + 15 = day 143

To find the Early Finish of TAU we compare the results of the two alternative equations, August 5 (workday 152) and July 25 (workday 143). We must select the latest of the early dates (largest workday number) because only this date will satisfy all of the predecessors. In our example TAU would have an Early Finish (EF) date of August 5 (day 152), which is both 40 days from the start of TAU and at least 15 days from the finish of RHO, satisfying both predecessor relationships. Graphically the completed forward pass of this model looks like Figure 4-7. (Note that the selected date is flagged with an asterisk.)

For the second example, let us change the problem slightly. Let task TAU start on May 16 (workday 95), leaving RHO as it was before. This new start date for TAU would give a new Early Finish of:

Figure 4-7

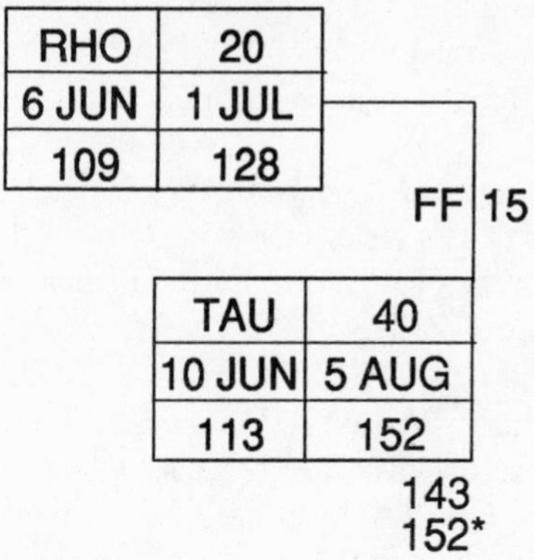

EF of TAU = ES of TAU + DU − 1 = 95 + 40 − 1 = day 134

This is equivalent to July 12. The alternative equation for the multiple predecessor relationship remains:

EF of TAU = EF of RHO + lag = 128 + 15 = day 143

As before, this is equivalent to July 25. Comparing these two values, we now find that day 143 is the latest early date; it is both 40 days from the start of TAU and 15 days from the finish of RHO. Figure 4-8 diagrams the completed forward pass.

Figure 4-8

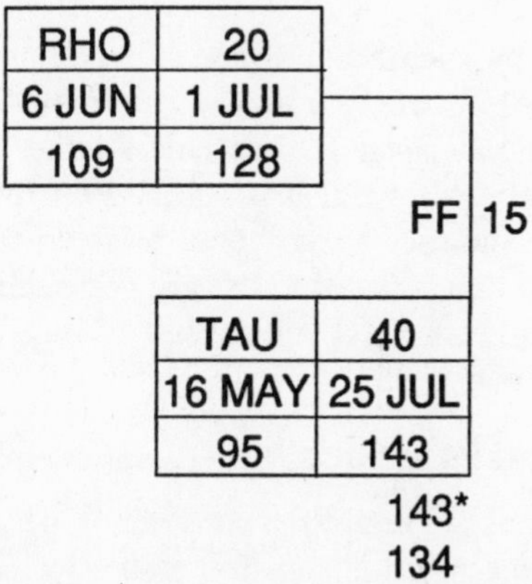

Now the Early Finish of RHO and the lag drive TAU's Early Finish date. It is important to understand what this means. The difference between TAU's Early Start and its Early Finish (143 – 95 + 1 = 49 days) no longer equals its duration (40 days). In effect, the finish-to-finish constraint breaks TAU into two parts (see Figure 4-9).

Task TAU2 is the last 15 days of TAU (the lag of the constraint). Task TAU1 includes the rest of the task (DU – lag: 40 – 15 = 25). RHO begins on day 109 and has a duration of 20 days, so it finishes on day 128 (109 + 20 – 1) just as it did in Figure 4-8. Task TAU1 starts on day 95 and has a duration of 25 days, it finishes on day 119 (95 + 25 – 1). TAU2 cannot start until both of these predecessors finish. When TAU1 finishes, the other predecessor, RHO, remains in process. Therefore task TAU2 must wait nine days (128 – 119 = 9) for RHO to finish. Once task TAU2 is initiated on workday 129, it finishes 15 days later or workday 143, which is also the EF of the combined task TAU. As a single task, TAU, the finish of the task appears extended or stretched.

Figure 4-9

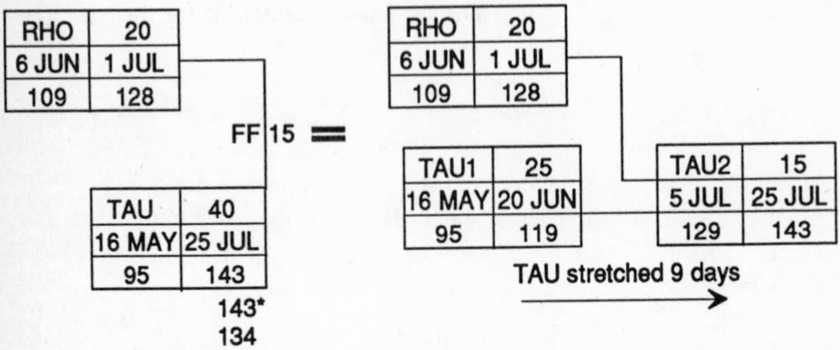

The task affected by such a constraint is called an *External Predecessor Stretched Activity* (EPSA) or simply a *stretched activity.* Such a task cannot start and finish in one continuous span of time unless either the start of the task is delayed or the FF predecessor's finish is speeded up (pulled to the left in the diagram) by nine days. The user must choose between these options.

This also results in two paths: the first goes through task TAU, the second through the entire span of task RHO and the last 15 days of TAU. The first path is 40 days long, the second is 35.

Start-to-Finish Forward Pass Equation

Although we characterized the last of the constraint types, the start-to-finish constraint, as impractical, we offer the equation associated with it:

$$\text{SF Constraint: EF of SN} = \text{ES of PN} + \text{lag} - 1$$

This states that the Early Finish of a successor equals the Early Start of the predecessor plus the lag (which is part predecessor and part successor). Since we are calculating a finish from a start, we must subtract one workday to satisfy the start-finish convention.

Figure 4-10

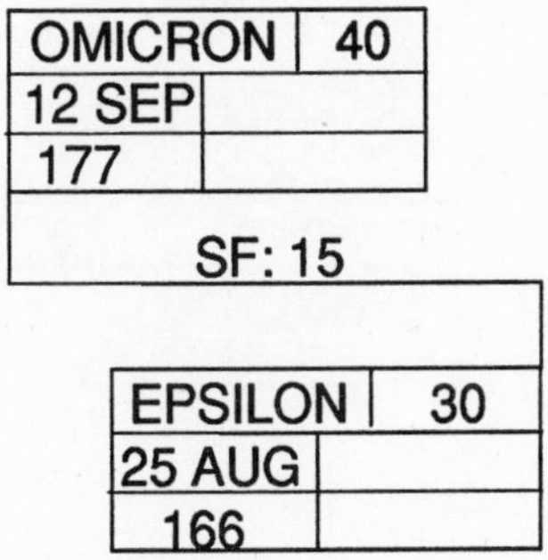

As an example, let us look at two activities, OMICRON and EPSILON, which are illustrated in Figure 4-10. OMICRON starts on September 12 (workday 177) and takes 40 days to accomplish. EPSILON starts on August 25 (workday 166) and takes 30 days to accomplish. A start-to-finish constraint with a 15-day lag binds OMICRON and EPSILON.

The Early Start of OMICRON, September 12, gives an Early Finish of:

$$\begin{aligned}\text{EF of OMICRON} &= \text{ES of OMICRON} + \text{DU} - 1 \\ &= 177 + 40 - 1 = \text{day } 216\end{aligned}$$

This is equivalent to November 4. The Early Start of EPSILON is August 25; its Early Finish depends on two predecessors: its own Early Start plus its duration (30 days) and the Early Start of OMICRON plus the lag on the constraint (15 days). This task's Early Finish date will be either:

$$\begin{aligned}\text{EF of EPSILON} &= \text{ES of EPSILON} + \text{DU} - 1 \\ &= 166 + 30 - 1 = \text{day } 195\end{aligned}$$

or

$$\begin{aligned}\text{EF of EPSILON} &= \text{ES of OMICRON} + \text{lag} - 1 \\ &= 177 + 15 - 1 = \text{day } 191\end{aligned}$$

The multiple predecessor rule now comes into play. The Early Finish of EPSILON would be the latest of the two dates, workday 195 (October 6). The finished forward pass in this situation looks like Figure 4-11.

Figure 4-11

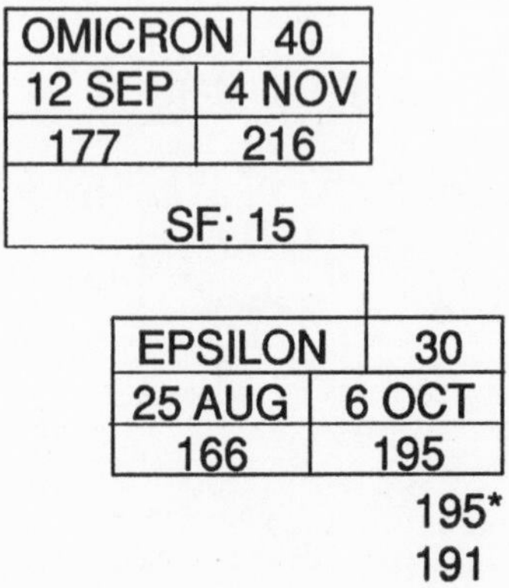

In this example, the Early Finish of EPSILON is driven by its Early Start date, but if the Early Start of OMICRON were five days later, or the lag were five days greater, or EPSILON were to start five days earlier, then the Early Finish of EPSILON would be driven by the Early Start of OMICRON. EPSILON would be an external predecessor stretched activity (EPSA).

Again, we introduce the start-to-finish equation for reference only. Remember that the lag value of this constraint type is a complex number as it involves a portion of both tasks, so it buries the interface of the two tasks in a vague reference. For this reason, modelers should avoid the start-to-finish constraint.

Predecessor Equation Summaries

This concludes our discussion of the normal predecessor types. The forward pass equations (that is, the second rule of the forward pass) can be summarized as follows:

DU Equation: (EF of Node) = (ES of Node) + DU – 1
FS Constraint: (ES of SN) = (EF of PN) + lag + 1
SS Constraint: (ES of SN) = (ES of PN) + lag
FF Constraint: (EF of SN) = (EF of PN) + lag
SF Constraint: (EF of SN) = (ES of PN) + lag – 1

MULTIPLE PREDECESSOR RELATIONSHIPS

Multiple predecessor relationships (MPRs) can complicate the forward pass through each of the constraint types as mentioned in the discussion of finish-to-finish and start-to-finish constraints. Apart from these examples, activities can have, and quite commonly do have, many predecessors of both start and finish. A task that has many predecessors can still have only one Early Start date and one Early Finish date, though. We must resolve these multiple predecessor relationships to determine an Early Start date and an Early Finish date for the successor of many predecessors.

This may appear much tougher than it really is. To accomplish it, we simply evaluate each predecessor relationship individually and then choose the earliest value that satisfies all of the constraints. The date that satisfies all of the predecessor relationships is the latest early date.

Since the Early Start of a task is always a predecessor of the Early Finish of that task, we must first calculate the task's Early Start date. Therefore, we need to separate the task's predecessors into two categories: the predecessors of its start and the predecessors of its finish. We then determine the task's Early Start by independently calculating the Early Start resulting from each predecessor and selecting the latest value. We determine the Early Finish by independently calculating the Early Finish resulting from each predecessor, including the task's Early Start plus its span, and selecting the latest value.

Figure 4-12 shows an activity HEA with two finish-to-start predecessors (SIN and ABU), two start-to-start predecessors (ANU and BEL), and two finish-to-finish predecessors (UTU and LER). From the data provided we need to determine the Early Start and Early Finish of HEA.

The predecessors of the start of HEA include:

1. The finish of SIN (FS)
2. The finish of ABU (FS)
3. The start of ANU (SS)
4. The start of BEL (SS)

These four predecessors give possible Early Start dates for HEA of:

ES of HEA = EF of SIN + lag + 1 = 190 + 0 + 1 = day 191

ES of HEA = EF of ABU + lag + 1 = 187 + 0 + 1 = day 188

ES of HEA = ES of ANU + lag = 180 + 10 = day 190

ES of HEA = ES of BEL + lag = 187 + 5 = day 192

Figure 4-12

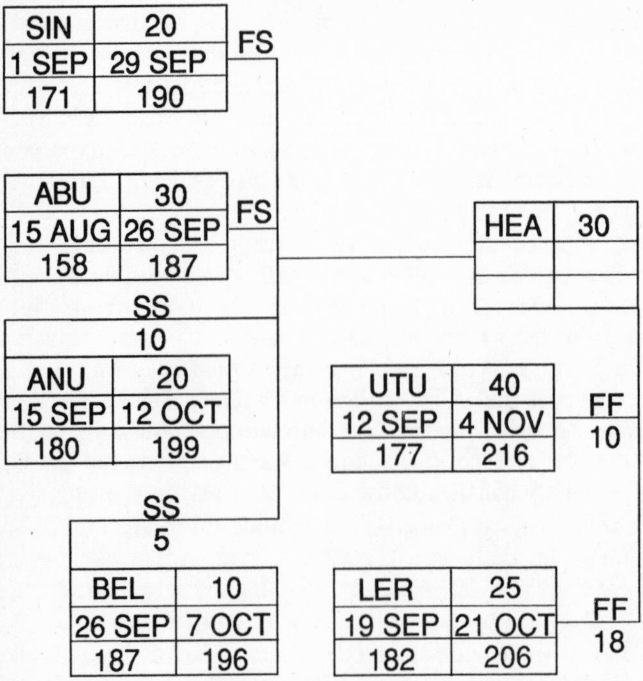

The latest of these four dates, day 192 or October 3, is the Early Start of HEA. This is the earliest date that satisfies *all* of the predecessor relationships. Once we know a task's Early Start, we can calculate its Early Finish. The predecessors of the finish of HEA include:

1. The start of HEA
2. The finish of UTU (FF)
3. The finish of LER (FF)

These three predecessors give independently computed Early Finish dates for HEA of:

EF of HEA = ES of HEA + DU − 1 = 192 + 30 − 1 = day 221

EF of HEA = EF of UTU + lag = 216 + 10 = day 226

EF of HEA = EF of LER + lag = 206 + 18 = day 224

The latest of these three dates, workday 226, November 18, is the Early Finish of HEA. This is the earliest date that satisfies *all* of the predecessor relationships. The finished forward pass of node HEA appears in Figure 4-13.

Figure 4-13

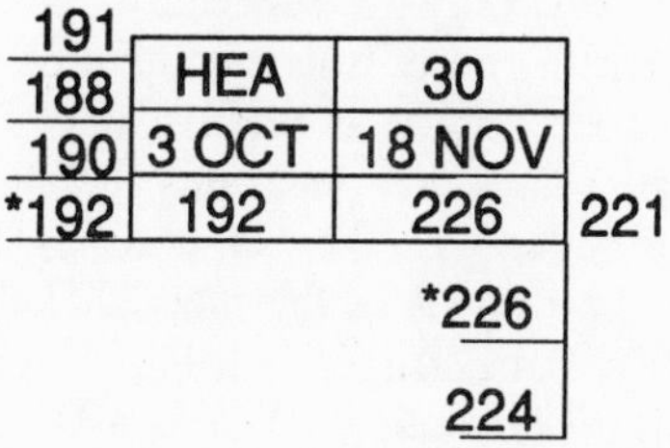

In this case, a finish-to-finish constraint drives the Early Finish date, which means that this task (HEA) is an EPSA or stretched activity. As modeled, task HEA would not start and finish in one continuous effort, unless:

1. The start of task HEA were delayed by at least five days, or
2. The finish of task UTU were advanced by at least five days and the finish of LER were advanced by at least three days. This is the amount of time each predecessor would have to be advanced to produce the same Early Finish date as the task's span.

We must choose whether to allow the task to remain stretched or to exercise either of these two options, and adjust the project and the model accordingly.

START NODES AND DATE PREDECESSORS

Start Nodes

So far, all our forward pass equations have dealt with numeric values only. In practice, we initiate the process from calendar dates, so we need a means of injecting dates into the model. The original example assumed that the ALPHA-OMEGA project started on February 1, but we did not explain how we introduced this date into the model. We need some means of associating a date with a start-up activity.

Since task ALPHA has no predecessor, the forward pass gives us no means to calculate its Early Start. This makes ALPHA a special node which we simply call a *start* or a *start node*. Since the forward pass cannot determine the Early Start of this node, we must do so outside the process. Prior to the forward pass process, we must identify all start nodes and then set appropriate Early Start dates for them. These dates establish the points of initiation of the forward pass.

We can easily see that node ALPHA is a start node in the ALPHA-OMEGA project, because it has no predecessors. The proper definition of a *start node*, however, is one for which the *forward pass cannot provide an Early Start date*. This condition means that only the start end of the node has no predecessors. It is possible for the finish end of a start node to have predecessors through FF constraints. The start node criteria is relative only to the start end. Let us examine the model in Figure 4-14 and determine the start nodes:

In this example, four nodes (A1, B2, C3, and F3) have no predecessors of their starts. These are the start nodes of the model. Nodes A1, B2, and C3 are rather obviously start nodes, but consider task F3. Although it is in the middle of the model and it has a predecessor (E2), it is a start node. The location of the node within the model has no bearing on whether it is a start node. Careful attention to its predecessor reveals that the finish-to-finish constraint (E2-F3) has no impact on the Early Start of F3; it only affects the node's Early Finish. Since the Early Finish of a node must always fall after its Early Start date, nothing in the forward pass process determines the Early Start of F3. This makes it a start node.

Figure 4-14

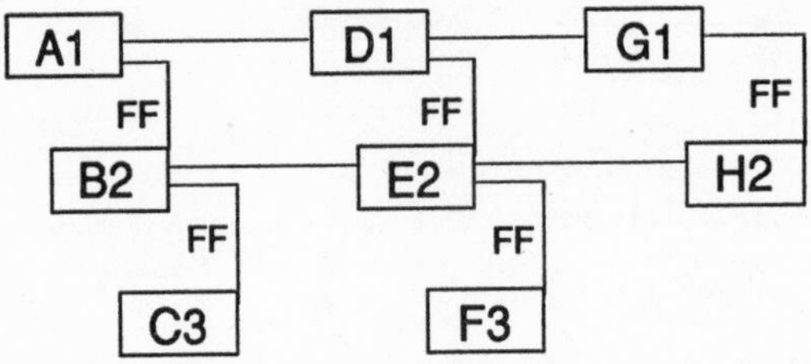

Let us examine each node in another project model, Figure 4-15, to determine which are the start nodes:

- NV is a start node; it has no predecessor of any kind.
- AZ is a start node; it has no predecessor of its start. The SF constraint CA-AZ establishes a predecessor of its finish.
- CA is a start node; it has no predecessor of any kind.
- OR is a start node; it has no predecessor of any kind.
- WA is a start node; it has no predecessor of its start, ID is a predecessor of its finish.

- TX is a start node; it has no predecessor of its start, OR and MT are predecessors of its finish.
- UT is a start node; it has no predecessor of its start, TX is a predecessor of its finish.
- NM is a start node; it has no predecessor of its start, UT and WA are predecessors of its finish. (NOTE, the final node in this project is a start node.)

The following nodes are not start nodes:

- WY is not a start node; NV is a predecessor of its start.
- ID is not a start node; CA is a predecessor of its start.
- MT is not a start node; OR and WY are predecessors of its start.
- CO is not a start node; ID is a predecessor of its start.

Figure 4-15

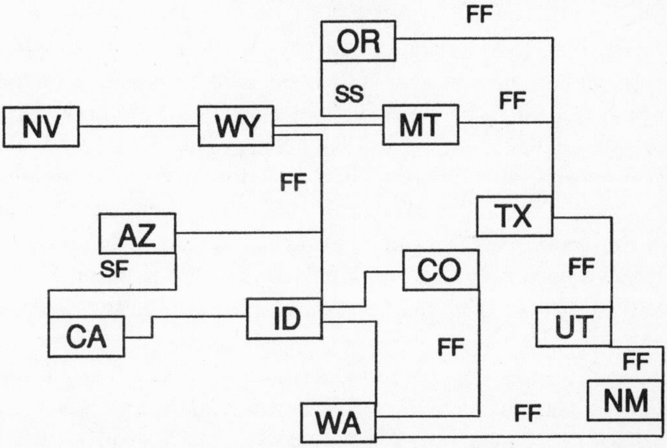

The examples in Figures 4-14 and 4-15 illustrate the process of identifying start nodes in a project model. We simply look at each node individually and determine whether it has a predecessor of its start. If it has such a predecessor, it is not a start node; if it does not have such a predecessor, it is a start node.

Setting Early Start Dates for Start Nodes. After identifying start nodes, we must determine their Early Start dates. We can do this by five techniques:

1. Apply a date predecessor to each start node

2. Set the Early Start date of start nodes equal to some common, predetermined date, like the project start date
3. Calculate the start nodes' Early Start dates from their Early Finish dates (the reverse of the forward pass process)
4. Set an Early Start date on each start node to establish a specified amount of time reserve (See Chapter 8)
5. Enter the date a task actually starts (See Chapter 9)

The modeler must choose among these techniques, and then supply the necessary information in the form of dates or numeric values.

Not all of these techniques are acceptable. As we examine them in detail, we will determine which are useful. The fourth technique requires an understanding of all the CPM processes, so we will defer its explanation until later (Chapter 8); it is, however, a valid method. Chapter 9 will explain the fifth technique, but it is enough at this point to state that it is also a valid method.

Start-No-Earlier-than Dates

A start-no-earlier-than (SNE) date, or start date predecessor, is simply an appropriate date a planner applies to a node. It is the date on which we intend to start the node or path. This date becomes another attribute of the node for use in the forward pass process to determine its Early Start date. We call this date a *start date predecessor* because it functions like a predecessor of the node's start. On a start node a start date predecessor is the only, and therefore the latest, Early Start date of the node. Every example so far has assumed the Early Start dates of start nodes. We need only add the initials *SNE* to the notation for designating date predecessors that we have used so far. The effort is in the determination of the date.

We can also specify start date predecessors for nodes that are internal to a model, that is, nonstart nodes. We do this to model delays on the starts of tasks caused by forces other than task interfaces. We cannot model these delays as constraints because constraints arise by definition only from task interfaces. Most, but not all, such external delays arise from some kind of resource problem, and we model them through the application of start-no-earlier-than (SNE) dates.

As an example, take our initial project model (ALPHA-OMEGA). Suppose that task GAMMA was scheduled to start on April 19, but, for some reason, we could not start it until May 2 (workday 85). (See Figure 4-16.) This delay does not arise from any task interface, so we cannot model it as a constraint.

Figure 4-16

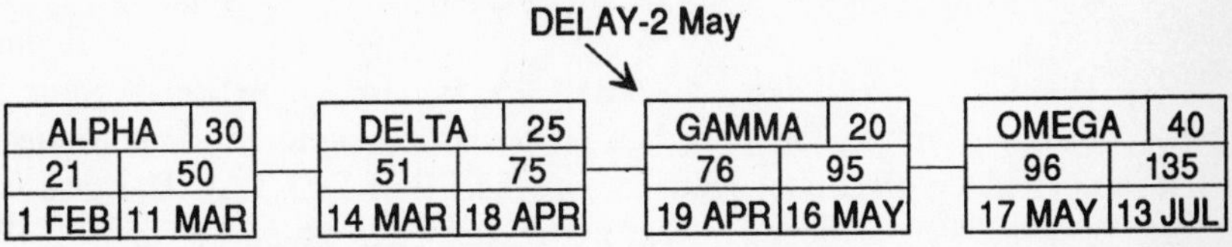

Modelers commonly make such a mistake and add a nine-day lag to the DELTA-GAMMA constraint. Unless something essential to these two tasks adds nine days of work, this is *invalid* modeling. The proper technique is to apply a start date predecessor or SNE date on GAMMA to represent the date of the delay, May 2 in our example. The SNE date acts as an additional predecessor of GAMMA's start. Thus, when the forward pass reaches the Early Start date calculation of GAMMA, we choose between:

ES of GAMMA = EF of DELTA + lag + 1 = 75 + 0 + 1 = day 76

or

ES of GAMMA = SNE date on GAMMA = day 85

The latest early date of these two choices, the 85th workday or May 2, becomes the Early Start of GAMMA. This new date will affect all downstream early date calculations, as shown in Figure 4-17. Notice that we attached SNE dates to both ALPHA and GAMMA.

Figure 4-17

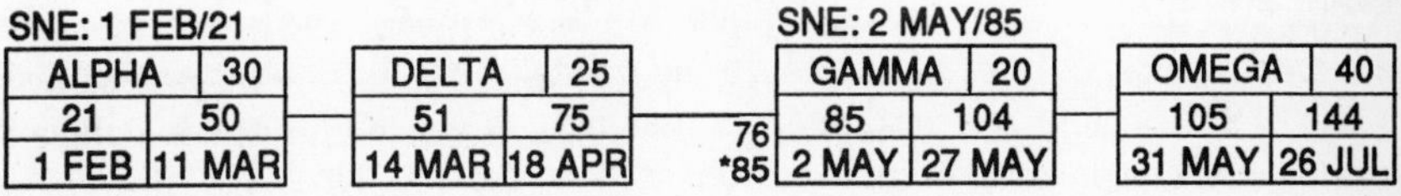

Many misinterpret SNE dates as a fixed dates. SNE dates will be the nodes' Early Starts only if they are the latest early dates. SNE dates function like any other predecessor. If, for example, the delay had moved the Early Start to April 11, or workday 70, the Early Start of GAMMA would be a choice of:

ES of GAMMA = EF of DELTA + lag + 1 = 75 + 0 + 1 = day 76

or

ES of GAMMA = SNE date on GAMMA = day 70

Now the latest early date, workday 76, becomes the Early Start of GAMMA. In this example, the start date predecessor is ignored, or more accurately it is overridden by a later predecessor (the EF of DELTA). The task can start later than the date predecessor, but not earlier (start-no-earlier-than: SNE). The SNE date does not automatically become the Early Start of the node. It will be the Early Start only if it is the latest date of all predecessors.

In a start node, the SNE date is the *only* predecessor so it is the Early Start of the node by default.

Finish-No-Earlier-than Dates

It may seem that if we model nonconstraint delays to a task's start as start date predecessors, then we should model nonconstraint delays to a task's finish as finish date predecessors, or finish-no-earlier-than (FNE) dates. In reality, however, all delays to a finish that are not constraint related are internal to the task. A resource problem that would affect the performance of a task could most likely be modeled as a function of the task's span. Such a delay would show up as an increase in duration. This leaves very little need for an FNE date, although Chapter 9 will explore this concept further.

Common Start Dates

In the second technique for setting the Early Starts of start nodes, the universal date method, we set all start nodes' Early Start dates to a common date such as the project start date or the funding release date. Though some starts in a project are often driven by a common date, many more are not. In a simplistic world with a single project start, this technique would obviously work, but this simplistic world exists only in examples. In the practical world, individual date predecessors (SNE dates) are preferable to assigning some common date. In fact, tasks that are driven by a common date, such as the project start date, should not be start nodes. Instead, an events node called the *Project Start* should precede them and list the appropriate start date as its SNE date. As an example, consider a project with three independent paths initiated at the project start date (January 4). Figure 4-18 shows an improper way to model this situation.

Instead, the model should show the initiation date in an events node that constrains all three paths. A proper model of this situation appears in Figure 4-19. The common events node predecessor gives the three paths (Task 1, Task 3, and Task 5) a common Early Start date of January 4. The events node, TaskA, measures starts, so the start date is reported while the finish date is ignored. It has no relevant meaning.

Figure 4-18

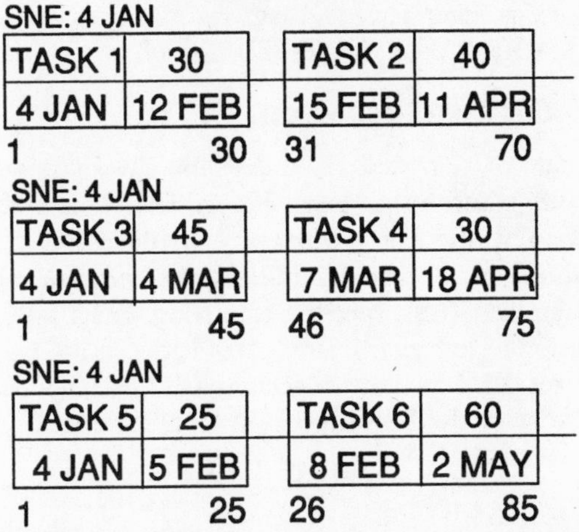

We use universal or global start node dates primarily as default values. CPM software programs commonly employ this method. When the user fails to establish an Early Start date for start nodes, some default date at least allows the accomplishment of the forward pass process. This assumes that any early date calculation is better than no early date calculation, a bad fix for poor modeling discipline.

We must understand and accept our responsibility to identify and provide all necessary model attributes in order for the CPM process to function properly.

Figure 4-19

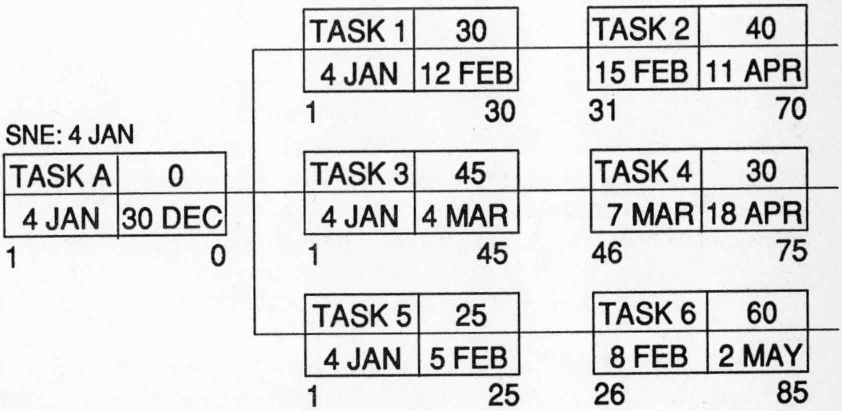

Appropriate Early Start dates for start nodes are part of this necessary data, just as are proper durations, interfaces, and lags for all nodes and constraints. Specifying some global default date violates this discipline.

Deriving Early Starts from Successors

As a third technique of start node date definition, we can calculate the Early Start of a start node based on its Early Finish date. This technique looks to a successor to determine the Early Start of a predecessor, searching the path downstream (to the right) of a node until some other path merges with it and establishes an early date (ES or EF). Subtracting spans and lags through the constraints from right to left yields early dates between the two points. The two examples in Figure 4-20 show how to accomplish this. The arrows in the figure indicate the directions of the forward pass date calculations.

Figure 4-20

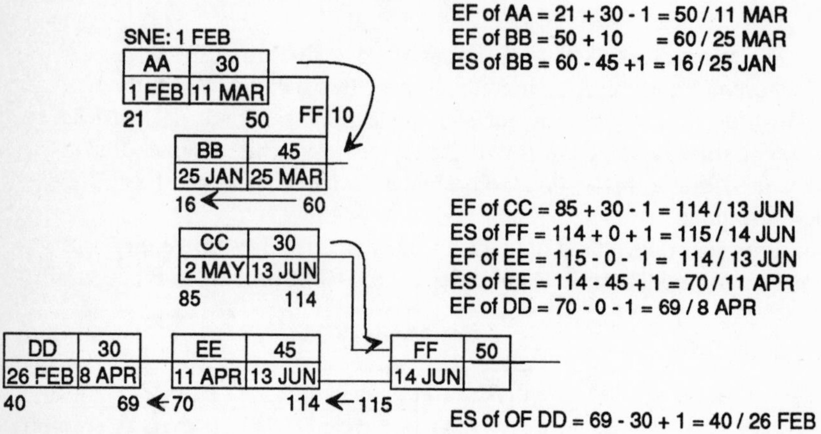

The weakness of this technique arises from its effort to determine the schedule of a task from the predecessors of a common successsor. Figure 4-21 and its associated date calculations show why this is a problem. The dates of the three example nodes would be determined by the following calculations:

$$\text{ES of X} = \text{SNE on X} = \text{day } 85$$

$$\text{EF of X} = \text{ES of X} + \text{DU} - 1 = 85 + 30 - 1 = \text{day } 114$$

$$\text{ES of Z} = \text{EF of X} + \text{lag} + 1 = 114 + 0 + 1 = \text{day } 115$$

EF of Y = ES of Z − lag − 1 = 115 − 0 − 1 = day 114

ES of Y = EF of Y − DU + 1 = 114 − 20 + 1 = day 95

In essence, the Early Start of task X establishes the Early Start of task Y through task Z. This does not adequately answer the original question, because the start of X is not a predecessor of the start of Y.

This technique is another poor way to overcome a lack of discipline, although the basic idea has practical merit. When to start an independent series of tasks is a reasonable question, but this method is an inferior way to answer the question. Chapter 8 will discuss the proper way to address this problem through a specified amount of time reserve, the fourth technique.

A specified time reserve addresses the question in a better way. We must present considerably more information before we can explain this technique in Chapter 8, though. At this point, however, we can see that the second and third techniques are inadequate for our purposes.

Figure 4-21

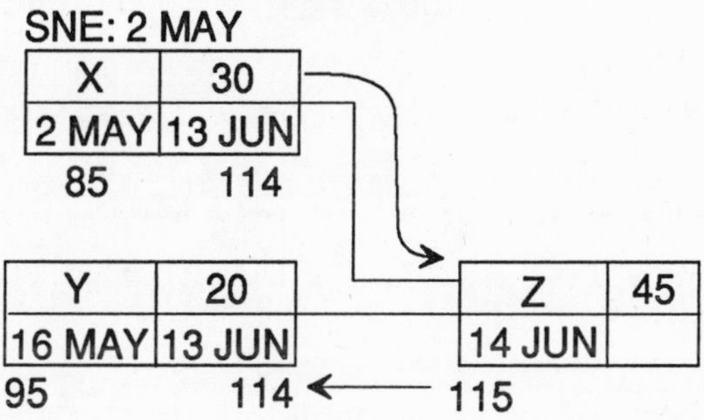

Section Summary

We can now summarize several salient points.

A. A start node is a node that has no predecessors of its start.

B. The user must provide the Early Start dates on all start nodes.

C. The user can do this by three practical techniques:

1. Apply a start date predecessor (SNE date) to each start node.
2. Set an Early Start date for the start node that establishes a specified amount of time reserve (see Chapter 8).
3. When activities begin, apply the actual start date of the task as its Early Start date (see Chapter 9).

D. A start date predecessor (SNE date) allows modeling of nonconstraint delays to the start of any task.

E. Adjusting the duration of a task allows modeling of nonconstraint delays on its finish, leaving little or no need for a finish-no-earlier-than (FNE) date predecessor. Such a date will, however, have other uses (see Chapter 9).

EVENTS NODES

A node with a zero duration is called an *events node*. Forward pass calculations involving such nodes need some explanation since their early dates can seem contradictory. We can now expand the brief discussion from Chapter 2.

Figure 4-22 shows an events node, AU, in the middle of a model. This node dictates that the paths AL-HG, AG-SN, and CO-TI cannot start until paths CU-PB, ZN-FE, and MO-NI have all completed. As such, it serves as a path funnel. We need to examine how to calculate the dates in this node and what they mean.

Figure 4-22

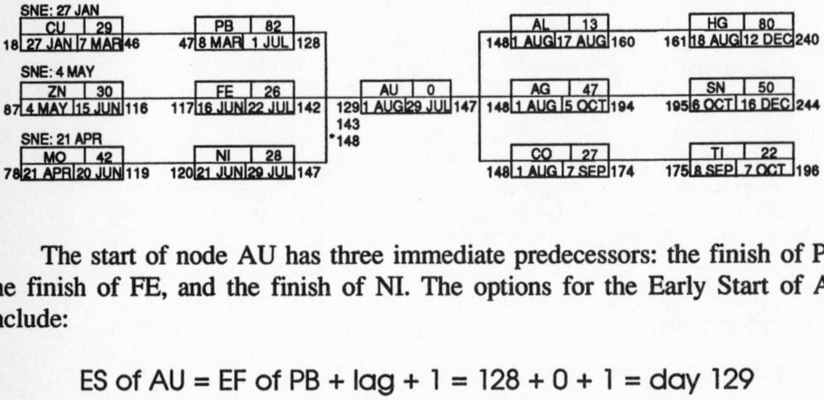

The start of node AU has three immediate predecessors: the finish of PB, the finish of FE, and the finish of NI. The options for the Early Start of AU include:

$$\text{ES of AU} = \text{EF of PB} + \text{lag} + 1 = 128 + 0 + 1 = \text{day } 129$$

$$\text{ES of AU} = \text{EF of FE} + \text{lag} + 1 = 142 + 0 + 1 = \text{day } 143$$

$$\text{ES of AU} = \text{EF of NI} + \text{lag} + 1 = 147 + 0 + 1 = \text{day } 148$$

The latest early date, workday 148 or August 1, is the Early Start of AU. Only on this date can node AU start and satisfy all of the predecessors.

The finish of AU has only one predecessor, the start of AU, so we calculate it:

EF of AU = ES of AU + DU − 1 = 148 + 0 − 1 = day 147 (July 29)

The Early Start and Early Finish of AU appear to be out of chronological sequence. The events node seems to finish before it starts. This is basically a perception problem, though.

Like all nodes in Precedence, AU has two associated events: one at the start and one at the finish. The finish of AU reflects the date when all of the node's predecessors finish (the latest finish—in this case, the EF of node NI). The start of AU reflects the date when its successors can start (the morning after node NI finishes). The finish of an events node measures its predecessors' finishes and the start measures its successors' starts.

Another way to understand this seeming contradiction uses the primary cause of the effect. Recall the start-finish convention. A start date (e.g., ES) reflects the beginning of the work period, whereas a finish date (e.g., EF) reflects the end of the work period. If we diagram this (as in Figure 4-23), giving each day length, we see that a start date of August 1 is equivalent to a finish date of July 29. An instant of work time separates the two points in time, since Saturday and Sunday are not workdays. That is what an event is—an instant of time. Our dates in the events node are simply portrayed as a start date (ES) and as a finish date (EF).

Figure 4-23

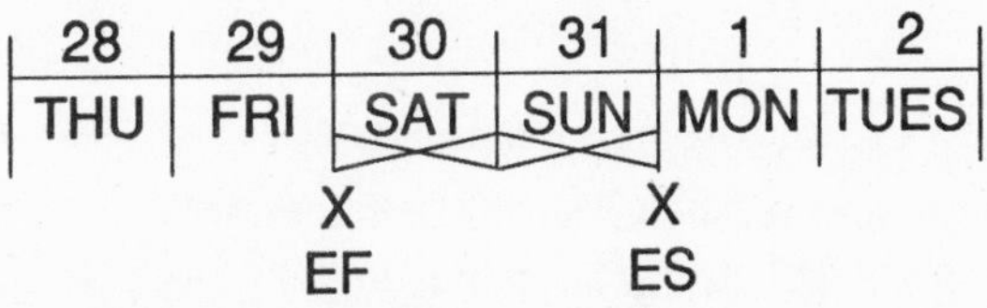

The dates in an events node look contradictory, but they are in fact equivalent points in time. The contradiction is a perception problem caused by the start-finish date convention. By understanding this, we can turn the contradiction into a very useful tool. Events nodes allow us to report events that measure starts or finishes relative to any activity or several independent paths.

Normally, an events node marks only one of the two events. We simply ignore the second unnecessary event. Therefore, we need a way to designate which end of the node to report, which is called an *event bias* (EB).

THE STANDALONE FORWARD PASS

By itself, the forward pass process can serve as a *set forward scheduler*. If we can determine the necessary parameters of a model (activities, constraints, durations, and lags) and understand when the project or segment of work will begin, then by the forward pass, we can determine a schedule of the activities that satisfies the model's conditions. If the resulting finish dates are not acceptable, we can alter the project model iteratively until we devise an acceptable plan.

This is most useful in the formative stages of a project, or when a project is undergoing major alterations. It gives a schedule by which to track the project's performance until new conditions force another round of changes. Then we dust the process off, and develop a new schedule.

This is the most common application of project modeling today, and no doubt it is a useful tool. By itself, however, this function does not justify all the inherent expense of project modeling. It does not provide enough value added to justify the expense of the modeling effort. Standalone forward pass modeling employs only one-third of the CPM capability. How can any process used in such a limited fashion justify its cost?

This is analogous to buying a luxury car and then just sitting in it and running the air conditioner. However great the air conditioner, eventually we must challenge the expense. As a result, we will probably get rid of the car without ever learning that the air conditioner was just an adjunct process of the car. Ignorance is and always has been the greatest obstacle to progress.

SUMMARY

A few rules govern the forward pass process:

First Rule: The process works left to right chronologically through a project, beginning at the model starts and culminating at the model finishes.

Corollary: We must calculate the early dates of a node's predecessor(s) before calculating the early dates of the node.

Corollary: We must calculate the Early Start of a node before calculating its Early Finish.

Second Rule: The appropriate predecessor equations for the forward pass include:

DU Equation: (EF of Node) = (ES of Node) + DU − 1
FS Constraint: (ES of SN) = (EF of PN) + lag + 1
SS Constraint: (ES of SN) = (ES of PN) + lag
FF Constraint: (EF of SN) = (EF of PN) + lag

SF Constraint: (EF of SN) = (ES of PN) + lag − 1

Third Rule: To choose among multiple predecessors (MPR) of a node's start or finish, calculate the date that would result if each predecessor were acting alone and then choose the latest early date. This is the earliest date that satisfies all predecessors.

Normally a node's Early Start (ES) date plus the duration drives its Early Finish. Any time a node has other external predecessors of its finish (FF and SF constraints), however, one of these relationships may drive its Early Finish. When this happens, the node is called an *external predecessor stretched activity* (EPSA) or simply a *stretched activity*. In response, we can:

1. Delay the start of the activity so that it can be accomplished in one continuous effort.
2. Pull in or advance the finish of the external predecessor that is causing the stretching problem.
3. Divide the task into two distinct activities.

Start nodes are nodes that have no predecessors of their starts. The user must provide Early Start dates for start nodes.

Start date predecessors or start-no-earlier-than (SNE) dates are one acceptable means of providing a start node's Early Start date. On nonstart nodes, they simulate delays to the nodes' starts.

Two other methods are acceptable for providing a start node's Early Start date:

1. Setting a date to provide a specified amount of time reserve (see Chapter 8).
2. Assigning the actual start date as the Early Start date (see Chapter 9).

CHAPTER 5

The Backward Pass

The backward pass has nothing to do with the line of scrimmage on a football field. Instead, it is a set-back subprocess in CPM that determines when tasks must be accomplished. It works like the forward pass, only in reverse.

For some reason, many people find the backward pass much more difficult to understand than the forward pass. This seems strange, because everyone uses set-back scheduling, the fundamental principle of the backward pass, quite commonly.

Most casual scheduling starts at an end point and works backward to determine the starting time necessary to reach the end point on schedule. To get to work on time, people perform miniature backward passes when they determine a schedule of all the tasks necessary to get ready for work that allows them to arrive on time. They set this schedule back all the way to the moment the alarm clock goes off.

The backward pass process is more complicated mathematically than the forward pass since it requires more subtraction than addition. The concept, however, is as simple as understanding that a series of tasks that takes 125 days cumulatively to accomplish must start at least 125 days prior to the required end date. For this reason, we label the backward pass the "must" schedule, as it defines when we must start and must finish a task in order to complete it and all its successors by the required time. We determine the schedule for every node in a model as a function of its successors.

The dates derived from the backward pass are called the *late dates.* By the backward pass process, we find the Late Finish (LF) and Late Start (LS) for every node in the project model.

Following the same graphic convention as the forward pass, we show the process flowing chronologically from right to left from the finishes of the model to the starts. For now, we define the term *finishes of the model* as tasks with no successors. Specifically, in the backward pass we determine when a task must finish based on the *successors of its finish* and when it must start based on the *successors of its start*. This leads to the first rule of the backward pass:

> The backward pass process begins at the finishes of a project and proceeds chronologically from the right to the left toward the start of the project model.

As in the forward pass, this rule has two corollaries:

1. We must calculate the late dates of a node's successor(s) before we can calculate the late dates of the node itself.
2. We calculate the Late Finish of a node before its Late Start because the finish of a node is always a successor of its start.

BACKWARD PASS EQUATIONS: DURATION AND FINISH-TO-START

The second rule of the backward pass takes the form of a set of five equations that calculate the specific dates from the different types of successors. We will explore each of these functions by solving sample problems. The first example (Figure 5-1) repeats the initial problem worked in the discussion of the forward pass. It illustrates (1) the duration equation and (2) the finish-to-start equation.

First, it is important to note that nothing in the definition of the model changes when we perform the backward pass. Predecessors and successors remain the same; ALPHA is a predecessor of DELTA, DELTA is a successor of ALPHA. The calculation process changes, however, reversing direction through the model. First, we start the backward pass at the finish or finishes of the model, task OMEGA in the example.

Figure 5-1

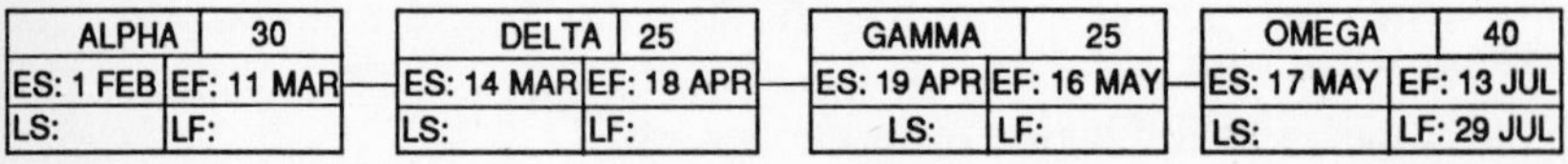

As in the forward pass, we initiate the backward pass with a date. The process seeks to determine *must* or *requirement* dates of the project, so we begin with such a date, for example a customer delivery date. A model finish is a completion of some series of effort and as such should have a requirement. Our example project must finish by July 29. Since the finish of task OMEGA is the finish of the project, the Late Finish date of OMEGA is July 29, which the calendar tells us is workday 147. (Note: We use the same five-day week calendar in the backward pass that we used in the forward pass.)

Since task OMEGA takes 40 days to accomplish, it must start no later than 40 days prior to the 147th day in order to finish on time. This introduces the first equation of the backward pass, which determines the Late Start (LS) of a task based on its Late Finish and duration. It is derived from the fundamental date math equation:

$$\text{DU Equation: LS of Node} = \text{LF of Node} - \text{DU} + 1$$

In our example, this would yield a Late Start of OMEGA of:

$$\begin{aligned}\text{LS of OMEGA} &= \text{LF of OMEGA} - \text{DU} + 1\\ &= 147 - 40 + 1 = \text{day } 108\end{aligned}$$

The 108th workday, June 3, is the Late Start (LS) of OMEGA. OMEGA must start no later than June 3 in order for the project to finish by its required date, July 29.

Since OMEGA cannot start until GAMMA finishes and OMEGA must start on June 3, GAMMA must finish no later than the end of the prior workday. Any lag on this constraint would back this finish off by the amount of the lag. This leads to the next equation, which calculates the Late Finish of a node based on the Late Start of its successor in a finish-to-start constraint:

$$\text{FS Constraint: LF of PN} = \text{LS of SN} - \text{lag} - 1$$

In our example, this would yield:

$$\begin{aligned}\text{LF of GAMMA} &= \text{LS of OMEGA} - \text{lag} - 1\\ &= 108 - 0 - 1 = \text{day } 107\end{aligned}$$

The 107th workday, June 2, is the Late Finish of GAMMA. The equation takes the type of constraint, the lag on the constraint, any nonwork days, and the start-finish convention into account. GAMMA must finish no later than June 2 in order for the project to complete on July 29.

Since the example features only finish-to-start constraints, we can calculate the remaining late dates by altering these two equations. Accomplishing this work should build an appreciation for the process flow through the entire path.

First, if GAMMA must finish by the 107th workday and it takes 20 days to accomplish, then it must start by:

$$\text{LS of GAMMA} = \text{LF of GAMMA} - \text{DU} + 1$$
$$= 107 - 20 + 1 = \text{day } 88$$

The 88th workday, May 5, is the Late Start of GAMMA. Since GAMMA cannot start until DELTA finishes, DELTA must finish:

$$\text{LF of DELTA} = \text{LS of GAMMA} - \text{lag} - 1 = 88 - 0 - 1 = \text{day } 87$$

The 87th workday, May 4, is the Late Finish of DELTA. The Late Start of DELTA must be 25 days earlier to accommodate its duration:

$$\text{LS of DELTA} = \text{LF of DELTA} - \text{DU} + 1 = 87 - 25 + 1 = \text{day } 63$$

The 63rd workday, March 30, is the Late Start of DELTA. Since DELTA cannot start until ALPHA finishes, ALPHA must finish:

$$\text{LF of ALPHA} = \text{LS of DELTA} - \text{lag} - 1 = 63 - 0 - 1 = \text{day } 62$$

The 62nd workday, March 29, is the Late Finish of ALPHA. Since ALPHA takes 30 days to accomplish, it must start no later than:

$$\text{LS of ALPHA} = \text{LF of ALPHA} - \text{DU} + 1 = 62 - 30 + 1 = \text{day } 33$$

The 33rd workday, February 17, is the Late Start of ALPHA. We have now calculated the dates by which each task must start and finish in order for the project to meet its schedule objective, July 29. These calculations are based on the model's nodes and constraints, the estimated durations and lags, and the known required finish date. The completed calculations for the total model (both forward and backward passes) appear in Figure 5-2.

Figure 5-2

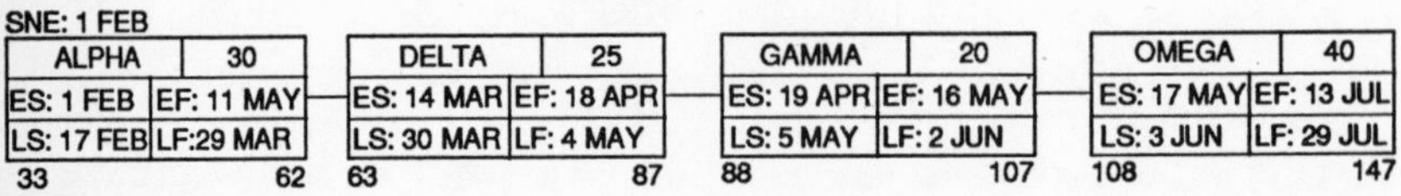

Adding up all the spans tells us that the project takes 115 days to complete, the same number we found from the forward pass. A project that must finish on July 29 and takes 115 workdays must then start on:

$$\text{LS of Path} = \text{LF of Path} - \text{DU} + 1 = 147 - 115 + 1 = \text{day } 33$$

Workday 33, February 17, is the same date derived from the backward pass. The total equals the sum of its parts in the backward pass, too. If, for some reason, the required finish date were pulled in to June 30, the project would have to start 115 days prior to this date, or:

$$\text{LS of Path} = \text{LF of Path} - \text{DU} + 1 = 127 - 115 + 1 = \text{day } 13$$

Figure 5-3

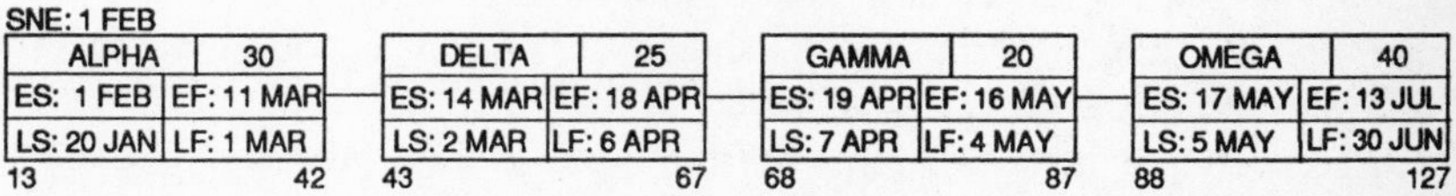

Workday 13 is January 20. Reworking the backward pass shows that this change will ripple backward all the way through the project, pulling in each date accordingly. This example is illustrated in Figure 5-3.

We have presented two of the five equations of the backward pass. Three other successor conditions remain to be explained: finish-to-finish, start-to-start, and start-to-finish constraints. We will now examine each of these successor types.

Finish-to-Finish Backward Pass Equation

Figure 5-4 gives an example of a finish-to-finish constraint where MU must finish 15 days prior to the finish of NU. Either a successor of NU or the requirement date dictates that NU must finish no later than July 15 (workday 137), again using the five-day week calendar.

Figure 5-4

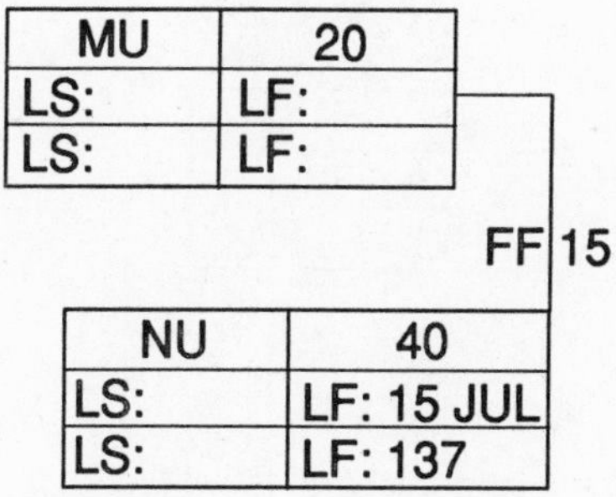

We can see that the finish date of the predecessor node (MU) could affect the finish of the successor. The Late Finish of the predecessor must reflect this relationship. This equation takes all the appropriate variables for this type of relation into account:

FF Constraint: LF of PN = LF of SN – lag

This equation determines the Late Finish of the predecessor in a finish-to-finish constraint based on the Late Finish date of the successor and the lag of the constraint. As both dates in the equation are finish dates, it is not necessary to make any adjustment for the start-finish convention. The equation requires the Late Finish date of the successor as input.

In our example, node MU must finish at least 15 days prior to NU's finish. Therefore, MU must finish by:

LF of MU = LF of NU – lag = 137 – 15 = day 122

The 122nd workday, June 23, is the Late Finish of MU. MU must finish by June 23 in order for NU to finish on July 15. Now we can calculate the Late Start dates of both nodes independently:

LS of NU = LF of NU – DU + 1 = 137 – 40 + 1 = day 98

LS of MU = LF of MU – DU + 1 = 122 – 20 + 1 = day 103

The 98th workday falls on May 19 and the 103rd workday on May 26. Figure 5-5 shows the completed backward pass of the model. We have determined when each task (MU and NU) must start and finish in order to complete task NU by July 15.

Figure 5-5

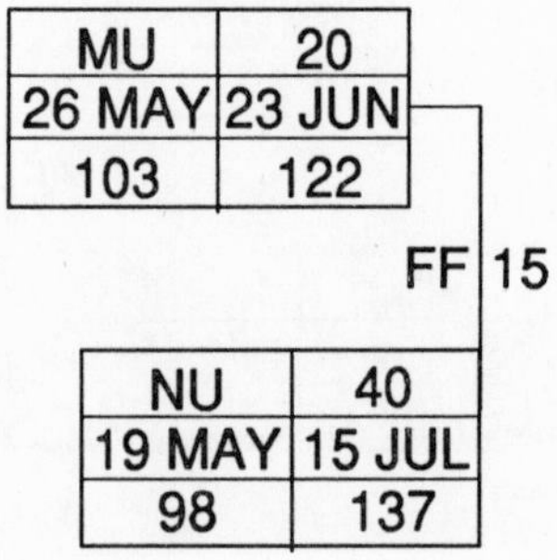

Start-to-Start Backward Pass Equation

The start-to-start equation resembles the finish-to-finish equation. In this type of relation we need to calculate the Late Start of a node based on the Late Start of its successor in a start-to-start constraint. The equation for this type of relation is:

SS Constraint: LS of PN = LS of SN – lag

This equation gives the Late Start of a predecessor in a start-to-start constraint based on the Late Start of its successor and the lag of the constraint. As both dates are start dates, we need make no adjustment for the start-finish convention.

Figure 5-6 shows an example of a start-to-start constraint where Task PHI must start 10 days before CHI starts. The Late Start of CHI has already been determined to be November 1 (workday 213). We still use the familiar five-day week calendar.

Figure 5-6

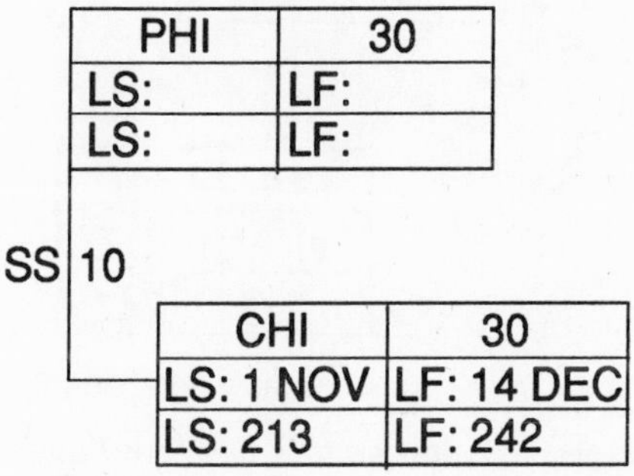

The start-to-start equation can tell us the Late Start of PHI in relation to the start of CHI, but PHI must also start no later than 30 days prior to its own Late Finish. Two dates affect the Late Start of PHI: the Late Finish of PHI and the Late Start of CHI. When two or more successors directly affect the calculation of a late date, we have a *multiple successor relationship (MSR)*. When this occurs, we must choose one of the dates that satisfies all of the successor requirements. Only the earliest date can do this. This leads to the third rule of the backward pass, the rule of the multiple successor relationship:

> To account for multiple successor relationships, calculate each independently, and then select the earliest of the late dates.

To see how this works, let us examine two examples. First let PHI's Late Finish be November 21 (workday 227). Its Late Start would be a choice of:

LS of PHI = LS of CHI – lag = 213 – 10 = day 203

or

LS of PHI = LF of PHI – DU + 1 = 227 – 30 + 1 = day 198

Workday 203 is October 18, and workday 198 is October 11. Of the two, workday 198, October 11, is the earliest late date so it becomes the Late Start of PHI. This date allows PHI to run for 30 workdays before it must finish on November 21, and also allows at least 10 workdays before CHI must start on November 1 (see Figure 5-7). It satisfies both successors.

Figure 5-7

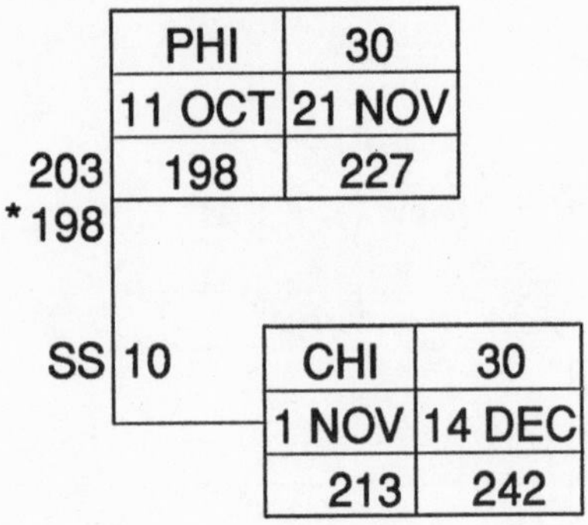

For the second example, suppose that the Late Finish of PHI is December 7, workday 237. Based on this late date, the Late Start of PHI would be a choice of:

LS of PHI = LS of CHI – lag = 213 – 10 = day 203

or

LS of PHI = LF of PHI – DU + 1 = 237 – 30 + 1 = day 208

Workday 203 is October 18 and workday 208 is October 25. In this case, the earliest late date would be workday 203, October 18 (see Figure 5-8). Only this date satisfies both successors.

In this example, the difference between the Late Finish and Late Start of PHI (237 – 203 + 1 = 35 days) no longer equals the duration of PHI (30 days).

Figure 5-8

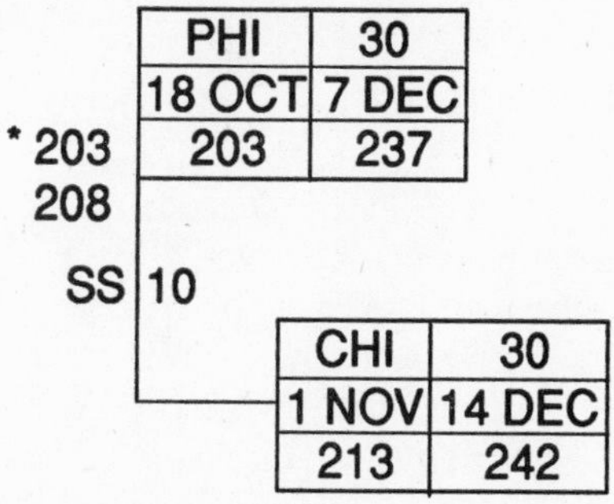

The start-to-start constraint (PHI-CHI) now drives the Late Start of PHI. As we saw earlier, the start-to-start constraint breaks the predecessor into two parts, as shown in Figure 5-9.

Figure 5-9

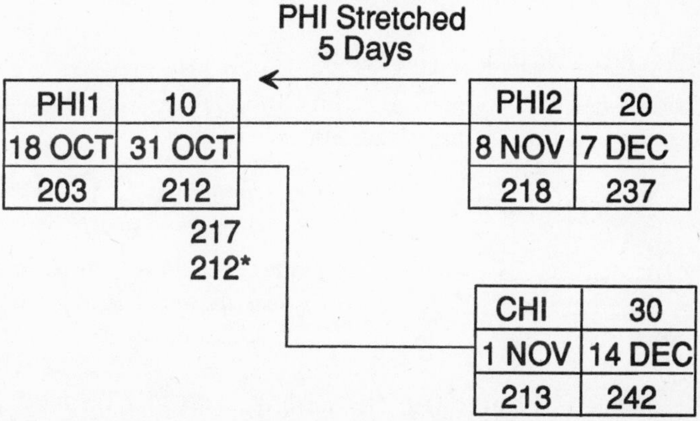

The first task, PHI1, covers the first 10 days of PHI, the length of the lag on the constraint. The second task, PHI2, covers the rest of PHI (DU – lag = 30 – 10 = 20 days). We know Task CHI must start on November 1 and PHI2 must finish on December 7. The duration equation tells us that PHI2 must start 20 days prior to its required finish, or:

$$\text{LS of PHI2} = \text{LF of PHI2} - \text{DU} + 1 = 237 - 20 + 1 = \text{day } 218$$

This is equivalent to November 8. To satisfy both of its successors, the start of PHI2 (November 8) and the start of CHI (November 1), PHI1 must finish in time to support the earliest of the two, in this case November 1. Thus,

PHI1 must finish by October 31. It must start at least 10 days prior to this date, or October 18:

$$\text{LS of PHI1} = \text{LF of PHI1} - \text{DU} + 1 = 212 - 10 + 1 = \text{day } 203$$

Thus, the start of PHI (and PHI1) depends on the need to start CHI on November 1. This successor does not, however, affect PHI2 or the finish of PHI.

When this occurs, the task is said to be an *External Successor Stretched Activity* (ESSA) or simply a *stretched activity*. This means that the task need not start and finish in one continuous span. This has a different effect than the external predecessor stretched activities we discussed when explaining the forward pass. Stretching in the backward pass process (creating ESSAs) means that the path through the initial portion of the task has a more stringent requirement than the path(s) through the entire task.

This will become important when we discuss paths in Chapter 6. At this point, however, it is enough to see that stretching can occur in both the forward and backward passes. Constraints can stretch the Early Finish of a task in the forward pass and the Late Start of a task in the backward pass.

Start-to-Finish Backward Pass Equation

We must examine one more type of constraint, the start-to-finish, impractical as it is. We give the equation for reference purposes only:

$$\text{SF Constraint: LS of PN} = \text{LF of SN} - \text{lag} + 1$$

This states that the Late Start of the predecessor in a start-to-finish constraint equals the Late Finish of the successor less the lag. Since we are calculating a start date from a finish date, we need to add one unit for the start-finish convention.

In the example in Figure 5-10, the finish of node THETA must occur at least 15 days after the start of node KAPPA. In the terms of the backward pass, the Late Start of KAPPA must occur at least 15 days prior to the Late Finish of THETA. The example assumes Late Finishes for KAPPA and THETA of November 30 (workday 232) and October 14 (workday 201), respectively.

The duration equation can supply the Late Start of THETA:

$$\text{LS of THETA} = \text{LF of THETA} - \text{DU} + 1 = 201 - 30 + 1 = \text{day } 172$$

Workday 172, September 2, is the Late Start of THETA.

The Late Start of KAPPA depends on two values: the Late Finish of KAPPA and the Late Finish of THETA. This constraint type creates a multiple

Figure 5-10

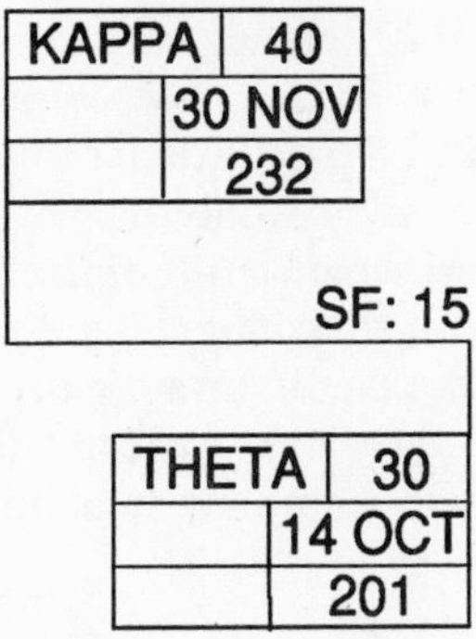

successor relationship (MSR), leaving a choice for the Late Start of KAPPA between:

LS of KAPPA = LF of KAPPA – DU + 1 = 232 – 40 + 1 = day 193

or

LS of KAPPA = LF of THETA – lag + 1 = 201 – 15 + 1 = day 187

Workday 193 is October 4 and workday 187 is September 26. The earliest late date, workday 187 or September 26, is the Late Start of KAPPA. The completed backward pass looks like Figure 5-11.

Figure 5-11

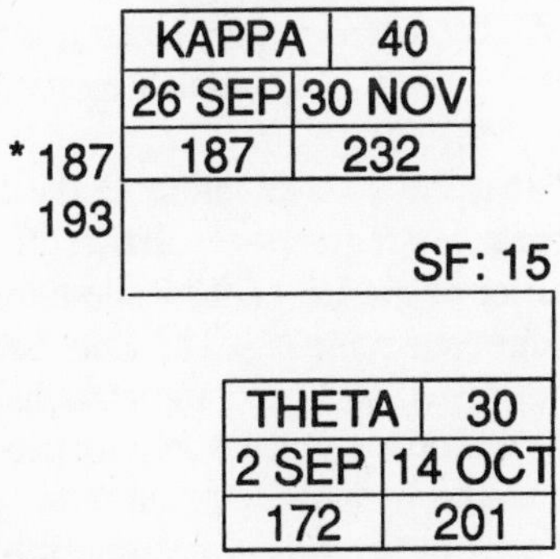

In this example, KAPPA is an ESSA, since the SF constraint pulls its Late Start back six days. If the Late Finish of KAPPA were six days earlier, or the Late Finish of THETA were six days later, or the lag value were six days less, the constraint would not affect KAPPA's Late Start. This combination of values determines whether the Late Start of an activity is stretched by a successor other than its own finish.

In the forward pass, external predecessor stretching can occur only on the Early Finish of a node and only in the presence of finish-to-finish or start-to-finish constraints. In the backward pass, external successor stretching can occur only on the Late Start of a node and only in the presence of start-to-start or start-to-finish constraints. If we avoid start-to-finish constraints, then only a finish-to-finish constraint can cause stretching in the forward pass and only a start-to-start constraint can cause it in the backward pass. Therefore, if we can determine at which end stretching occurs, we can also determine the specific constraint that causes it. We will learn how to do this in Chapter 6.

Section Summary

We conclude the discussion of the backward pass for normal successor types with a summary of the equations that make up the second rule of the backward pass:

DU Equation: (LS of Node) = (LF of Node) – DU + 1
FS Constraint: (LF of PN) = (LS of SN) – lag – 1
FF Constraint: (LF of PN) = (LF of SN) – lag
SS Constraint: (LS of PN) = (LS of SN) – lag
SF Constraint: (LS of PN) = (LF of SN) – lag + 1

MULTIPLE SUCCESSOR RELATIONSHIPS

Having explained the backward pass for each type of successor, we can combine them into complex relationships. In the forward pass, these were called multiple predecessor relationships (MPRs), because predecessors drive the forward pass. Successors drive the backward pass, so its complex relationships are called multiple successor relationships (MSRs).

We have already seen that the determination of the Late Start of a predecessor in a start-to-start constraint requires resolution of a multiple successor relationship. We find the earlier of the LF of the node minus the duration or the LS of the successor minus the constraint lag. The Late Start of the predecessor in a start-to-finish constraint created similar complications.

A task may have several other successors, increasing this complexity. Nevertheless, we resolve all multiple successor relationships the same way. If a node end (event), either the start or the finish, has multiple successors, we must find the late dates for each independently and then choose the one that satisfies all successors, that is, the earliest late date.

Figure 5-12 shows a model fragment in which node AFA has multiple successors: TCU, BYU, LSU, CSU, and USC. This example assumes that the late dates of the successors have already been calculated, so the next step is to determine the late finish of AFA. Looking at the constraints, we see that the successors of AFA's finish are:

1. The start of TCU (FS)
2. The start of BYU (FS)
3. The finish of LSU (FF)

These three successors give three Late Finish dates for node AFA:

LF of AFA = LS of TCU − lag − 1 = 133 − 0 − 1 = day 132

LF of AFA = LS of BYU − lag − 1 = 143 − 5 − 1 = day 137

LF of AFA = LF of LSU − lag = 139 − 10 = day 129

The earliest of these late dates, workday 129 or July 5, is the Late Finish of AFA. Resolving multiple successor relationships is no more difficult than that.

Figure 5-12

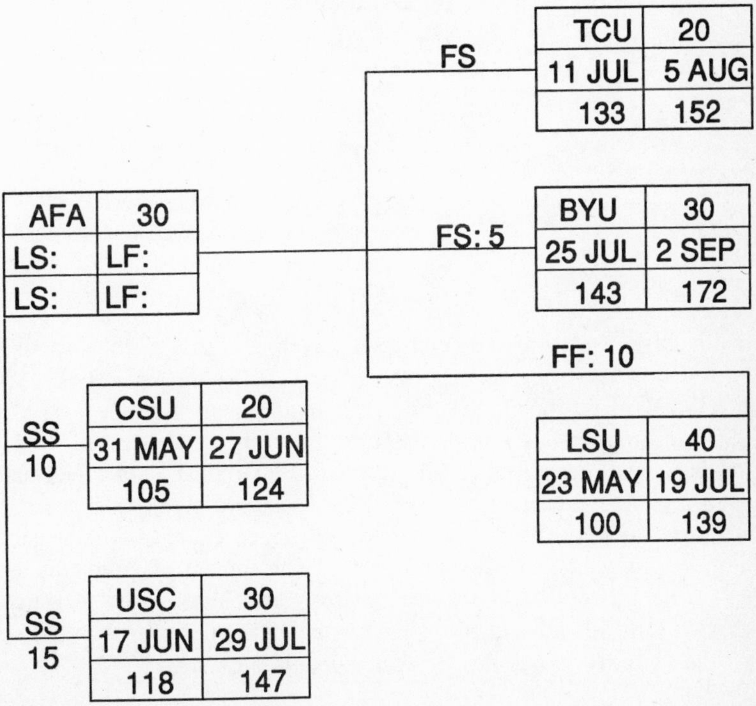

Having calculated the Late Finish of AFA, we can determine its Late Start. First, we determine the successors of its start:

1. The finish of AFA
2. The start of CSU (SS)
3. The start of USC (SS)

These three successors give three Late Start dates for AFA:

LS of AFA = LF of AFA – DU + 1 = 129 – 30 + 1 = day 100

LS of AFA = LS of CSU – lag = 105 – 10 = day 95

LS of AFA = LS of USC – lag = 118 – 15 = day 103

The earliest late date among these three values, workday 95 or May 16, is the Late Start of AFA, as shown in Figure 5-13.

Figure 5-13

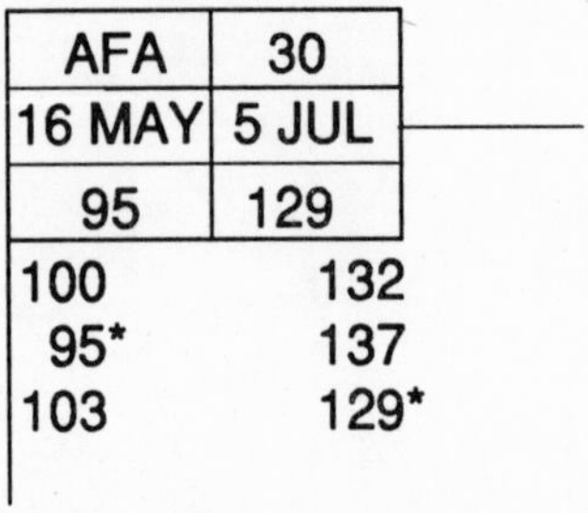

In this case, a start-to-start constraint determined the Late Start of node AFA, so this task is a stretched activity (ESSA) (LF – LS + 1 = 129 – 95 + 1 = 35 days > DU = 30 days). Therefore, the requirement through the path . . . AFA-CSU . . . imposes a more stringent requirement on the start of AFA than any path that goes through its finish. Since different constraints affect tasks in the forward pass (FF predecessors) and in the backward pass (SS successors), it would be very unlikely that a task would experience equal stretching simultaneously in the forward and backward passes. The impact of stretching is also different in the forward and backward passes. Stretching in the forward pass may require a schedule adjustment, while stretching in the backward pass simply indicates the path that exerts the greatest schedule pressure.

FINISH NODES AND DATE SUCCESSORS

Finish Nodes

Just as we needed to inject dates into the forward pass process, we must also inject dates into the backward pass process. As the predecessor-driven forward pass process requires date predecessors, the successor-driven backward pass process needs date successors. Thus, we call the dates with which we begin the backward pass *date successors.*

In our examples, we have already initiated the backward pass from Late Finish dates imposed on the nodes at the far right of the model. In our initial sample project, ALPHA-OMEGA, we imposed this date, July 29, on the Late Finish of OMEGA. OMEGA has no successors so we cannot calculate its Late Finish by the backward pass process. This makes OMEGA a special node called a *finish* or a *finish node.* Since the backward pass cannot determine the Late Finish of any finish node, we must do so as one of our model construction and maintenance chores.

These dates are the proper points of initiation for the backward pass. Any date generated from the backward pass process descends directly from these dates. Any invalid or missing late dates taint the entire process.

It is easy to see that OMEGA is a finish node: it has no successors and lies at the far right of the project model. This is not the criteria for a finish node, though. Instead, ask whether or not the Late Finish of the node can be calculated by the backward pass process. If not, the node is a finish node.

This means that finish nodes have no successors of their finishes. A finish node may appear anywhere in the model, and it may have successors with SS constraints. It cannot, however, have successors of its finish. To test this understanding, let us examine the simple model in Figure 5-14 and determine its finish nodes.

Figure 5-14

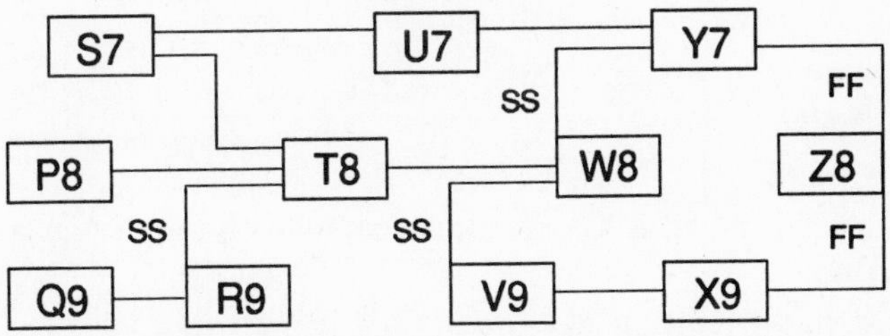

Three nodes (Z8, W8, and R9) have no successors of their finishes. These are the finish nodes of this model. Z8 has predecessors of its finish, but no successors, so it is a finish node. W8 has a successor (Y7), but it is a successor of its start. Since W8 has no successors of its finish, it is also a finish node. Node R9 near the beginning of the model has a successor (T8), but it has no successor of its finish, so it is also a finish node. The backward pass process gives us no way to calculate the Late Finish of R9, although we can calculate its Late Start through the start-to-start constraint R9-T8. Remember the finish of any node is always a successor of its start, and the backward pass process always flows right to left (from the finish of a node to its start).

Examine the other nodes in this example. Each has a successor that allows us to determine its Late Finish date through the backward pass. Node Y7 has Z8, V9 has X9, T8 has W8, etc. We must, however, provide the Late Finish dates on nodes Z8, W8, and R9.

For another exercise, examine Figure 5-15 to determine its finish nodes.

Figure 5-15

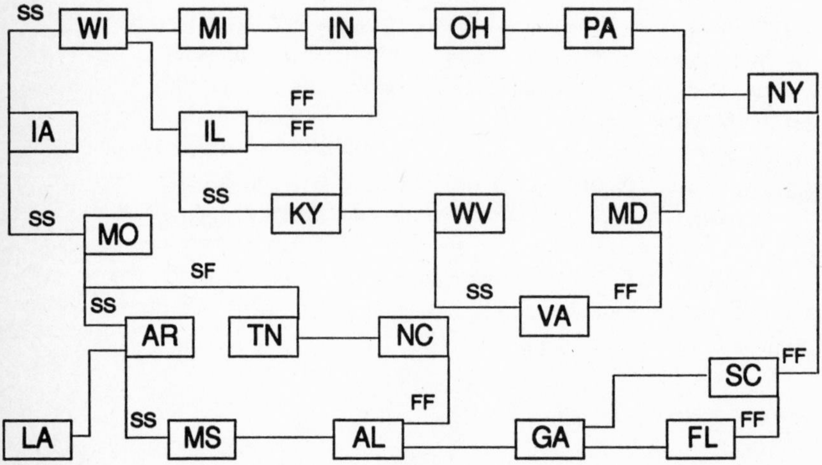

Finish Nodes

- NY is a finish node; it has no successors.
- WV is a finish node; it has no successors of its finish. VA is a successor of its start.
- NC is a finish node; it has no successors. AL is a predecessor of its finish.

- AR is a finish node; it has no successors of its finish. MS is a successor of its start.
- MO is a finish node; it has no successors of its finish. AR and TN are successors of its start.
- IA is a finish node; it has no successors of its finish. WI and MO are successors of its start.

Nonfinish Nodes

- PA is not a finish node; NY is a successor of its finish.
- OH is not a finish node; PA is a successor of its finish.
- IN is not a finish node; OH is a successor of its finish.
- MI is not a finish node; IN is a successor of its finish.
- WI is not a finish node; MI and IL are successors of its finish.
- IL is not a finish node; IN and KY are successors of its finish.
- KY is not a finish node; WV is a successor of its finish.
- MD is not a finish node; NY is a successor of its finish.
- VA is not a finish node; MD is a successor of its finish.
- TN is not a finish node; NC is a successor of its finish.
- LA is not a finish node; AR is a successor of its finish.
- MS is not a finish node; AL is a successor of its finish.
- AL is not a finish node; GA and NC are successors of its finish.
- GA is not a finish node; SC and FL are successors of its finish.
- FL is not a finish node; SC is a successor of its finish.
- SC is not a finish node; NY is a successor of its finish.

These two models (Figures 15-14 and 15-15) illustrate the process of identification of finish nodes. We must determine whether each node has a successor of its finish. Those that do not are finish nodes. Once we have identified all the finish nodes in the model, then we can proceed to establish their Late Finish dates to allow the backward pass to function properly. *All finish nodes must have a late finish date determined by the user.*

Setting the Late Finish Dates for Finish Nodes

We can provide Late Finish dates for finish nodes by five techniques:

1. Apply a date successor to each finish node.

2. Set the Late Finish of finish nodes equal to some common, predetermined date, like the project expected completion date.
3. Calculate the finish nodes' Late Finish dates from their Late Start dates (the reverse of the backward pass process).
4. Set the Late Finish of a finish node to its Early Finish as a default value.
5. Set a Late Finish date on each finish node to establish a specified amount of time reserve (see Chapter 8).

We must examine these options for setting late dates thoroughly. Some of the techniques are not acceptable, and we must determine which to use and which to avoid. The fifth technique requires understanding of all the CPM processes, so we will defer its explanation until Chapter 8. It is, however, one of the valid means of providing the Late Finish date on a finish node.

Finish-No-Later-than (FNL) Dates

In defining a finish date successor, we simply apply an appropriate date to a node. It is the required finish date of the end event. This date becomes an attribute of the node for use only in the backward pass process to determine its Late Finish date. We call this a *finish date successor* because it acts on a node like a successor of its finish. Such dates are more commonly called *finish-no-later-than* (FNL) *dates.*

When we attach a finish date successor (or FNL date) as an attribute of a finish node, it becomes the Late Finish date of the node. We did this in all of the backward pass examples by providing Late Finish dates on all finish nodes. When we impose these dates on a node, we identify them with the label *FNL.*

We can apply finish-no-later-than (FNL) dates to any node in a model, not just finish nodes. Often, a task that is a predecessor to another effort in the project must be completed on a certain date, adding an external requirement to the node. We model such a requirement through the application of an FNL date.

Recall the initial project, the ALPHA-OMEGA model. If we were required to deliver the result of task DELTA on April 29, two requirements would apply to its finish: to satisfy its delivery date of April 29 and to satisfy a path requirement to complete task OMEGA on July 29. We solve this problem just like a multiple successor relationship, as both requirements must be satisfied.

Figure 5-16

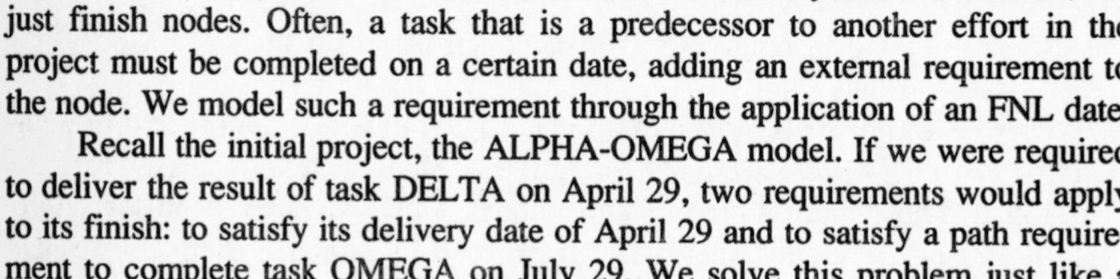

When the backward pass reaches the Late Finish calculation for DELTA, we must choose between:

LF of DELTA = LS of GAMMA – lag – 1 = 88 – 0 – 1 = day 87

or

LF of DELTA = FNL date on DELTA = day 84

We select the earliest late date, workday 84 or April 29, as the Late Finish of DELTA. This date provides for both the delivery of DELTA on April 29 and the path requirement that OMEGA finish by July 29. It is as if Task DELTA had another successor.

Figure 5-17

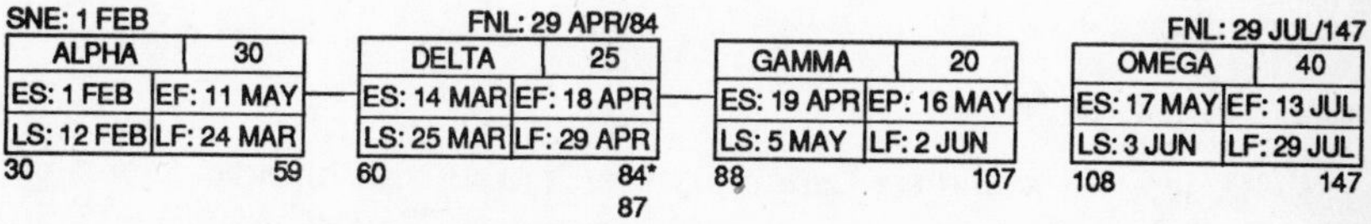

This new Late Finish date on DELTA would affect all late date calculations upstream (to the left) of DELTA, as illustrated in Figure 5-17, which shows the revised completed backward pass. Notice that we attached finish-no-later-than dates to both OMEGA and DELTA.

A finish-no-later-than date may not always be the Late Finish of a node. It becomes the Late Finish of a node only if it is the earliest date of all the successors. When we apply an FNL date to a finish node, it is the only successor and therefore, by default, the earliest late date. This may not be true, however, of an FNL date imposed on a node that is internal to a model. If, for example, the delivery requirement of DELTA specified May 20 or workday 99, then the Late Finish of DELTA would be a choice of:

LF of DELTA = LS of GAMMA – lag – 1 = 88 – 0 – 1 = day 87

or

LF of DELTA = FNL date on GAMMA = day 99

In this case, the earliest late date is workday 87 or May 4 and the model remains unchanged from its original values. The Late Finish date of DELTA (May 4) will still satisfy both successors. It just happens that the downstream requirement on DELTA is currently more strict. A node might have to finish earlier than its FNL date, but it cannot finish later (finish-no-later-than date).

The backward pass process sorts out from all successors the most demanding schedule requirement of each task to find the late dates.

Start-No-Later-than Dates (SNL)

It would seem that external date requirements could be applied to a task's start as well as its finish, giving start date successors or start-no-later-than (SNL) dates. In practice, however, external demands very rarely affect the starts of activities. Few tasks deliver anything at their starts. It could conceivably happen, though. As an example, a contract might require that certain tasks begin by certain dates. We would need SNL dates to model such requirements.

Generally, however, we find little need for this type of date. Instead, we generally determine Late Start dates by subtracting tasks' durations from their Late Finish dates. We could also model these rare occurances with an events node and an FNL date one day earlier than the demand to account for the start-finish convention.

Common Finish Dates

In one technique, we would set Late Finish dates of all finish nodes to a single predetermined date, such as a project completion date. Undoubtedly, we would treat some finish nodes in a project this way, but many more would have independent requirements. Thus, this technique has very limited practical value.

This method is used principally to supply default values. If the user fails to establish a Late Finish date on a finish node, the process defaults to this date to make sure a backward pass remains possible. The value of the technique in this application is questionable. If, say, 1 out of 10 late dates is arbitrary, how accurate can any late date in the model be? This is a bad fix for poor modeling discipline.

Activities that are driven by a common date should be modeled together rather than individually. Correct modeling technique defines an events node that is constrained by the last activity in each path. Only this node needs to carry the requirement date. Figure 5-18 illustrates an example of three paths that share a finish requirement date (LF = December 23). They have been modeled as three independent finishes although it does not seem likely that required finishes on the same date would be independent.

To properly model this situation, we should create an events node as a successor to all three paths which would then impose the common date requirement on each path. Only this node should have the finish date successor of December 23. This technique is illustrated in Figure 5-19, which funnels all three paths from Figure 5-18 into a common events node. Notice that the Late Finish date of the activity at the end of each path is identical to the end node requirement date (December 23).

Figure 5-18

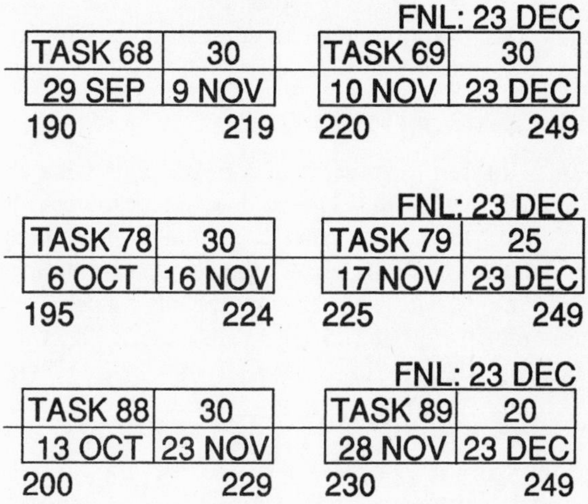

Figure 5-19

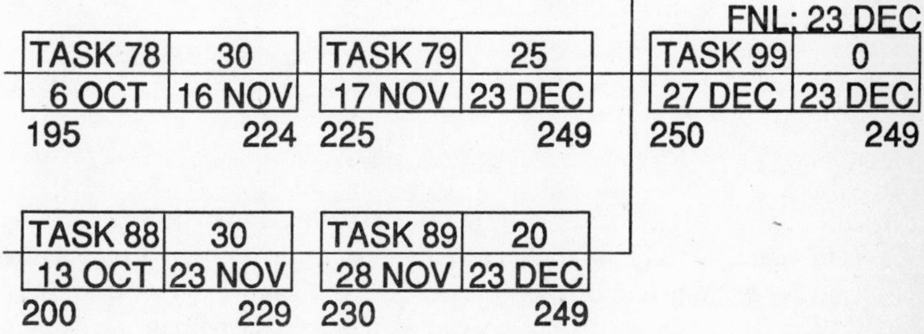

If the FNL date were to change, we would have to note the change in only one place: the end node. If, however, we had modeled this requirement in each independent path, as in Figure 5-18, we would have to note the change everywhere we applied the FNL date. This increases the chance for error. In Figure 5-19, we let the backward pass process do some of the work for us.

Task99 is an events node by which we find finish dates, so its finish date (LF) is pertinent. The start date (LS) refers to the start date of any nodes subsequent to Task99. Our example contains no such nodes, so this date is superfluous and should be ignored.

Deriving Late Finishes from Predecessors

We could search to the left (upstream) of a finish node with an unknown date successor to find a late date with a known date successor calculated from some other finish. We could then work from the point of intersection back to the finish node with the unknown Late Finish date, similar to the forward pass. This allows us to calculate late dates, but are they of any value?

In the simple model in Figure 5-20, finish node B3 lists a finish date successor (November 18), whereas finish node L3 shows no known date successor. To find a common point from which to calculate a late date, we search through nodes L2 to L1 to X2. We can generate the Late Finish of X2 from the date successor on B3 through the path X2-B1-B2-B3. Taking this date (May 17), we can work back through the path L1-L2-L3 to calculate late dates. This means that the requirement to complete L3 is driven by the requirement to complete B3, which contradicts the relationships in the model. Nowhere does B3 appear as a successor of L3; their only relationship is a common predecessor, X2.

Figure 5-20

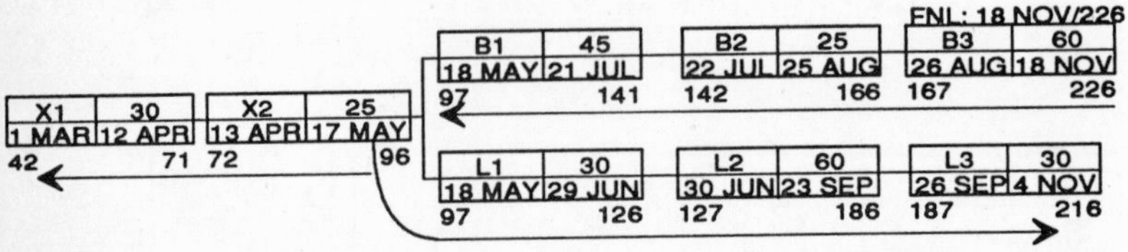

In addition to this inconsistency, the successor path to node X2: L1-L2-L3, exerts no influence on its Late Finish calculations, or on any of its predecessors. This makes the technique an invalid means of establishing requirement dates on these finish nodes since it reflects no causal relationship among the activities. This is another bad fix for poor modeling discipline.

Defaulting a Finish Node's Late Finish to its Early Finish

In the normal CPM process, the backward pass always follows the forward pass. As a result, every finish node has an Early Finish date calculated before the process determines its Late Finish. If a finish node has no established late date, the Early Finish is available. When using this technique, the Late Finish of the finish node is set equal to the tasks Early Finish and the backward pass commences. The finish node and its predecessors have zero (No) Path Time Reserve and only convergence with a path with negative Path Time Reserve will override it.

This is simply a variation of the default date method, assuming that a backward pass based on bogus dates is better than no backward pass. Furthermore, the technique defines the requirement of a path equal to its schedule. When we can finish is when we must finish. All of this is contrived, with little basis in practical modeling. The technique is one more example of a bad fix for sloppy modeling discipline.

Section Summary

It may help to review the salient points in this section:

A. Finish nodes are nodes that have no successors of their finishes or nodes for which the rules of the backward pass can supply no Late Finish date.

B. The user must provide Late Finish dates for all finish nodes.

C. Two techniques are practical for providing Late Finish dates for finish nodes:

1. Apply a finish date successor (FNL date) to each finish node
2. Set a Late Finish date for the finish node that provides a specified amount of time reserve (see Chapter 8).

D. A finish date successor (FNL date) allows us to model external schedule requirements (or successors) on the finish of any task.

E. A start date successor (SNL date) allows us to model external schedule requirements on the start of a task. This need rarely arises and negates the requirement for this date parameter.

EVENTS NODES

The backward pass process also reverses the forward pass through events nodes. Its end result is a start node with a start date that is later than its finish date, but this reflects the same perception problem as in the forward pass caused by the start-finish convention.

Figure 5-21 shows a model fragment with an events node embedded in the path logic. The finish of node AU has three successors: the start of AL, the start of AG, and the start of CO. To find the Late Finish of AU, we choose between:

$$\text{LF of AU} = \text{LS of AL} - \text{lag} - 1 = 138 - 0 - 1 = \text{day } 137$$

Figure 5-21

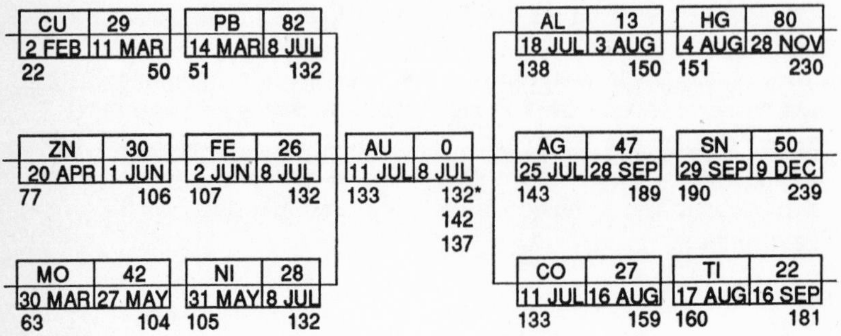

$$\text{LF of AU} = \text{LS of AG} - \text{lag} - 1 = 143 - 0 - 1 = \text{day } 142$$

$$\text{LF of AU} = \text{LS of CO} - \text{lag} - 1 = 133 - 0 - 1 = \text{day } 132$$

The earliest late date, workday 132 or July 8, is the Late Finish of node AU. Only this date satisfies all of the successors.

The start of AU has one successor, the finish of AU, so the Late Start of node AU is:

$$\text{LS of AU} = \text{LF of AU} - \text{DU} + 1 = 132 - 0 + 1 = \text{day } 133$$

The Late Start of AU is workday 133 or July 11.

The finish end of this node measures finishes and it supplies the Late Finish date of any immediate predecessor of the node, except for predecessors for which another successor imposes an earlier requirement. This Late Finish date imposes the strictest limit of all the events node's successors and it establishes a requirement for all of its predecessors. Note the LF of tasks PB, FE and NI is 8 Jul, the same as node AU. The start end of node AU measures starts. It imposes the most demanding start date of all the node's successors.

All constraints into and out of an events node should be finish-to-start, in which case, the events node takes its Late Start date from the Late Finish date of the node. In the example, the Late Start of AU, July 11, is also the Late Start of node CO. Therefore AU takes its late dates from the path CO-TI and the date successor at the end of this path. When we bias the desired end of an events node to report, only the corresponding late date (LS or LF) will be shown.

THE STANDALONE BACKWARD PASS

The backward pass in a standalone mode serves as a *set-back scheduler*. If we can determine the necessary parameters of a model (activities with durations and constraints with lags) and know when the project or segment of work must finish, then the backward pass can determine a schedule for the activities that satisfies all of the conditions of the model. A countdown to a missile launch is a good example of a set-back schedule. The MRPII (Material Requirements Planning) process is another.

The backward pass helps us determine a schedule based on a known end date. If the end date changes, we can quickly derive a new schedule based on the work and its preferred sequence. But a set-back scheduling process can only determine a schedule; it contributes nothing to project tracking and control. It does not let us project slips or changes forward to determine their impacts.

As a standalone process, the backward pass has a very limited, or at least infrequent, application. Only when we mate the backward pass with the other CPM processes can we thoroughly exploit it.

SUMMARY

The backward pass is a CPM set-back scheduling process defined by the following rules:

First Rule: The process works right to left chronologically through a project, beginning at the model finishes and culminating at the model starts.

Corollary: We must calculate the late dates of a node's successor(s) before calculating its own late dates.

Corollary: We must calculate the Late Finish of a node before its Late Start.

Second Rule: The backward pass includes an equation for each constraint type:

DU Equation: (LS of Node) = (LF of Node) – DU + 1
FS Constraint: (LF of PN) = (LS of SN) – lag – 1
FF Constraint: (LF of PN) = (LF of SN) – lag
SS Constraint: (LS of PN) = (LS of SN) – lag
SF Constraint: (LS of PN) = (LF of SN) – lag + 1

Third Rule: To choose among multiple successors (MSR) of a node's finish or start, calculate the dates that would result from each successor acting alone, then choose the earliest late date.

Normally, we find the Late Start of a node by subtracting its duration from its Late Finish (LF). Any other external successors of a node's start (with SS and SF constraints) may drive the node's Late Start date. When this happens, the node is called an *external successor stretched activity* (ESSA), or simply a *stretched activity*. This does not mean that the schedule of the task must change, but only that the requirement for the first portion of the task imposes a stricter schedule than the latter portion of the task.

Finish nodes are nodes that:

1. Have no successors of their finishes
2. Require the user to provide their Late Finish dates

Finish date successors or finish-no-later-than (FNL) dates provide one means of setting a finish node's Late Finish date. We can apply them to non-finish nodes to simulate external schedule requirements on the nodes' finishes.

One other method is acceptable for providing a finish node's Late Finish date: Setting a Late Finish that provides a specified amount of time reserve (see Chapter 8).

CHAPTER 6

Project Time Reserve

The calculation of time reserve is the simplest of the CPM processes. Understanding the meaning of the time reserve value and its relationship to other model data and processes is much more difficult, though.

Of the several formats for expressing time reserve, two provide especially rich information inputs for project management:

1. The difference between the cumulative span of work along a path and the time available to the path. This value, called *Path Time Reserve* (PTR), is the primary subject of this chapter.
2. The time that the working calendar excludes. This could include weekends, holidays, and off-shifts, any time excluded by the project calendar. This value is a finite resource potentially available to a project. It is commonly called overtime, but we refer to it as *Calendar Reserve* (CR). Chapter 7 will discuss it.

Chapter 1 defined the concept of Path Time Reserve as the difference between the cumulative duration of work and the time available to accomplish the work (two parameters). This is time available in excess of that committed to accomplish the work of the project, so management can apply it to problems to accommodate unexpected delays and extensions of work. This is important to keep in mind as we examine the calculation of Path Time Reserve through the CPM process.

The original definition of PTR yields a method for its calculation that becomes impractical when activities lie along more than one path, that is, when tasks have more than one predecessor or successor. Since most projects involve multiple paths, we need another way to calculate PTR. This alternate process is the Critical Path Method (CPM).

This method still derives the Path Time Reserve from the difference between path length and path time available, but the CPM techniques for calculating Path Time Reserve tend to confuse this fact. CPM calculates time reserve at each node, creating the perception that it is a property of the node rather than the path. This is an erroneous perception and we must maintain the understanding that Path Time Reserve is still a property of the paths of a project.

The calculation of PTR in CPM is simple. Subtract the early date of a node from its late date. We could subtract ES from LS, EF from LF, or both. If both operations always gave the same answer, if, that is, we could always be sure that:

$$LS - ES = LF - EF$$

then we could choose either value. This equation is not always true, however. The PTR at the beginning of a node can differ from that at the end. How and why this occurs will be shown later, but, because these two values are not always identical, we must calculate both values and compare them.

Subtracting ES from LS gives the Start-end Path Time Reserve (SPTR). In equation form:

$$SPTR = LS - ES$$

Subtracting EF from LF gives the Finish-end Path Time Reserve (FPTR). In equation form:

$$FPTR = LF - EF$$

To make these calculations, we do not work with calendar dates such as:

June 10 – June 1 = 9 days

Instead we work with equivalent numerical values derived from the project calendar. If June 1 were day 106 and June 10 were day 113 on a five-day week calendar, we would write:

day 113 (June 10) – day 106 (June 1) = 7 days

The PTR calculations depend on the calendar for the node, leaving out nonworkdays or Calendar Reserve (in this case a weekend). Simply stated, we express the calculated values of SPTR and FPTR in work units (days) pertinent to each task.

It is important to keep the plus or minus sign of the PTR values correct. Using the equations exactly as shown (subtracting early dates from late dates) will do that. If the value of the late date is smaller than the value of the early date, that is, if the late date occurs before the early date, then the PTR value will be negative. This indicates that a path includes more cumulative work than time available to accomplish it, something of a problem. This is referred to as *negative PTR* or *negative time reserve*. If the value of the late date of a node is larger (the date occurs later) than the value of the early date, then the PTR value will be positive. This indicates that the path has more time available to accomplish work than it has cumulative work to complete. A sufficiently large difference of this kind indicates a healthy project.

In summary, the process for calculating time reserve values includes the following steps:

1. From the appropriate calendar, find the numerical equivalents of the four early and late dates (ES, EF, LS, and LF) on every node in the model. This should be residual from the forward and backward pass processes.
2. Enter these values into the following equations to determine the Path Time Reserves (PTRs) at each end of every node:

$$SPTR = LS - ES$$

$$FPTR = LF - EF$$

Figure 6-1 shows a model with these steps accomplished. The math for calculating the time reserves in each of the nodes appears below:

TWINK
SPTR = LS – ES = 21 – 26 = –5
FPTR = LF – EF = 45 – 50 = –5

NERDD
SPTR = LS – ES = 21 – 26 = –5
FPTR = LF – EF = 49 – 45 = +4

TWITT
SPTR = LS – ES = 50 – 51 = –1
FPTR = LF – EF = 79 – 84 = –5

Figure 6-1

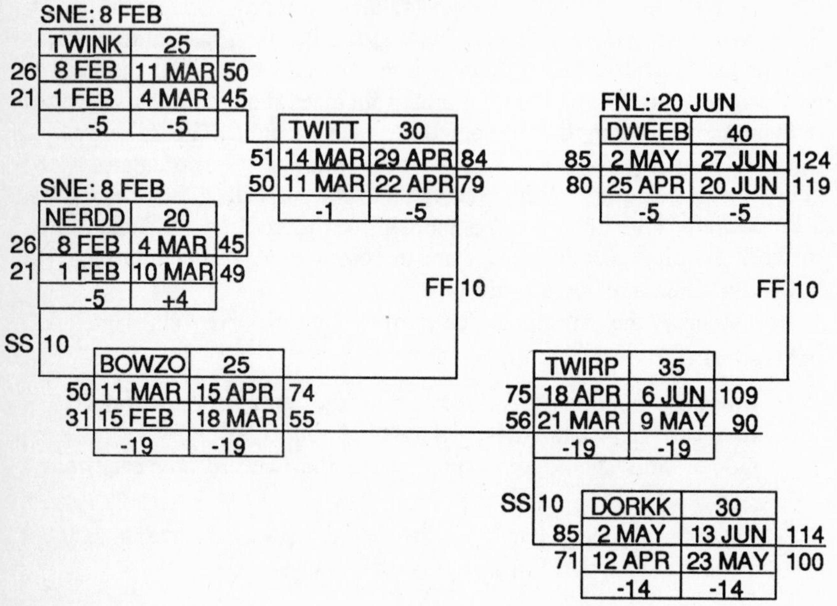

BOWZO
SPTR = LS – ES = 31 – 50 = –19
FPTR = LF – EF = 55 – 74 = –19

DWEEB
SPTR = LS – ES = 80 – 85 = –5
FPTR = LF – EF = 119 – 124 = –5

TWIRP
SPTR = LS – ES = 56 – 75 = –19
FPTR = LF – EF = 90 – 109 = –19

DORKK
SPTR = LS – ES = 71 – 85 = –14
FPTR = LF – EF = 100 – 114 = –14

This is a straightforward, simple process. Understanding the data derived from it, however, becomes much more complex.

PATH VERSUS NODE TIME RESERVE

As stated earlier, the most common misconception regarding time reserve arises because CPM calculates the values within each node, so it is easy to interpret it as a property of the node. Whether it is or not is very simple to prove. If time reserve belonged to each node individually, a node could consume some time reserve without affecting its successors' time reserves. Therefore, if values in successor nodes change with changes in one node's time reserves, then time reserve belongs to the path as a whole. It is a simple test to determine whether time reserve belongs to a node or to the path.

Figure 6-2

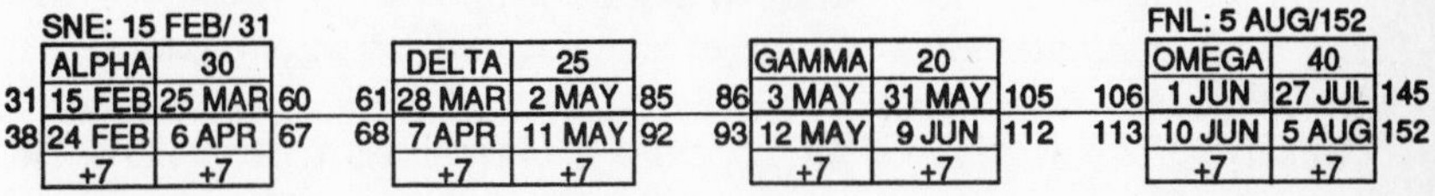

Figure 6-2 shows the conditions of a project as of February 1. The time reserve reported at each node end is +7, so if time reserve belonged to each task, then ALPHA, DELTA, GAMMA, and OMEGA would each have seven days. Let Task ALPHA use some of its time reserve by slipping its finish five days. To model this condition, we increase the span of Task ALPHA by five days and then recalculate the early and late dates and time reserves of each activity. The result appears in Figure 6-3.

Figure 6-3

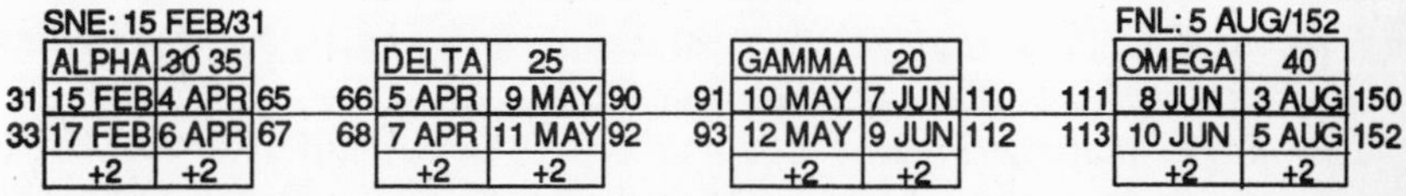

The five-day slip on one task reduced time reserve reported on all of the tasks by five days. Task ALPHA's slip diminished not only its own Path Time Reserves, but the reserves reported on all other tasks on the path, as well. These results demonstrate that the PTR calculated at the ends of each node do not belong to the node. *Time reserve is a property of the path.*

To understand this better, let us further examine what actually happened in this simple project. In the original example (Figure 6-2) the project had 115 days of work to accomplish (30 + 25 + 20 + 40) in 122 days [152 (Aug. 5) – 31 (Feb. 15) + 1 = 122]. This leaves time reserve of +7 days (from the Path Time Reserve equation of Chapter 1: PTR = TA – CSP = 122 – 115 = +7). In the updated example (Figure 6-3), the cumulative duration of work increased by 5 days to 120, while time available remained at 122 days. This reduced the time reserve of the path to +2 (PTR = TA – CSP = 122 – 120 = +2). The CPM process produced the same figure, even though it calculated the time reserve values within each node (at every node end).

Project time reserve that is calculated at both ends of each node belongs, not to any one task in a path, but to the entire path or paths on which the task lies. It therefore belongs to the project and should be managed by the project management. PTR values are like flow meters on a river. The characteristics of the river, its length, volume, rate of descent, etc., control the flow. We can measure the flow at various points to understand what is happening to the river. PTR calculations accomplish the same thing in that they allow us to measure what is happening to the path at various points (every node end) along the path.

As its end result, the CPM process determines the path's time reserves at every event (node end) in the project.

Since time reserve is always measured relative to the remaining work and remaining time available, it can give us the best perspective on how the current condition affects the remaining project.

PATHS

To completely understand project time reserve, we must understand paths and their relationship to the CPM process. We must fundamentally change our thinking when we begin to incorporate project modeling into our management processes. *We must begin to treat a project not as a series of independent activities, but rather as a series of paths of related activities.*

The single path in Figures 6-2 and 6-3, ALPHA-DELTA-GAMMA-OMEGA, shows up clearly. Conversely, the project in Figure 6-1 has many paths: TWINK-TWITT-DWEEB-, NERDD-TWITT-DWEEB-, NERDD-BOWZO-TWIRP-, NERDD-BOWZO-TWITT-DWEEB-, -BOWZO-TWIRP-DORKK-, etc. Let us examine the time reserve of each of these paths. The nodes of path NERDD-TWITT-DWEEB contain three different time reserve values: -5, +4, and –1. This seems to contradict the principle that time reserve is a property of a path, since each node appears to have a different value. Realize, however, that these three nodes also lie on other paths in the model. The time reserves reflected at each node end could pertain to any one of the other paths. To understand this, we must analyze this situation in greater detail.

The time available to path NERDD-TWITT-DWEEB is 94 days [119 (June 20) – 26 (Feb. 8) + 1 = 94]. The cumulative length of the path is 90 days (20

+ 30 + 40 = 90), which leaves a time reserve for the basic path of +4 days (94 – 90 = +4). This value appears only in the FPTR of node NERDD. All other PTR values on this path are affected by other paths. For instance, the start of TWITT must wait five days for the finish of TWINK, increasing the cumulative path length at this point to 95 days. The time available to the path is still 94 days, so the time reserve of the path at this point is –1 (94 – 95 = –1). This value appears in the SPTR of TWITT, the point of the delay.

The finish (EF) of TWITT is delayed an additional four days because the finish (EF) date of BOWZO (plus the lag) makes it an external predecessor stretched activity (EPSA). This further increases the path length at this point to 99 days, leaving a reserve of –5 (94 – 99 = –5), as the FPTR of TWITT indicates. As TWITT alone drives the date calculations of DWEEB, the SPTR and FPTR at the ends of this task reflect the same value (–5). A similar analysis of every path and every node end in the model gives insight into the effects of paths upon one another, helping us to identify any problems.

Paths influence one another's time reserve values at their points of intersection. Such influence always imposes the lesser of the possible time reserve values because the earliest late date in the backward pass and the latest early date in the forward pass determine time reserves. The choice among values in multiple predecessor and successor relationships determines the early and late dates for each node as a result of the worst case of all the paths on which it lies. Stated differently:

> A node that lies on more than one path will always reflect the Path Time Reserve of the path with the least reserve.

EXTERNAL CONSTRAINT STRETCHED DATES

In Figure 6-1, the following nodes have SPTR values that equal their FPTR values: TWINK, DWEEB, BOWZO, TWIRP, and DORKK. Two of the nodes have unequal time reserve values: NERDD and TWITT. What difference between the nodes in the first list and the nodes in the second list could cause this to happen?

The two nodes in the second list are stretched by external constraints in either the forward pass (TWITT) or the backward pass (NERDD) while none of the nodes on the first list show such stretching. Any stretching due to external constraints causes the value of time reserves to differ from the start end to the finish end of a node. This should come as no great surprise, because if a task is not externally stretched, the difference between its ES and EF is its span, which is also the difference between the LS and LF. If, however, an external predecessor stretches the EF or an external successor stretches the LS, the time between start and finish dates exceeds the span of the task. This means that the differences between the early and late dates at each end of the node will not be the same value.

Different forces cause stretching in the forward pass (FF constraints) than in the backward pass (SS constraints). Therefore the odds are extremely remote that a node would show the same amount of external constraint stretching in the forward pass that it shows in the backward pass. When one end or the other of a node is stretched, the time reserve values at the ends of the node (SPTR and FPTR) will differ. This gives us a means of determining when, at which end, and by how much external constraint stretching occurs in a node. We can determine when a node experiences stretching through a pair of equations:

$$\text{SPTR} = \text{FPTR: No external constraint stretching}$$
$$\text{SPTR} \neq \text{FPTR: External constraint stretching}$$

The same principle allows us to determine which pass caused the stretching. If SPTR exceeds FPTR (as in node TWITT), stretching occurred in the forward pass (the node is an EPSA). When FPTR exceeds SPTR (as in node NERDD), we know that stretching occurred in the backward pass (the node is an ESSA). Expressing this in equations:

$$\text{SPTR} > \text{FPTR: Forward pass stretching}$$
$$\text{SPTR} < \text{FPTR: Backward pass stretching}$$

The amount of stretching (AOS) in a node equals the absolute value of the difference between the two time reserves. To express this principle in an equation:

$$/\text{AOS}/ = \text{SPTR} - \text{FPTR}$$

Let us examine all of these relationships through an example. We can tell that node TWITT is stretched since SPTR does not equal FPTR (–1 >< –5). We know also that the Early Finish of this node is stretched in the forward pass since SPTR exceeds FPTR (–1 > –5). Furthermore, the node end is stretched by four days [SPTR – FPTR = –1 – (–5) = –4]. Likewise, we know that node NERDD is stretched because SPTR does not equal FPTR (–5 < +4) and that the Late Start of the node is stretched in the backward pass because SPTR is less than FPTR (–5 < +4) by nine days [SPTR – FPTR = –5 – (+4) = –9].

Once we have identified and quantified any stretching, we can take the necessary actions. If SPTR is less than FPTR, indicating external successor stretching in the backward pass, we normally do nothing other than noting the more demanding successor on the first portion of the task and ensuring that it is satisfied. If SPTR exceeds FPTR, indicating external predecessor stretching in the forward pass, we have several options that affect the schedule of the task.

1. Prevent the stretching by splitting the schedule of the task into two segments.
2. Delay the start of the task so it can proceed in one continuous span. The length of the necessary delay is the difference between the SPTR and the FPTR, the AOS.
3. Attempt to pull in or advance the finish of the predecessor in the FF constraint that causes the stretching by the amount of stretch (AOS).

This whole concept of externally driven Early Finish and Late Start dates may seem unimportant, but these are real forces that impact project effort. It also points out that paths with finish-to-finish and start-to-start constraints do not always pass through entire activities. Tasks become divided at points determined by the lags on the constraints and paths can pass through only one end of such nodes. This can complicate the analysis of the paths of our project.

CRITICAL PATH

Definition

In project management mythology, the *critical path* is the sequence of work in all projects which, if identified and managed, always leads to success. Such wishful thinking has little to do with the practical CPM process. If any such path existed, project managers could probably find it only through hindsight while musing, "If only we had . . ." Unfortunately, by then they can do nothing about it. Still, many seek the magical critical paths of their projects in hopes of making management simple.

The word *critical* seems to cause this misconception about the critical path. It implies pivotal importance, something crucial or indispensable. Certainly a manager must attend to anything critical in this sense, but this definition applies only loosely to the critical path of a project. In CPM, the critical path can be defined in two ways:

1. The critical path in a project is the longest path relative to the time available to complete it.
2. The critical path in a project is the path with the least project time reserves.

Unfortunately, understanding which activities and constraints lie on the path with the least amount of time reserve does not advance the project at all. It is generally an exercise in the obvious, anyway. Further, identifying the work along this path does not mean that we can manage it in isolation. Defining the critical path is just part of time reserve analysis.

In the model fragment shown in Figure 6-1, the critical path lies along: -BOWZO-TWIRP-. This path comes from an unseen predecessor of BOWZO and continues to an unseen successor of TWIRP. This path's negative 19 days of time reserve certainly needs management attention, but we cannot isolate on it because several other paths are also in trouble. TWINK-TWITT-DWEEB-etc. has negative reserve, as does the branch through DORKK. Subsequent analysis may reveal new and different problems. Further, the rate of time reserve erosion is often even more revealing of our problems. We simply cannot limit the focus of our analysis to the critical path.

Our quest is not to find one important path, but rather to understand the overall health of our project's schedule. If we understand what is happening to the project's time reserves, we can assess that health, isolate any problems, and begin resolving them.

With this understanding, we can learn to identify the critical path from the proper perspective, as just one of the paths that we must examine. Also, we can learn much about analyzing any path from our study of the critical path.

In Figure 6-1, we isolated the critical path by simple observation. Basically this is how we find the critical path in any project after completing the forward and backward passes and time reserve calculations. We must, however, develop a disciplined process for accomplishing this work.

Of all the paths to any given node in a project, one has the *least time reserve* (LTR). The critical path is the LTR path that leads to the node with the smallest time reserve value in the entire project. In a simple model with target dates only on start and finish nodes, the LTR path will flow from a start node to a finish node, maintaining a constant value throughout.

In a more typical model that has many internal target dates (FNL dates on nonfinish nodes and/or SNE dates on nonstart nodes), the LTR path may:

1. Terminate at a nonfinish node
2. Exhibit variation in the PTR value through the model

In this kind of model, the process of identifying the critical path becomes more complex. Fundamental principles of identifying this path still apply, however.

First, we limit our search to paths that terminate at finish nodes. This allows us to focus on the LTR that affects a project objective—the finish node of a path. We could select a specific finish node or look for the finish node in the project with the smallest FPTR. To do this, we arrange the finish nodes in ascending order relative to time reserve. The finish node with the smallest FPTR will top this list. Next, we examine all of this node's predecessors and select the one with the smallest time reserve. The constraint binding the two nodes is part of the critical path as well. We repeat this process until we trace a path all the way back to a start node.

An example will demonstrate how this process works. Figure 6-4 shows a model with the results of the forward and backward passes and time reserve calculations. To isolate the critical path of this project, we follow this process:

Step 1 We identify two finish nodes: WHOZ and THIS. The time reserve (FPTR) on node WHOZ (–5) is less than that on node THIS (+5). This tells us that the critical path has a PTR of –5 and it ends at WHOZ.

Step 2 The predecessors of node WHOZ (WHAT, WHAR, and WHEN) have time reserve values of +5, 0, and –5, respectively. The smallest value is –5, so the path moves through the FS constraint (WHEN-WHOZ) to node WHEN.

Step 3 The predecessors of node WHEN (WHYY and WICH) have time reserve values of 0 and –5, respectively. The smallest value is –5, so we see that the path moves through the FS constraint (WICH-WHEN) to node WICH.

Step 4 Since node WICH is a start node (it has no predecessors), we have found the entire critical path of the model: WICH-WHEN-WHOZ with a PTR value of –5.

This path is 60 days long (10 + 20 + 30) and has to be accomplished in 55 days [226 (Nov. 18) – 172 (Sept. 2) + 1 = 55]. This indicates that the path includes five more days of work than it has time available. This negative time reserve means that this path is in trouble. It is more important at this point, however, to see that the time reserve values calculated in the CPM process agree with this analysis. In this example only the critical path is in trouble as it has negative time reserves, there are other paths with marginal positive or zero reserve.

Notice the impact of time reserve at the points of path intersections. When node WHYY intersects with node WICH at the start end of WHEN, the least time reserve value (–5 versus 0) carries forward. When the path WHYL-WHAT intersects with path WICH-WHEN, WHOM-WHAR, and WHOO-WHAR at the

Figure 6-4

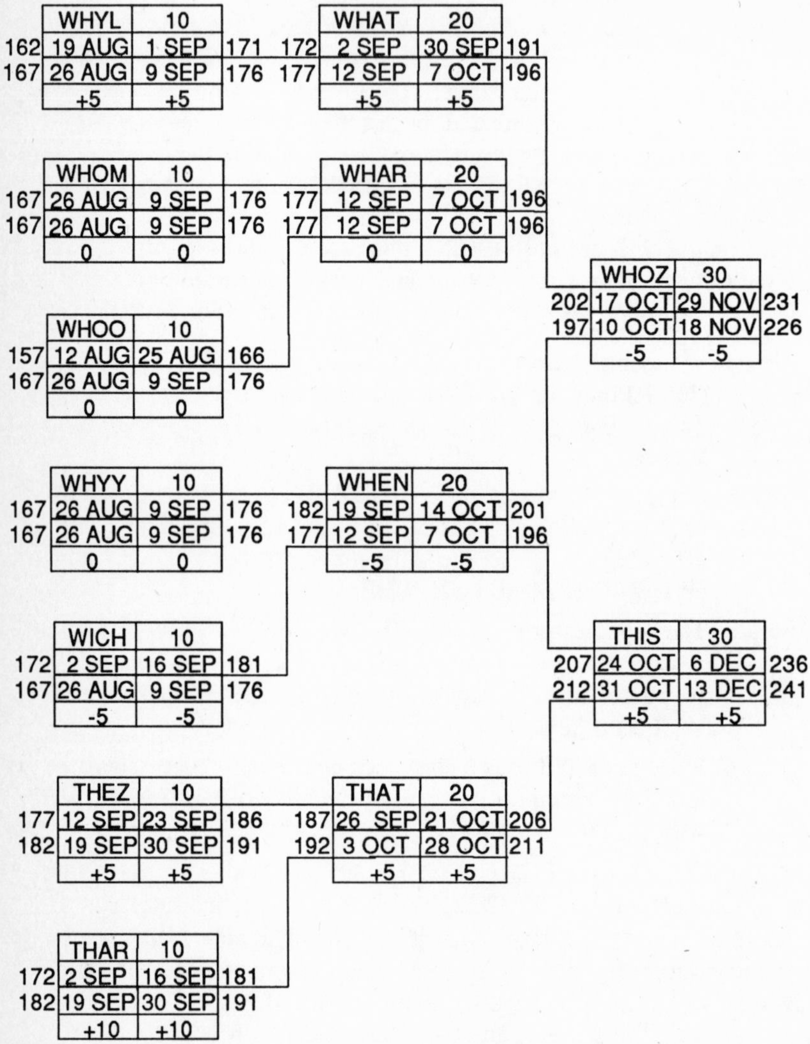

start end of WHOZ, the least time reserve value (–5 between +5, 0 and –5) carries forward. As paths merge, the least time reserve value always prevails because in the forward and backward passes we must choose the strictest among the multiple predecessor and successor relationships. The dates calculated on each node in both passes result from the worst-case path into and out of the node.

FF Constraints and the Critical Path

A second example reveals a new twist on critical path determination. To find the critical path in Figure 6-5, we follow our simple process:

Step 1 Node F16 is the only finish, so its time reserve of +5 is the lowest finish end value in the model by default. This makes node F16 the end of the critical path with +5 days of reserve.

Step 2 The predecessors of node F16, (nodes F15 and F106) have PTR values of +5 and +10, respectively. The path moves through the node with the lesser value, +5, following the FS constraint (F15-F16) to node F15.

Step 3 The predecessors of node F15, (nodes F4D, F105, and F101) have PTR values of +5, +10, and +15. The path follows the smallest value, +5, through the FF constraint (F4D-F15) to node F4D.

Step 4 Node F4D's only predecessor, node F100, has a PTR of +5. The path moves through the FS constraint (F100-F4D) to node F100.

Step 5 Since F100 is a start node, we have identified the entire critical path: F100-F4D-F15-F16, with a PTR value of +5.

Figure 6-5

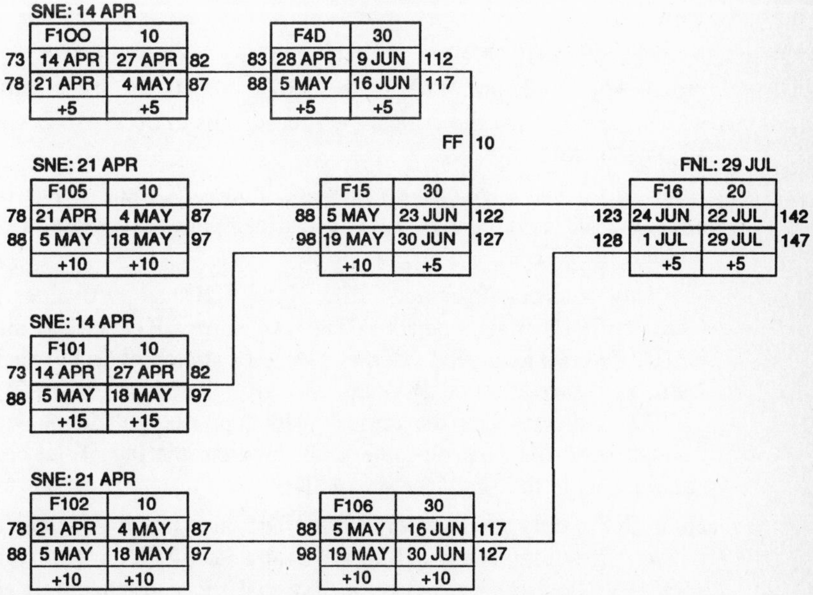

The finish-to-finish relationship between nodes F4D and F15 differs from any interface in the first example in a couple of ways. The PTR at the start end of node F15 (+10) exceeds the PTR at its finish end (+5). This indicates that node F15 was stretched by an external predecessor in the forward pass by 5 days. It also indicates that the critical path does not go all the way through node F15; it goes only through the finish end. If we examine the predecessors of the finish of F15, we find that it picks up the PTR value of +5 from the end of node F4D. This means that the critical path actually goes through the finish-to-finish constraint to include only the last 10 days of F15, the stretched portion of the task.

The critical path of this model is really F100-F4D-the last 10 days of F15-F16. This path includes 70 days of work (10 + 30 + 10 from F15 + 20) to be completed in 75 days [147 (July 29) – 73 (April 14) + 1 = 75 days]. This leaves five more days of time available to the path than its work requires. The PTR calculated at the end of each node along this path agrees with the difference between the cumulative length of the path and the time to accomplish the path (+5). This also verifies that the lag value on a constraint is part of the cumulative path length (CSP).

SS Constraints and the Critical Path

Another example takes this a step further. Let us find the critical path of the model in Figure 6-6 through our simple process.

Step 1 Nodes LBJ, GRF, MND, and GNB are finish nodes with PTR values of +5, 0, –5, and +10. In this group, –5, is the least value, which indicates that the critical path terminates at node MND with –5 days of reserve.

Step 2 Node MND has only one predecessor, node JEC, and it has the expected PTR value of –5. The critical path, therefore, runs through node JEC.

Step 3 The only predecessor of node JEC, Node IKE, has two values of PTR. Its SPTR is –5, and its FPTR is 0. Since SPTR is less than FPTR, we have identified external successor stretching in the backward pass through node IKE amounting to five days [0 – (–5) = 5]. This indicates that the critical path does not go all the way through node IKE, but only the initial portion, the lag of the constraint. This is the first 10 days of IKE.

Step 4 Node IKE's only predecessor, Node FDR, also has two values of project time reserve. Its SPTR is –5 and its FPTR is +5. Thus, node FDR is also stretched by an external successor in the backward pass, but by 10 days [+5 – (–5) = 10]. The critical path does not go all the way through node FDR, but only through the initial portion, the lag of the constraint (FDR-IKE) or the first five days of FDR.

Step 5 Node FDR has no predecessors, so it is the start node at which the critical path terminates. The critical path of the model is: the first 5 days of FDR-the first 10 days of IKE-JEC-MND.

Figure 6-6

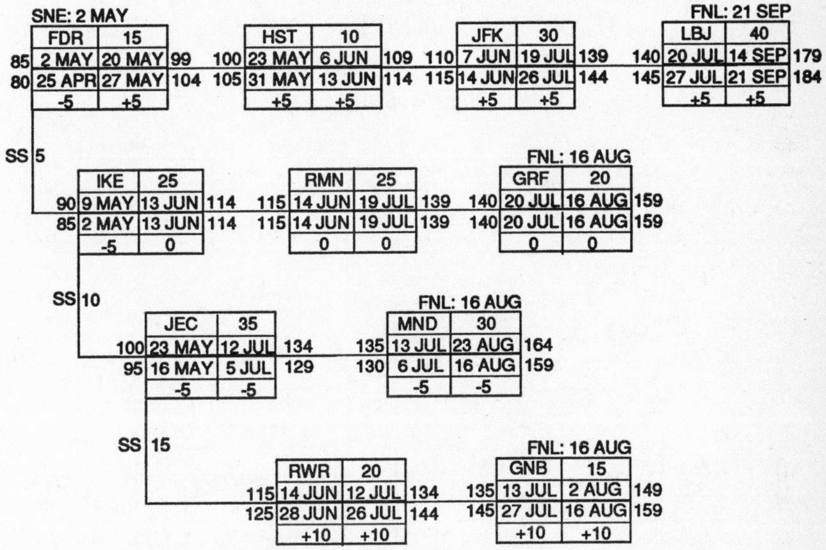

This yields a total path length of 80 days (5 from FDR + 10 from IKE + 35 + 30 = 80) to be accomplished in 75 days [159 (Aug. 16) – 85 (May 2) + 1 = 75 days]. The resulting shortfall of five days corresponds to the PTR of the node ends along this path and is the problem that we must address. We must also pay close attention to the zero and +5 reserve paths. Besides identifying the critical path, more importantly we have assessed the schedule health of our paths. These examples of path analysis lead to some conclusions:

1. A path can go through a portion of a task in a finish-to-finish or start-to-start constraint relationship.
2. The portion equals the lag on the constraint.
3. The lag therefore is part of the cumulative length of the path just like any node duration along the path.

4. Thus, the CPM process treats lags just like node durations as part of the cumulative sequence of work, so they must represent the accomplishment of work. This was stated as a rule in Chapter 2; it has now been proven.

The Longest Path versus the Critical Path

Another common, but erroneous notion regards the critical path as the longest path in the model. This would require that critical path determination depend on cumulative path length alone. Our definition states, however, that the critical path is a function of (1) path length and (2) time availability. Figure 6-7 will help us understand this better.

Figure 6-7

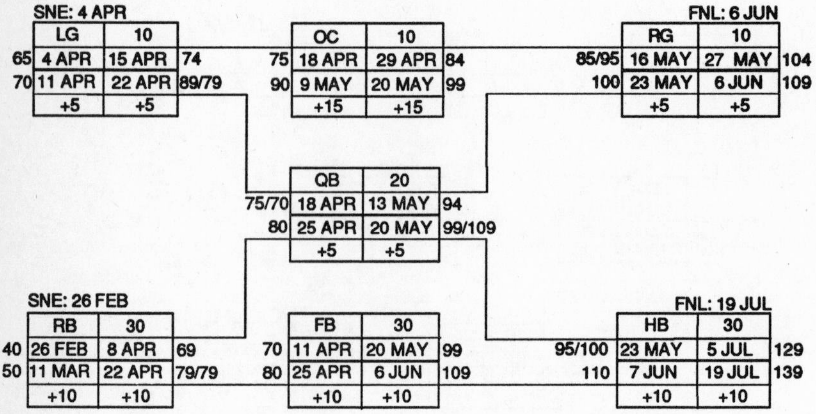

The several paths in this model have varying cumulative lengths:

LG-OC-RG: 30 days
LG-QB-RG: 40 days
RB-FB-HB: 90 days
RB-QB-HB: 80 days
RB-QB-RG: 60 days
LG-QB-HB: 60 days

The longest path, RB-FB-WR, includes 90 days of work to be completed in 100 days [139 (July 19) – 40 (Feb. 2) + 1 = 100 days]. This yields a reserve of +10 days. Path LG-QB-RG includes less work, 40 days, but it must be accomplished in only 45 days [109 (June 6) – 65 (April 4) + 1 = 45 days], leaving a +5 day time reserve. Path LG-QB-RG has less reserve (+5) than the longest path, RB-FB-WRG (+10). Stated differently, LG-QB-RG is the longest path *in terms of the time available for its accomplishment.* It is the critical path, although there are several other paths that take longer to complete, because it has the least time reserve.

If a project has only one start and one finish, then all paths would have the same time availability, that of the longest path, so the longest path would be the critical path. This is a special case, however. We rarely encounter a project having only one independent start and one independent finish in the real world, where we need to understand and use the CPM process.

This exercise further emphasizes the fallacy of concentrating only on the critical path. In the example, the path with the greatest concern to the project could be the longest path, the critical path, or any other path, for that matter. The critical path is only a mathematical parameter of the project; identifying it contributes *nothing* to our management of the outcome of the project. Controlling a project is just not as simple as identifying and managing one path of tasks. We must come to grips with this truth if we are ever to get adequate value for our efforts. Perhaps it would be better to use the term *path with the least time reserve* and cease calling it the critical path.

PATH LENGTH DETERMINATION

We often need to know the length of a path, but it is very cumbersome to add all the spans, delays, and lags to determine it. Fortunately, this arithmetic is unnecessary, as a simple algebraic equation alteration will yield this value based on easily available parameters. The basic path time reserve equation is:

$$PTR = TA - CSP$$

Stated verbally, path time reserve equals time available to a path minus its cumulative sequence. Simple algebraic rearrangement gives an equation for the cumulative sequence of a path:

$$CSP = TA - PTR$$

The cumulative sequence of a path determined in this way would include task durations, constraint lags, and delays along the path. We determine Path Time Reserve for any path through the CPM process and our subsequent analysis. To find the time available to accomplish a path, we simply find the Early Start date of the path start node (ESP) and subtract it from the required or Late Finish date of the end node of the path (LFP):

$$TA = LFP - ESP + 1$$

This is the band of time in which a path is to be accomplished. We have used this equation to determine time available (TA) throughout previous chapters. A path that begins on February 1 and must end on December 23 has 229 available work days (using the five-day week calendar with holidays):

$$\begin{aligned} TA &= LFP - ESP + 1 \\ &= \text{day } 249 \text{ (Dec. 23)} - \text{day } 21 \text{ (Feb 1)} + 1 = 229 \text{ days} \end{aligned}$$

If we had determined the PTR of this path to be -25, then the cumulative length of the path would be:

$$CSP = TA - PTR = 229 - (-25) = 254 \text{ days}$$

Taking this process further, we could quickly examine the impact of switching to a six-day week calendar to see if working Saturdays could alleviate the time reserve problem indicated by the –25 day figure. Using the six-day week calendar, the time available between February 1 and December 23 is:

$$\begin{aligned} TA &= LFP - ESP + 1 \\ &= \text{day } 297 \text{ (Dec. 23)} - \text{day } 25 \text{ (Feb. 1)} + 1 = 273 \text{ days} \end{aligned}$$

$$\begin{aligned} \text{No. of Sat.} &= TA \text{ (6-day week)} - TA \text{ (5-day week)} \\ &= \text{day } 273 - \text{day } 254 = +19 \text{ days} \end{aligned}$$

From this we can determine that there are 19 days more time available in the six-day week versus the five-day week. This is then the number of Saturdays between the two dates. Our preliminary analysis indicates that working Saturdays (19 extra days) would not solve the problem. There is still a short fall of six days (25 – 19 = 6). This analysis can only be considered cursory though, because the influence of another path (or any constraint on the path) could affect our solution. Only a full CPM time analysis would account for these possibilities. Nevertheless, we see how path length, the time available to a path, and Path Time Reserve affect one another.

TIME RESERVE AS A MANAGEMENT PARAMETER

The critical path gives the project manager no magical powers, but identifying and understanding the worst-case or least-time-reserve path within a project is still important. This cannot overshadow or even approach the importance of the true purpose of the CPM process, however, which is to identify and understand all Path Time Reserves (PTRs) in a project, so that we can manage them. This is a basic process; we monitor the left side of the PTR equation to spot problems, then we manipulate the right side of the PTR equation (TA – CSP) to correct them.

We must use more than a single method to monitor the left side of this equation (PTR) to identify time reserve problems. First, we need to recognize any path with insufficient reserves. Certainly any path with a negative reserve value falls into this category, but a path with no reserve or a small positive value could very well need attention, as well. Therefore we aspire to maintain in all paths some positive value that we deem adequate to meet schedule objectives.

How much reserve is adequate depends upon several variables, one of which is the length of the path. A path with a year of cumulative duration could need a month of reserve, whereas a three-year path would certainly need more. There are no hard and fast rules for these quantities, but certainly as the length of the path of work increases, so does the need for reserve.

When we have completed half the work of a path, we probably need less reserve than we needed at the beginning of the path. If we are halfway through a path and have used 50 percent of the time reserve, we probably need not worry. If we have used 75 percent of the reserve by the same point in the path, we may be in trouble. We must watch the rate of time reserve erosion (ROE) in relation to the rate of progress (ROP) of the project along with the value of the time reserves. We will explore this important concept more thoroughly in Chapter 12.

Risk also affects the need for time reserve on a path. A pure CPM process may not evaluate time reserve in relation to risk, because task durations are independent of risk. We can still evaluate the impact of risk in two ways. We could use a sophisticated risk analysis tool to add quantitative values to the basic CPM figures. There is, however, an easier way to derive some very useful analytical data. We can simulate risk conditions in the base CPM model by increasing durations of any task(s) in proportion with perceived risk.

Suppose, for instance, a vendor has committed to accomplish a task in a month. In the base model, the task's duration would equal the work days in a month. This may create no time reserve problems, but we fear that the vendor will actually take six months to perform the task. This creates the risk. It is a simple matter, though, to change the duration of this task to six months and then assess the impact of such a slip. If this evaluation were to reveal a severe problem, we could plan some way to avoid or mitigate the potential impact.

As one strategy to mitigate risk, we should build extra Path Time Reserve into the concerned paths to absorb the potential delay. Whatever the fix, though, we must first identify potential problems and assess their impacts through modeling techniques. This is called what-if analysis, and we will discuss it more thoroughly later.

MANAGING PROJECT RESERVES

With the calculation and understanding of Path Time Reserve through the CPM process, we have come full circle. Chapter 1 introduced the concept of PTR as a project management parameter. It has taken five additional chapters to explain the derivation of this value from the project modeling process thoroughly.

We coined the term *time reserve* because of its analogy to cost reserve, a percentage of a contract's value that project managers set aside to solve unexpected problems. Path Time Reserve provides a similar crisis fund, but in the form of time rather than dollars. We can determine cost reserve through straightforward calculations, though. Determining time reserve requires a much more complicated project modeling process. Yet we must bear the burden of this work to develop this key project parameter.

To solve a project problem, we should always evaluate time versus cost. We must assess the cost to the project reserves of the additional effort of detecting and solving problems. In this kind of analysis, we often find that we can trade one of the reserves to preserve the other. Working overtime spends cost reserve to solve schedule problems; in essence, we trade cost reserve for time reserve. We could just as well decide to sacrifice some time reserve and not work the overtime, preserving cost reserve and ultimately profit at the potential expense of the schedule. This is one of the primary roles of project management, making decisions on the utilization of project reserves. Time reserve calculations better equip the project management for such decisions.

Full project modeling provides essential information of this kind that conventional scheduling techniques do not. In making the forward and backward passes, we in essence evaluate every task, determining its relationship to the total project through its predecessors (in the forward pass) and its successors (in the backward pass). We no longer consider each task as an isolated, independent effort, but rather as an integral part of a total project. In essence, our evaluation shows us how each task fits into the project scheme, so we no longer view the project as a series of activities, but rather a series of paths of activities.

Monitoring PTR values allows us to understand the impacts of our current situation on the balance of the project. As the project evolves and changes with the completion of tasks, we can determine the effect of each of these on the remainder of the project. This makes PTR a proactive parameter, as it relates to the future, not the past.

In any modeling process, however, we must keep a certain perspective. We are dealing with a forecast of the future and there are no absolutes in forecasting. The CPM process is not an infallible crystal ball. We cannot take the values derived from this process as absolutes. Rather, we must treat them as relativistic values. This certainly does not mean that CPM data is invalid, or that the result is not worth the effort. It does mean that we must view the information in the proper perspective.

There is no magic in this process. Used properly, it allows us to measure and understand potential impacts. It will not tell us when a test will fail, a requirement will change, or a vendor will deliver late, but it can help us understand the impacts of all of these factors on the rest of the project. Used properly, time reserve management allows us to detect and therefore react to adverse impacts earlier than through conventional scheduling processes.

TIME RESERVE AND THE PROJECT MANAGEMENT PARAMETERS

The discipline of the project modeling process must be as good as its intentions. The penalties of poor discipline are useless data and a great deal of effort for very little returned value. The fundamental principles of this discipline are generally uncomplicated, provided we keep the foremost objective, identifying PTR, firmly in mind. Successful model analysis relates all model parameters to the time reserve equation.

$$PTR = TA - CSP$$

The various parameters of the model determine the two values on the right of this equation (TA and CSP).

Time available to a path (TA) depends on two model parameters. First, the date values imposed on the model as either date predecessors (SNE dates) or date successors (FNL dates) affect TA by specifying the amount of time available to a path. Basically the time available to a path is the difference between the SNE date on the path start node (the Early Start of the start node or ESP) and the FNL date on the path finish node (the Late Finish of the finish node or LFP). These dates form the time bounds within which the path must be performed, according to the equation:

$$TA = LFP - ESP + 1$$

SNE dates on nonstart nodes or FNL dates on nonfinish nodes can alter the time available, as well. This establishes subpaths with different values for time available. Thus, entering date targets into a model requires thorough knowledge and rigid discipline.

The other parameter that controls time available to a path is the calendar against which we measure durations and lag values. Variations in workday patterns (e.g., a six-day week versus a five-day week) alter the time available to a path. Therefore the appropriate calendar must be properly established and related to each activity duration and constraint lag.

Two sets of parameters control the other term on the right side of the PTR equation, the cumulative sequence of the path. These are the sequence of the path and its cumulative span. The sequence of a path depends on the constraints of the project (FS, SS, FF), which reflect the interrelationships of specific tasks. Every constraint ties two nodes into a path. By continuously stringing predecessors and successors together with the appropriate types of constraints, we define the sequence of each path and the serialization of durations (activities) and lags (constraints). Every path extends from a node with no predecessor (a start node) to a node with no successor (a finish node). Many paths may flow out of each start node, into each finish node, and through all other nodes. The defined constraints of the model control all of this.

The length or cumulative duration of a path depends on the durations of nodes and lags of constraints. Through these numeric values, we model the work to be accomplished by the project, then we sum the values through the sequence of the path to determine CSP. In calculating this value, we treat task durations and constraint lag values exactly the same way. Both together determine the expected span of work. Each must therefore represent the expected time it will take to accomplish the effort of the path.

In summary, the key to proper project modeling techniques is understanding the relationship between the model parameters and the PTR equation:

$$PTR = TA - CSP$$

Time available to a path (TA) is a function of:

1. The calendar on which we measure each node's duration and each constraint's lag value
2. SNE dates on the start nodes
3. FNL dates on the finish nodes
4. Any internal date target (predecessor or successor)

The cumulative sequence of a path (CSP) is a function of:

1. Specific constraints that relate two tasks to one another. A string of consecutive predecessors and successors bound together by constraints of appropriate types (FS, SS, or FF) define the sequence of a path.

2. Durations on tasks and lags on constraints sum to the cumulative work of a path.

These are the variables that project planners must determine and maintain in a project model. The basic purpose of the CPM process, the determination of Path Time Reserves, requires that the parameters of the model follow these rules. Each parameter determines schedule values in the forward or backward pass, and each parameter also affects the calculation of time reserves.

If the rules of the parameters are properly followed, the modeling process will yield not only the desired schedule, but valid time reserve results, as well. If these rules are not followed, regardless of intent, the process will yield information with only minimal value.

Meticulous modeling also requires that we remember that every value in the model is an estimate—it reflects our best understanding of our project. Because we cannot perfectly understand the future, errors are introduced into the process. This means only that results are relativistic rather than absolute. It does not in any way diminish the value of the effort. Imperfect forecasts of the future are certainly far better than closing our eyes and leaving the project to fate. The choice is up to each project manager. They can attempt to control the future, or allow the present to control them.

SUMMARY

We calculate Path Time Reserve at both ends of each node by the following equations:

$$\text{Start end: } SPTR = LS - ES$$
$$\text{Finish end: } FPTR = LF - EF$$

These calculated values are expressed in work units (e.g., workdays) relative to the appropriate calendar.

Although we calculate Path Time Reserve (PTR) within each node, the value is relative to the path or paths on which the node lies. We measure PTR at node ends, or events, just like flow meters in a river. A node that lies on more than one path will always reflect the PTR of the path with the lowest value.

PTR indicates when a node is stretched by an external constraint, in which pass it is stretched, and by how much. When SPTR does not equal FPTR, the node is stretched. When SPTR exceeds FPTR, the node is stretched in the forward pass. When SPTR is less than FPTR, the node is stretched in the backward pass. Subtracting FPTR from SPTR gives the amount of stretch (AOS).

The project manager must decide how to respond to external constraint stretching. Backward pass external successor stretching typically requires no response. To respond to forward pass external predecessor stretching, the user has three choices:

1. Prevent the stretching by splitting the schedule of the task into two segments.
2. Delay the start of the task to allow it to proceed in one continuous span. The length of the delay necessary to do this is the AOS.
3. Advance the finish of the predecessor in an FF constraint that causes the stretching by the AOS.

The critical path is:

1. The longest path in a project relative to the time available for its completion
2. The path in the project with the least project time reserves (the LTR path)

In a model with internal target dates, the most critical path may terminate in a nonfinish node. Its PTR value may also vary through the model.

We identify the critical path in a model through the following process:

1. Select the finish node with the lowest PTR value
2. Examine its predecessors and isolate the one with the lowest PTR value
3. Trace the path back through the linking constraint
4. Repeat steps 2 and 3 until reaching a start node

A path passes through the initial portion of a predecessor task in a start-to-start relationship and the final portion of a successor in a finish-to-finish relationship. The portion of the task the path passes through equals the lag on the constraint, so we add lag values onto the cumulative length of a path. The CPM process treats constraint lags and node durations exactly the same; therefore both must represent the accomplishment of work.

Searching for the critical path is generally an exercise in the obvious. The effort may very well isolate a path that is not the most critical to the project. At any rate, it is only part of time reserve analysis.

The following equation provides the cumulative length of a path:

$$CSP = TA - PTR$$

The time available to a path (TA) is derived from the equation:

$$TA = LFP - ESP + 1$$

where LFP and ESP are the Late Finish of the path and the Early Start of the path, respectively.

The project manager must take action regarding any path that has:

1. Negative, zero, or marginal values of PTR
2. An excessive rate of time reserve erosion (ROE) relative to the project's rate of progress (ROP) (see Chapter 12)
3. A source of risk that could result in excessive time reserve erosion

Time reserve and cost reserve are both management parameters. Often one can be traded for the other. We must assess the impact of project progress and any changes on either parameter.

The path activities and constraints are essential parameters that directly affect the Path Time Reserves of a path. Time available to a path depends on:

1. Date predecessors and date successors injected into the model on any node
2. The calendars on which we measure each activity's duration and each constraint's lag

The sequence of a path depends on:

1. The types of constraint relationships, which reflect the interrelationships among tasks
2. The cumulative length of a path depends on the durations of activities and the lags of constraints.

The model must maintain the principles of these relationships or the CPM process will yield invalid time reserve values.

CHAPTER 7

Calendar Reserve and Free Reserve

Besides Path Time Reserve, another significant form of time reserve, *calendar reserve,* may provide additional time to apply to project problems. It measures the periods of work excluded from the project calendar, and therefore from the calculation of Path Time Reserve (PTR). This reserve is commonly called *overtime.*

In the year displayed in the five-day, eight-holiday work calendar, the calendar reserve would include:

1. 53 Saturdays
2. 52 Sundays
3. 8 holidays
4. All off-shift time (potential 2nd and 3rd shifts, assuming an eight-hour workday)

This amounts to 113 days of holidays and weekends and 253 second and third shifts on weekdays.

In the six-day week calendar with eight holidays, the calendar reserve would include:

1. 52 Sundays
2. 8 holidays
3. 4 Saturdays associated with holidays
4. All off-shift time

This amounts to 64 days of holidays and weekends and 302 potential second and third shifts on workdays.

Both of these calendars assume that the labor force works only one shift. If they work multiple shifts, the principles remain the same, but calculating the off-shift reserve is more complex. It could be zero, or each shift could work off-shift overtime, provided adequate facilities were available.

There is a considerable amount of overtime available. An average month with five-day work weeks, eight-hour shifts, and one holiday includes:

1. 152 working hours
2. 520 hours of calendar reserve

Calendar reserve in a month exceeds straight work time by almost three and a half times. Much of this *overtime* is practically unusable, but we could reasonably count on every Saturday and half of a second shift as sustained overtime in a one-shift operation. This amounts to 112 hours of sustained overtime per month, or about 70 precent of the straight time. Simply through the use of overtime we can increase the work accomplished in a month by 70 percent, an enormous asset. Is it any wonder that overtime is such a common method of solving schedule problems?

Quantifying available calendar reserve is more involved than simply counting hours. First, we divide it into categories: Saturdays, Sundays, holidays, half-shifts on workdays, etc. Second, we define the period of time within which to calculate the calendar reserve by simply specifying a start date and a finish date. Then we calculate the available overtime periods in each category between the specified dates.

For instance, on the five-day week, eight-holiday calendar, how many Saturdays and half-shifts fall between June 1 and December 30? The number of available second shifts would equal the number of workdays between the two dates. This would be a simple computation using the appropriate calendar and the basic date math equation.

$$\begin{aligned} DU &= FIN - ST + 1 \\ &= \text{day } 253 \text{ (Dec. 30)} - \text{day } 106 \text{ (June 1)} + 1 \\ &= 148 \text{ days} \end{aligned}$$

The 148 workdays between June 1 and December 30 correspond to 148 half-shifts (or 74 equivalent workdays) of available overtime during this same period. Simply stated, available workday extension periods between two dates equals the number of workdays between the two dates.

Determining the number of Saturdays between two dates is not as straightforward. We could divide the number of workdays by 5 to get a rough estimate of the number of weeks, and therefore the number of Saturdays. In the example, this is 29.6 (148/5), giving an estimate of 29 Saturdays. The calendar shows 30, but that includes the Saturdays after Thanksgiving and just prior to Christmas. The actual number available for overtime is 28. Our first answer, though close, is not accurate.

We can accurately calculate the number of Saturdays between two dates by comparing two calendars: in this case the normal five-day week versus the six-day week, which excludes only Sundays and Saturdays following holidays. We first calculate the available workdays between the two dates in both calendars:

Five-day week: day 253 (Dec. 30) – day 106 (June 1) + 1
= 148 days

Six-day week: day 301 (Dec. 30) – day 126 (June 1) + 1
= 176 days

These two calendars differ only in that the six-day week calendar includes the Saturdays that do not immediately follow or precede a holiday. Therefore, subtracting the number of days in the five-day week calendar from the number of days in the six-day week calendar will leave the number of Saturdays that can be worked between the two dates: 176 - 148 = 28. In this way, we can calculate nonwork periods like Saturdays, Sundays, and/or holidays between two dates. In our example, we found 148 additional half-shifts and 28 Saturdays available for overtime between June 1 and December 30. This amounts to an extra 102 workdays or 20+ work weeks.

To summarize, we calculate the various categories of calendar reserve by one of two means:

Extensions of Workdays: To determine the number of potential overtime periods (either half-shifts or full shifts) between any two dates, we calculate the number of workdays between the selected dates from the basic date math equation:

DU = Finish Date – Start Date + 1

Specific Nonworkdays: To quantify the overtime periods on nonworkdays, such as the number of Saturdays, Sundays, and/or holidays between two dates, we find the difference between the number of workdays between the two dates in the normal work week calendar and the number of days between the same two dates in the calendar that includes the nonwork periods:

Overtime Calendar: (Finish Date – Start Date + 1) minus
Normal Calendar: (Finish Date – Start Date + 1)

By these methods, we can calculate the various categories of calendar reserve available to a project, a path, or any segment of work.

Once we have determined how much calendar reserve or overtime is available, we must properly manage it. If our decision about utilizing overtime is strictly based on schedule nonperformance, we are managing reactively, not proactively. We are reacting to historical parameters that only indirectly affect the future. Only when we make our decision based on the future condition of the project does control become proactive. Since we want to efficiently affect the outcome of the remaining project with the use of overtime, we must make proactive decisions about its use.

Path Time Reserve (PTR) is the figure by which we measure downstream conditions, so it affects our management of calendar reserve. In fact, calendar reserve is an element of the Path Time Reserve equation in the form of time available (TA). As concerns arise about Path Time Reserves, we can selectively apply available calendar reserve as one potential solution.

Careful management of calendar reserve requires understanding that it diminishes with time. It is at its greatest in the beginning of a project or path and disappears steadily until the finish. For instance, a year-long project starts with 52 available Saturdays, but six months into the project only 26 Saturdays remain before the project end. Every week that passes takes with it any unused overtime.

Unfortunately, without careful management the need for overtime is almost always the greatest at the end of a project. This is another sign of reactive management. Reactive managers only monitor and control project status, so they tend not to see beyond the near term. Further downstream, where the project objectives await, they see little threat. They wait, they delay action until deadlines are imminent, but the availability of overtime to meet the problems is at its lowest point in the project and only drastic measures can correct the project course.

A proactive manager would simply use calendar reserve earlier in the project, when it is more abundant. There is no panic, just a controlled, methodical adjustment to the project's course. It is not used wholesale either, but only along the paths that indicate PTR problems. Full understanding and careful management of all time reserves of a project makes for judicious and efficient use.

Cost Considerations

A cost impact analysis should accompany every use of calendar reserve because overtime almost always increases the cost of the project. The cost of a task is basically the duration of the task multiplied by the amount of resource needed to accomplish the task multiplied by the cost of the resource. A 10-day task that takes three people for its entire duration at a cost of $50 an hour per person costs $12,000:

10 days x 3 people x $400/day/person
($50 per hour x 8 hours) = $12,000

We will call this amount the *baseline cost* of the task and use it for comparison as we examine some cost variations that could occur in the course of this task.

Example 1. Suppose that the total span does not change, but 2 of the 10 days' work are performed on weekends at time and a half. The total cost becomes $13,200, an increase of $1,200.

(8 days x 3 people x $400/day/person) +
(2 days x 3 people x $600/day/person) = $13,200

We altered the cost per person for the portion of the task worked on overtime. The higher labor rate for that portion increases the overall cost of the task.

Example 2. Suppose that the task is performed on 10 normal workdays, but the workers must work an extra half-shift each day to get the job done. This increases the duration of the task by 10 half-days or 5 days. This impact affects the cost per day of the resource by increasing the hours worked: $600 (12 hours X $50 per hour, the standard pay rate for all hours worked) versus $400. The total cost becomes $18,000, an increase of $6,000.

10 days x 3 people x $600/day/person = $18,000

The total cost of the task increased with the increase in its span.

Example 3. Suppose that the task takes two additional Saturdays to finish on schedule. This increase in task duration of two days would boost the total cost to (a) $14,400 if the workers received a straight-time pay rate for the Saturdays, or (b) $15,600 if they were paid time and a half for the Saturdays.

a. 12 days x 3 people x $400/day/person = $14,400

b. (10 days x 3 people x $400/day/person) +
(2 days x 3 people x $600/day/person) = $15,600

Again, the total cost of the task increases with the increase in its span.

All three examples applied overtime or calendar reserve to accomplish the task, and in each case this had an adverse cost impact. This will always be true unless the original cost parameters (task duration, labor rate, number of workers, and length of the workday) remain unaltered or are altered to offset any increase with an equal decrease. If any of the three parameters change, costs will change unless the workers donate the overtime free of charge. Therefore, as a general rule:

> As a project uses its calendar reserves, it also uses its cost reserves.

Whenever we use calendar reserve, we must understand the impact on cost reserves because this ultimately affects the profit of our project.

FREE RESERVE

Free reserve *measures the amount of time a particular task can slip without affecting any of its successors.* Although the concept of free reserve seems useful, it has serious flaws in practice. To benefit a project, it must be understood and kept in proper context.

Whereas Path Time Reserve measures the dependence of each task on the other tasks in a project, free reserve measures the independence of each task from the other tasks in the project. While PTR belongs to the paths on which the task lies, free reserve (if any) belongs to individual tasks. We make this statement with some reservations.

Turning the definition of free reserve into a value requires some translation and assumption. In a project model, schedule slippage would appear in the early dates (ES and EF) moving to the right. To find free reserve, we must determine how far the Early Start and/or Early Finish of a task can slip without affecting the Early Start and/or Early Finish of a successor. Unlike Path Time Reserve, the backward pass and its late dates have no effect on free reserve; it exists strictly in the forward pass.

A slip of the Early Start of a task could affect the Early Start of a start-to-start (SS) successor. A slip of the Early Finish of a task could affect the Early Start of a finish-to-start (FS) successor or the Early Finish of a finish-to-finish (FF) successor. (We have deliberately ignored start-to-finish relationships.) Thus we need to evaluate the free reserve at the start end (SFR) and the finish end (FFR) of a task separately. We determine the SFR of each task by evaluating all successor constraint relationships of the start of each task. We determine the FFR of each task by evaluating all successor constraints of the finish of each task. If a task has more than one successor relationship, we must choose the one with the smallest FFR value, the one that would be the first successor affected.

A task has start-end free reserve (SFR) only if the node has a start-to-start successor. If not, no free reserve exists. If the task has at least one SS successor, then the value of SFR is the smallest quantity calculated from the following equation for all SS successors:

$$\text{SFR of Node} = \text{ES of SN} - \text{ES of Node} - \text{lag}$$

The Early Start of a task can slip by this amount of time and not disturb any of its start-to-start successors' Early Start dates. We do not include the Early Finish of the node as a successor for purposes of SFR determination.

Likewise, a node has finish-end free reserve (FFR) only if it has either a finish-to-start successor or a finish-to-finish successor; in other words it cannot be a finish node. If the task has a successor of its finish, it has no finish-end free reserve. If the task has at least one successor of its finish, then the FFR value is the smallest result of the following equations for every FF or FS successor:

$$\text{FF constraint: FFR of Node} = \text{EF of SN} - \text{EF of Node} - \text{lag}$$
$$\text{FS constraint: FFR of Node} = \text{ES of SN} - \text{EF of Node} - \text{lag} - 1$$

The smallest value resulting from these equations gives the amount of time the Early Finish of a task can slip without affecting any of its FS successors' Early Start dates or its FF successors' Early Finish dates.

To understand better how we employ these equations to measure the free reserves of a task (SFR and FFR), we need to work a sample problem. Figure 7-1 shows a model fragment with the forward pass completed (ES and EF dates are filled in). We have omitted the backward pass or late dates since they are not needed. The free reserve values for each node end appear at the lower left and right of each node.

The start ends of nodes IN, OZ, LB, YD, PT, and QT have no constraint successors (SS relationships) so they have no start-end free reserve (SFR). Only node FT can have SFR. Finish nodes LB, YD, and QT have no successors of their finishes, so they have no finish-end free reserve (FFR).

Nodes IN, OZ, FT, and PT have successors so they can have FFR. Let us examine each of these in detail. The finish of node IN has a finish-to-start successor (node FT), so its FFR is:

$$\begin{aligned}\text{FFR of IN} &= \text{ES of FT} - \text{EF of IN} - \text{lag} - 1\\ &= 118 - 117 - 0 - 1 = 0\end{aligned}$$

Node IN has no finish-end free reserve, that is, its EF cannot slip even a day without affecting a successor.

Figure 7-1

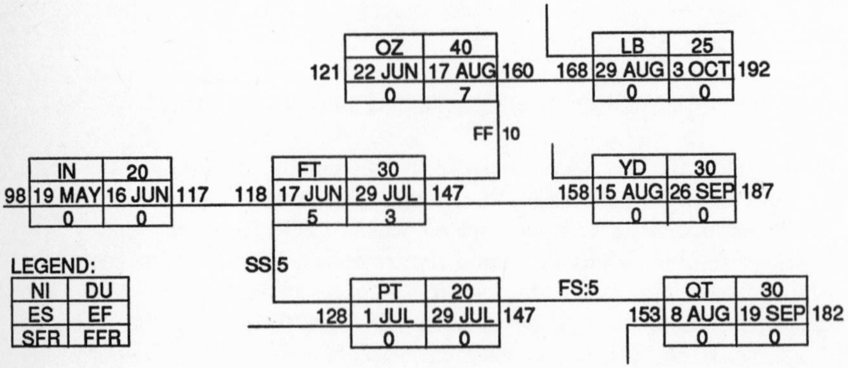

The finish of node OZ also has a finish-to-start successor (node LB), and its FFR is:

$$\text{FFR of OZ} = \text{ES of LB} - \text{EF of OZ} - \text{lag} - 1$$
$$= 168 - 160 - 0 - 1 = 7$$

Node OZ has seven days of finish-end free reserve, so its EF can slip to the right up to seven days before it affects any successor. To show this, suppose that The EF slipped seven days to August 26 (day 167). The ES of LB is August 29 (day 168) is not affected. If the EF were to slip eight days to August 29 (day 168), it would bump the ES of LB forward to August 30 (day 169). The successor is impacted.

The finish of node PT has a finish-to-start successor (node QT) and its FFR is:

$$\text{FFR of PT} = \text{ES of QT} - \text{EF of PT} - \text{lag} - 1$$
$$= 153 - 147 - 5 - 1 = 0$$

Node PT has no FFR, so it cannot slip at all without affecting a successor. If the EF of PT were to slip one day to August 1 (day 148), the ES of QT would also slip one day:

$$\text{ES of QT} = \text{EF of PT} + \text{lag} + 1$$

$$= 148 + 5 + 1 = \text{day } 154 \text{ (Aug. 9)}$$

Node FT has potential free reserve at both ends of the node (SFR and FFR). The start end of node FT has a start-to-start successor, and its start-end free reserve (SFR) would be:

$$\text{SFR} = \text{ES of PT} - \text{ES of FT} - \text{lag}$$
$$= 128 - 118 - 5 = 5$$

The Early Start of task FT can slip up to five days before affecting a successor, in this case the ES of PT.

The finish end of FT has two successors, so its finish-end free reserve (FFR) would be a choice between:

$$\text{FFR} = \text{ES of YD} - \text{EF of FT} - \text{lag} - 1$$
$$= 158 - 147 - 0 - 1 = 10$$

or

$$\text{FFR} = \text{EF of OZ} - \text{EF of FT} - \text{lag}$$
$$= 160 - 147 - 10 = 3$$

We select the smallest of these values, so the FFR of task FT would be three days. The Early Finish of task FT can slip up to three days before it affects any successor, in this case the EF of OZ.

A comparison of the two free reserve values of node FT (its SFR and FFR) shows that its start can slip five days, whereas its finish can slip only three days. If the start of the task slipped five days, however, the finish of the task would probably also slip five days, exceeding the free reserve at the task's finish. Adjusting the free reserve at the start end of a node to correspond to the free reserve at its finish end would keep the SFR within the bounds of the FFR value. If we make this adjustment, start-end free reserve (SFR) really provides no additional information—it is either zero or some value equal to or less than the FFR. We can either eliminate it altogether or calculate it to only for external successors (SS constraints). We will follow this rule, calculating SFR only for a node that is the predecessor in a start-to-start constraint. A node will have SFR if it is the predecessor in an SS constraint and the successor has another, more demanding predecessor.

Node FT in Figure 7-1 meets these conditions. It is the predecessor in a start-to-start constraint (FT-PT) and the successor in this relationship (PT) has a more demanding predecessor that compels it to start no earlier than workday 128. At any rate, the initial portion of task FT has five days of free reserve.

Free reserve is always a positive value or zero; it is never negative. This is because the early dates of a task's successor can never precede the early dates of the task itself. A node can have a free reserve value greater than zero only when its successor or successors have other, more demanding predecessors. In our example the finish of node FT has free reserve because its successors (the start of YD and the finish of OZ) have more demanding predecessors. Another unseen predecessor drives the Early Start date of Node YD to August 15 (workday 158), and the Early Start and duration of OZ drive its Early Finish to August 17 (workday 160). The finish of task IN, on the other hand, can have no free reserve because it is the only predecessor of its successor.

External predecessor stretching in the forward pass can also affect free reserve calculations. When this effect occurs, the Early Start of a task could slip by the amount of stretch (AOS) before disturbing its Early Finish date. The effects of stretching already appear, however, in the AOS and ASF parameters, so we need not include their effects in the SFR calculations. The additional variable would only complicate free reserve calculations.

Even so, it is useful to study the effects of external predecessor stretching on free reserve calculations. Figure 7-2 modifies our first example, stretching the Early Finish dates of two nodes (FT and PT) by 10 days each in the forward pass. All other relationships remain the same.

Figure 7-2

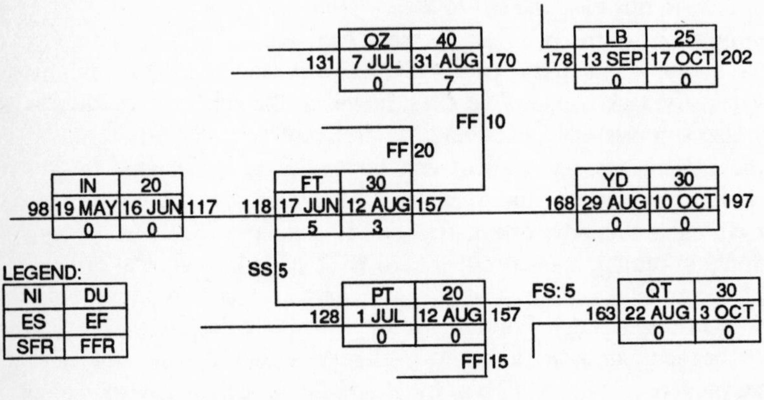

In this example the SFR of FT would be:

$$\text{SFR} = \text{ES of PT} - \text{ES of FT} - \text{lag}$$
$$= 128 - 118 - 5 = 5$$

This shows that our SFR calculations are unaffected by the external predecessor stretching of either FT or PT. The SFR of FT relates solely to its relationship with successor PT. The EFs of FT and PT are stretched (EPSA) but has no effect on their FFR. This is solely dependent on the early dates of the task's successors. The effects of the stretching appear instead in the AOS and ASF fields.

Free reserve can occur in one other instance. When date predecessors determine the Early Start of a node (latest early date), they create free reserve between the node's start and the finish of its predecessor. Figure 7-3 illustrates this. SNE dates (date predecessors) affect the starts of four nodes, although they are not always the most demanding predecessor. (This case is only an example; few real-world models have date predecessors on all nodes.)

Figure 7-3

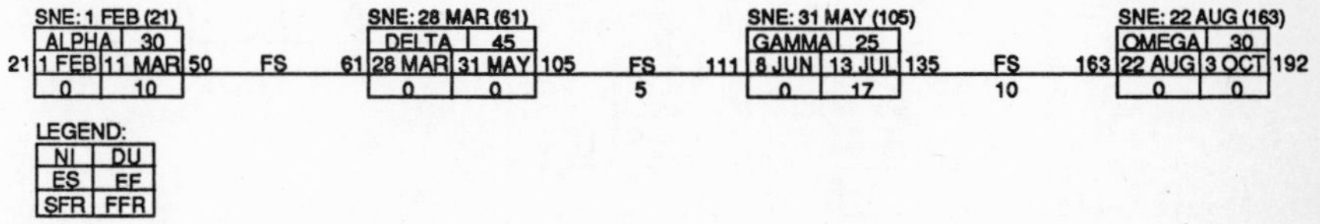

Since none of the nodes have successors of their start ends (that is, SS constraints), all have SFR values of zero. Node ALPHA has FFR because of the SNE date on its successor, Task DELTA. The FFR equation gives an FFR for ALPHA of 10:

$$\begin{aligned}\text{FFR of ALPHA} &= \text{ES of DELTA} - \text{EF of ALPHA} - \text{lag} - 1\\ &= 61 - 50 - 0 - 1 = 10\end{aligned}$$

Task GAMMA has a 17-day FFR for the same reason. The SNE date of its successor, Task OMEGA, delays its start.

$$\begin{aligned}\text{FFR of GAMMA} &= \text{ES of OMEGA} - \text{EF of GAMMA} - \text{lag} - 1\\ &= 163 - 135 - 10 - 1 = 17\end{aligned}$$

This equation also accounts for a 10-day lag on the constraint. Remember that lags on FS constraints are not very common and must represent the accomplishment of effort.

The last example, Task DELTA, also has a successor with an SNE date. In this case, however, the predecessor constraint overrides the date predecessor so the successor's Early Start date comes from the Early Finish of DELTA. The FFR calculation yields the following result:

$$\text{FFR of DELTA} = \text{ES of GAMMA} - \text{EF of DELTA} - \text{lag} - 1$$
$$= 111 - 105 - 5 - 1 = 0$$

A node's SNE date creates free reserve (FFR) in a task's predecessor only when it takes precedence over all other constraints.

Figure 7-4

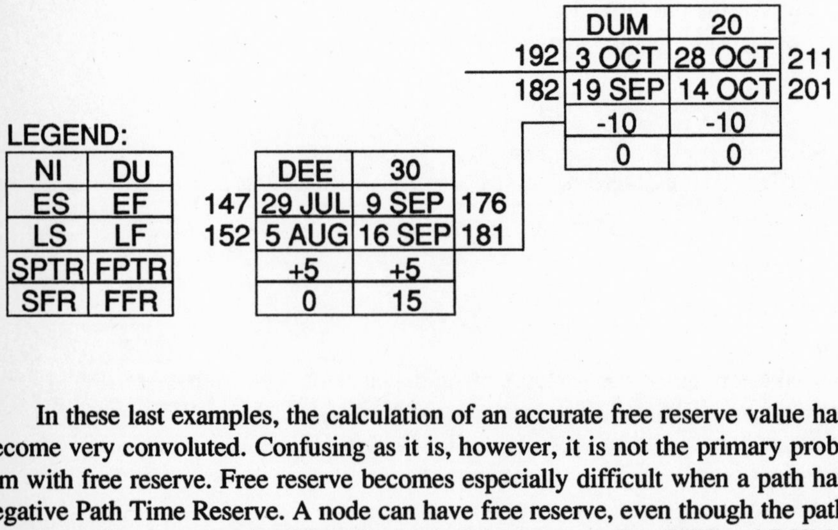

In these last examples, the calculation of an accurate free reserve value has become very convoluted. Confusing as it is, however, it is not the primary problem with free reserve. Free reserve becomes especially difficult when a path has negative Path Time Reserve. A node can have free reserve, even though the path on which it lies has negative PTR. Figure 7-4 shows two tasks (DEE and DUM), with their early (ES and EF) and late (LS and LF) dates, Path Time Reserves (SPTR and FPTR), and free reserves (SFR and FFR). Task DEE has 15 days of finish-end free reserve:

$$\text{FFR of DEE} = \text{ES of DUM} - \text{EF of DEE} - \text{lag} - 1$$
$$= 192 - 176 - 0 - 1 = 15$$

This free reserve comes from the Early Start of DUM on October 3. DUM, however, is on a path with a negative PTR value (–10). The project management should be trying to pull this date to the left (finish the path earlier), but if task DEE consumes its FFR, it could very well negate these problem resolution efforts. Suppose the managers had arranged to pull the Early Start of DUM in by 10 days to September 19 (workday 182) to correct the negative PTR problem.

In the meantime, however, task DEE consumed its finish-end free reserve, slipping its Early Finish 15 days to September 30. This delays task DUM to an October 3 start, wasting all the efforts (overtime, extra resources, etc.) expended to pull in DUM's Early Start date.

The free reserve at node DEE's finish, or at least a portion of it (10 days), is a phantom. Failure to understand and control it could cause serious damage to the project. Even though task DEE has some FFR, project management cannot allow it to be used, or at least they would have to subtract the negative time reserve on the successor from DEE's FFR value.

To calculate valid and accurate values for start-end free reserve, we must consider the following factors:

- The ES and lag of any SS successor
- Any negative SPTR values on any SS successor

These factors affect the validity of finish-end free reserve:

- The ES and lag of any FS successor
- The EF and lag of any FF successor
- Any negative PTR value on any finish successor

All of these factors complicate the calculation of free reserve beyond that for any other parameter derived from the CPM process. Despite all this effort, its value is minimal. It only measures the independence of the tasks, which cannot help us solve any pathwide problems. It is much simpler to ignore free reserve and we will not include its calculation in our models.

SPAN ALTERED FLOAT

Some schools of thought calculate a time reserve value from the duration of the task. We will call this type of reserve *Span Altered Float* (SAF). It is generally calculated by the following equation:

$$SAF = LF - ES - DU + 1$$

If the Early Finish of the task is determined by the Early Start date plus the duration, this equation gives the same answer as either of the PTR equations. To show this, substitute the equivalent value for the ES.

$$
\begin{aligned}
SAF &= LF - ES - DU + 1 \\
&= LF - (EF - DU + 1) - DU + 1 \\
&= LF - EF - DU + DU - 1 + 1 \\
&= LF - EF
\end{aligned}
$$

This relationship does not hold true, however, if an external predecessor determines the EF of the node. Stretching affects either the EF or LS of the task. The LF and ES of a task are never affected, though, so the SAF equation gives a value for Path Time Reserves that ignores stretching. We cannot ignore stretching, however. We need to know when it occurs, on which pass, and in what amount. To find this information, we must calculate two values of Path Time Reserve for each node with the following equations:

$$
\begin{aligned}
SPTR &= LS - ES \\
FPTR &= LF - EF
\end{aligned}
$$

The concept and definition of Span Altered Float is therefore meaningless.

PADDED SPAN RESERVE

There is yet another value of time reserve called *padded span reserve (PSR)* which is the difference between the estimated span of a task and the actual span. Clearly these two values will not always be equal. We cannot determine the actual span of a task until it is completed, however, at which point the reserve has been lost. In the on-going project, this value is generally useless.

Sometimes, however, activities occur more than once in a project. In large-scale manufacturing or another repetitive environment, this value could help determine future spans of similar activities. The actual span of the task, rather than this time reserve, is the important value in such situations.

Some seek to calculate this relationship because they believe estimates add a great deal of positive time reserve to every activity. This is commonly called *sandbagging*. If this were true, detail tasks would almost always finish on schedule, which is generally not the case. This would tend to indicate that most values of PSR are negative.

SUMMARY

Calendar reserve is an important project control parameter. Free reserve, span altered reserve, and padded span reserve are either too much trouble to calculate for the value they return or meaningless.

Two kinds of time reserve values are useful to a project: Path Time Reserves (SPTR and FPTR) and calendar reserve (CR). We can divide calendar reserve into three categories: Extensions of normal workdays, nonworkdays, and holidays.

Extensions of the workday include overtime periods appended onto normal workdays. They increase the number of hours in the workday. The number of such periods available to a project equals the number of workdays between the start and finish of the project:

Finish Date – Start Date + 1

Nonworkdays include overtime periods that are normally not workdays such as Saturdays or Sundays. The number of these periods that are available to a project equals the difference between the start and finish dates of the project on the normal calendar and on the calendar that includes the nonwork periods:

Normal Calendar: (Finish Date – Start Date + 1) minus
Overtime Calendar: (Finish Date – Start Date +1)

Holidays are special days on which no work is performed. They are listed separately from the calendar. Normally these are *not* calculated separately from the normal nonworkdays, so the equations for normal nonworkdays can give values for both holidays and nonworkdays.

Managers can apply calendar reserve values in conjunction with Path Time Reserve to the resolution of problems. Calendar reserve, or overtime, increases the time available to a path. It is one means of correcting a PTR problem.

To determine the impact of employing calendar reserve, managers quantify it by category. A 30-day PTR problem can be corrected by working 30 Saturdays, but this is not possible if only 10 Saturdays are available. Working 10 Saturdays along with an extra half-shift per day on normal workdays during the same 10 weeks (giving 25 equivalent workdays) will solve the problem.

Calendar reserve is an important, finite project resource, and one that should be managed proactively.

CHAPTER 8

Modeling Variants: The Good, The Bad, and The Ridiculous

Over time, many different techniques have been developed to model various project situations. These techniques generally start with valid objectives, but sometimes they try to achieve them by invalid methods. As they run the full gamut of validity, we classify these modeling techniques in various categories: some are good, some are bad, and some are outright ridiculous.

THE GOOD

Date Targets Determined by Time Reserves

Sometimes we are uncertain as to the end dates for finish nodes or the start dates for start nodes. Conditions may seem to allow some model finishes to occur independently. Likewise, some project starts need only meet the needs of the rest of the project. In these situations, we might establish these dates based on the paths with which these nodes interface. The best method to accommodate this is through time reserves. We could determine a finish node required date (LF) by adding a time reserve to the expected completion date (EF date) determined by the node's predecessors. Likewise, we could set the expected initiation (ES) of a start node by subtracting a time reserve from the date by which the node must start to meet its successor's requirements (its LS date). These techniques establish a schedule of a start or finish node that satisfies the requirements of the project and leaves a specified time reserve.

This technique requires two additional parameters of the activity record called *specified time reserves*. In Figure 8-1, node SOU is a finish activity with no date successor. This is called a *hanging finish* (from Chapter 5). In lieu of an FNL date, we have applied a *finish specified time reserve* (FSTR) of 30 days. After completing the forward pass, we set the Late Finish of a finish node with such an FSTR value as its Early Finish date plus the FSTR value, or:

LF = EF + FSTR
= day 137 (July 15) + 30 = day 167 (Aug. 26)

This value becomes the Late Finish date of the node for use in the backward pass calculations. Any subsequent CPM processing that changes the Early Finish date of SOU will similarly change the node's Late Finish date through the above calculation, keeping the Early Finish and the Late Finish 30 days apart. If we wished to maintain any calculated Late Finish date, we would have to remove the FSTR value and impose an FNL date. We would then use this value in all subsequent processing. Note that the 30 days of reserve moves upstream through node BIT, and overrides the 40 day reserve of path YEN-ATT at the common node LEK.

Figure 8-1

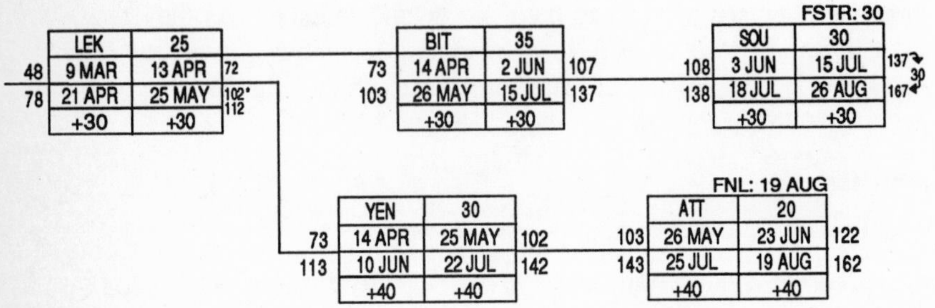

We could handle project starts similarly. Figure 8-2 illustrates the start end of node MAYA, which has no expected start date (SNE date). This is normally called a *hanging start* (from Chapter 4). For the initial forward pass, we take a hanging start default value (normally the project start date) for the Early Start date of MAYA. After completing the backward pass, we process the entire project through a second forward pass, recalculating the Early Start date of node MAYA and any other like start nodes to add a start specified time reserve (SSTR) of 30 days. We do this by subtracting the SSTR value from the node's Late Start date:

ES = LS − SSTR
= day 67 (April 6) − 30 = day 37 (Feb. 23)

We have now established a schedule for this node based on its successor requirements.

Node MAYA lies along a *side path,* that is, a small path outside the mainstream of the project. In such case, we need to ensure that we complete this work without interfering with any other effort. The process we have outlined sets an adequate schedule that satisfies the balance of the project and includes a reserve. Its main drawback is that we must process data through an additional forward pass, increasing the data processing time.

If subsequent backward passes change the Late Start of our start node, the Early Finish date of the node will change along with it. If we wish to retain any date, we must replace the SSTR value with the desired date as an SNE date.

Figure 8-2

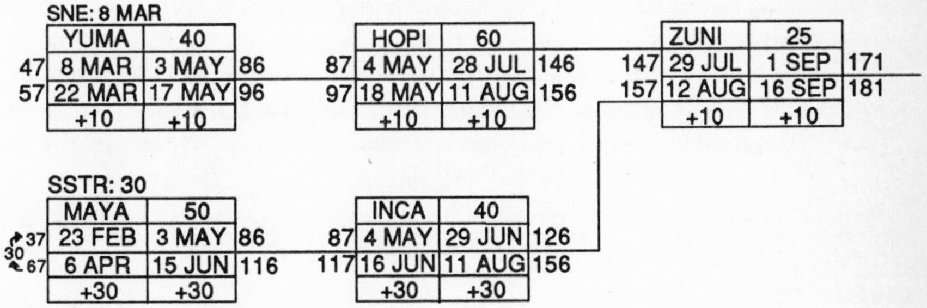

These two mechanisms are sound ways to introduce time reserves into a model and establish schedule requirements for certain nodes. They could be useful, although their application is limited to setting start and finish dates that are not imposed by project requirements.

Lag Values on Finish-to-Start Constraints

As a general rule, finish-to-start (FS) constraints should not have lag values because all lag values must represent the accomplishment of work. In some situations, however, a lag between tasks can meet this requirement. Certainly in modeling a process like the curing of concrete, the curing period could be entered as a lag value on the constraint between the task FINISH CONCRETE PAD SURFACE and its successor with a seven-day week calendar. This lag does represent project activity of sorts.

We can model this situation just as easily by defining a CONCRETE CURING node with the same duration. The effect on date calculations and time reserve is identical in either case. The user must choose between giving this chemical process visibility or merely accounting for it in the model.

Additionally, we could model trivial tasks like PACK AND SHIP, RECEIVE ON DOCK, or any transportation time in this way. We may not want to give these efforts the visibility of a node, but we must nevertheless account for their span. Establishing lags on finish-to-start constraints that equal the duration of the effort enables us to do this.

In these examples we did not violate the proper procedure for time reserve calculation. We can verify this by testing to be sure we can replace any lag value on a finish-to-start constraint with a node having the same span. We must always maintain the principle that:

All lag values must represent the accomplishment of path effort.

A lag value on an FS constraint satisfies this principle if we can replace it with a node having the same duration.

To verify that CPM processes treat lag values the same as task durations, examine Figure 8-3. It illustrates a simple three-node path with lag values on both finish-to-start constraints. The nodes represent 90 days of cumulative work (30 + 30 + 30), to be accomplished in 145 days:

$$TA = LFP - ESP + 1 = 239 - 95 + 1 = 145$$

If we only consider the tasks' durations, our reserve would be 55 days:

$$PTR = TA - CSP = 145 - 90 = 55$$

Figure 8-3

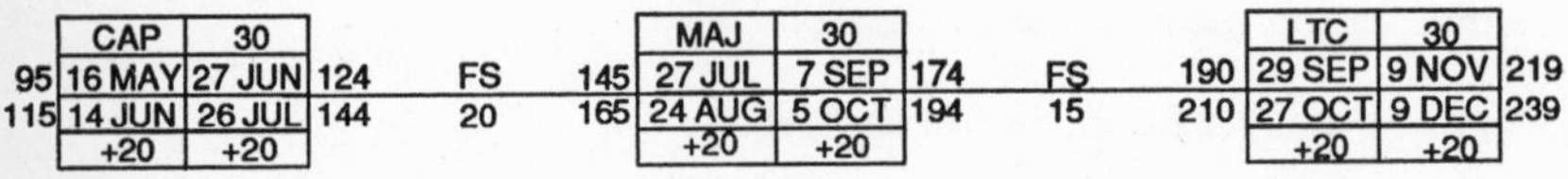

Examining the SPTR and FPTR values calculated at each node, however, reveals only 20 days, 35 days less than we figured. The 35 days of lag values (20 + 15) in this path account for the shortfall. Lags are added in the forward pass and subtracted in the backward pass and in fact contribute to the path's cumulative length through the CPM equations. In effect, we do not have 90 days of work, but rather 125 (90 + 35). This total length accounts for the calculated time reserves:

$$PTR = TA - CSP = 145 - 125 = 20$$

In our calculations, there is no difference between a duration on a node and a lag on a constraint; they both contribute to the cumulative length of the path. Therefore both values must represent the same thing in our model, namely the expected length of time necessary to accomplish detail effort. All numeric values in our model (duration and lag) represent spans that make up the cumulative length of the paths.

Lags as a Percentage of Effort

So far, we have expressed lag values on finish-to-finish and start-to-start constraints as numerical values that are either a portion of the predecessor's or successor's duration. The lags on these constraint types quantify the relationship between two tasks and also determine the extent to which the two tasks can overlap. We could express this kind of task relationship just as well, however, as a percentage of one of the related tasks. For example, a successor, task B, can start after 50% of its predecessor, task A, is done or in another fashion, 25% of the work of a successor, task D, remains to be completed once the predecessor, task C, has finished.

Figure 8-4a shows an example of a start-to-start constraint between nodes WISHY and WASHY. Instead of stating the lag value that defines the relationship between the starts of these two tasks as a specific amount of time, we state that the successor (WASHY) can start after completion of one-third (33 percent) of its predecessor (WISHY). As the duration of WISHY is 30 days, a 10-day delay separates the two nodes (.33 x 30 = 10). We expect WISHY to start on July 25 (its ES date), so WASHY could start on August 8 (10 working days later).

Figure 8-4

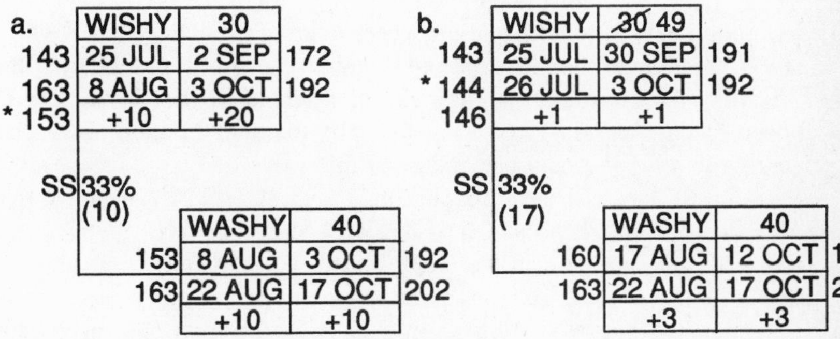

This lag value also affects the backward pass. In this case, it drives the Late Start calculation of WISHY, making it an external successor stretched activity (ESSA). But the amount of the lag still depends on the predecessor's duration. Suppose that our predecessor task, WISHY, overruns its expected duration of 30 days by 14 days, pushing its finish date (EF) forward to September 30. It is reasonable to expect that this would have some comparable effect on the start of WASHY. Since we expressed the lag as a percentage of the predecessor, we now multiply 33 percent by the new 49-day duration. This equates to a 17-day lag (.33 x 49 = 17, rounded up to the nearest integer), so WISHY's successor will start August 17 (workday 160). The delay in the predecessor causes a delay in the start of the successor by seven days (see Figure 8-4b).

It is interesting to note the effect of this slip on the backward pass. In our initial situation, the Late Start date of WISHY depended on the SS constraint from WASHY. The Late Finish of WISHY (on day 192) less its duration (49 days) gives an earlier date (192 – 49 + 1 = 144) than the Late Start date of WASHY (day 163) less the lag (17 days: 163 – 17 = 146). Now the Late Start of WISHY depends on the successors of its finish, no longer on the SS constraint to WASHY. The schedule is driven by a different path as a result of the delay. However, we must understand that this is peculiar to the situation expressed in the example.

Finish-to-finish constraints could be expressed similarly. In the example shown in Figure 8-5a, node HEM is the predecessor of node HAW in an FF constraint. The lag between these two activities is expressed as a percentage (20 percent) of the duration of the successor (HAW). Since HAW is 25 days long, the lag is 5 days (.20 x 25 = 5). As any finish-to-finish constraint creates a multiple predecessor relationship, we find that the Early Finish date (day 129) of HEM plus the lag (5 days) gives a later finish (129 + 5 = 134) than the Early Start of HAW (day 100) plus the duration (25 days: 100 + 25 – 1 = 124). This means that node HAW is an external predecessor stretched activity (EPSA).

If HAW runs longer than its expected duration, the slip in the finish of the task could affect the constraint as well. As an example, if the finish (EF) of HAW slipped 15 days to July 19 (see Figure 8-5b), the total duration of the task would increase to 40 days. As a percentage of this value, the lag increases to 8 days (.20 x 40 = 8). Processing the forward pass with this set of values, we find that HAW's duration (40 days) added to its Early Start (day 100) gives a later finish (100 + 40 – 1 = 139) than the Early Finish of HEM (day 129) plus the lag (8 days: 129 + 8 = 137). The task is no longer stretched by its predecessor constraint; instead, it is driven from its own start. A different path drives the finish of HAW due to the task's delay. This is also relative to this particular set of circumstances.

Figure 8-5

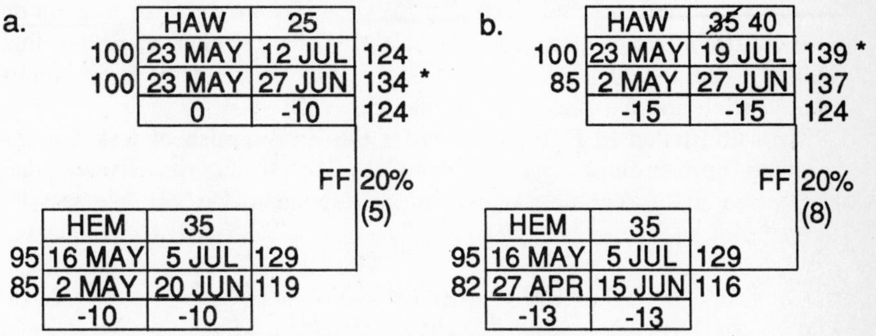

Expressing constraint lags as percentages of activity durations gives valid estimates of the relationship between two activities. It is fluid in that variations in the spans of the underlying predecessors or successors can affect the lag. This better expresses some work interrelationships.

Summary

These are only three examples of modeling variations that are consistent with the principles of time reserve management. There are undoubtedly others. However, as we will see, there are also many variations that violate or diminish these principles. We must carefully examine any modeling technique that we use to be sure that it preserves valid and accurate values of Path Time Reserve or other related parameters. In most cases, this is how we determine the validity of a modeling technique.

THE BAD

Many common techniques adversely affect the calculation of PTR or otherwise convolute the CPM path analysis processes. These processes exhibit some merit, but carry with them certain flaws. We need to examine some of these and understand how and why they violate the model. We may still elect to use them, but only with a full understanding of their ill effects.

Dates Altered to Compensate for Stretching

One common modeling mistake is to contend that the duration of a task is inviolable. This causes problems when a constraint relationship delays either the finish (EF) or the start (LS) of a task, when, that is, a task undergoes external constraint stretching. In order to maintain a task's duration when this effect occurs, we must adjust the Early Start or the Late Finish. We call this *date altered compensation for stretching* (DACS). For instance, if a predecessor in a finish-to-finish constraint stretches a task's finish date (EF), creating an EPSA, then this DACS process recalculates the start of the task (its ES date) to accommodate the duration and the stretched EF date (ES = EF – DU + 1).

This is illustrated in Figure 8-6, where the Early Finish of task SW depends on the finish-to-finish constraint from BD. The DACS process would then recalculate the Early Start date of SW to correspond to the EF date less the duration:

$$\begin{aligned}\text{ES of SW} &= \text{EF of SW} - \text{DU} + 1\\ &= 222 - 40 + 1 = \text{day } 183 \text{ (Sept. 20)}\end{aligned}$$

This maintains consistency between our date calculations and the task's duration. It also creates 20 days of free reserve between the finishes of nodes JF and SW.

A backward pass causes a similar effect when a start-to-start constraint drives the Late Start date of a predecessor. When a constraint pulls a Late Start date to the left (creating an ESSA), the DACS process recalculates the task's Late Finish date from the LS date and the task's duration (LF = LS + DU – 1). In Figure 8-6, the Late Start date of task SB depends on the Late Start of AH minus the lag on the constraint. The DACS process would recalculate the node's LF date as:

$$\begin{aligned}\text{LF of SB} &= \text{LS of SB} + \text{DU} - 1\\ &= 153 + 20 - 1 = \text{day } 172 \text{ (Sept. 2)}\end{aligned}$$

Two effects result from the application of the DACS process. First, as stated earlier, the difference between a task's early dates (ES to EF) or late dates (LS to LF) will always equal its duration. Constraints affect dates, but they never stretch tasks. As a practical consideration, we certainly do not want tasks to be stretched or delayed. This is inefficient. We should remember, however, that we have three options to respond to external predecessor stretching in the forward pass:

1. Delay the start of the task
2. Improve the Early Finish date of the delaying predecessor
3. Perform the task in two parts

Figure 8-6

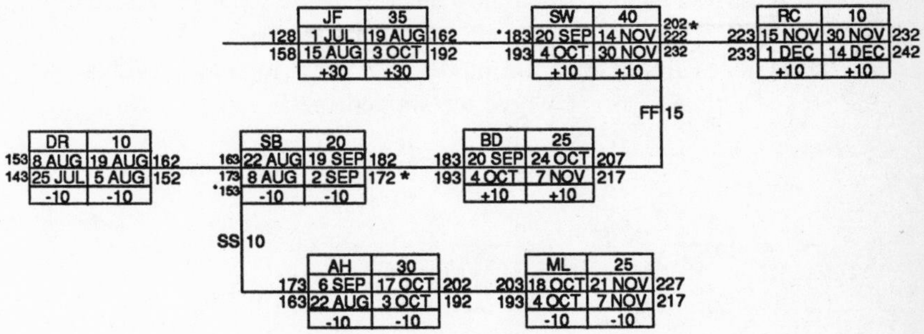

The DACS process simply makes the choice automatically and removes the responsibility for this decision from the user. Unfortunately, it makes the worst choice possible for the accomplishment of the project. It automatically delays the entire schedule of a task and denies management the opportunity to improve it through one of the other two choices.

Management must decide what to do on a case-by-case basis. The CPM process should inform us of any external constraint stretching so that we can evaluate the circumstances and decide on a response. It should not make the decision for us.

DACS also assures that the Path Time Reserve calculated at one end of a node will always equal that at the other, giving a single value of time reserve for each node. This results from the persistance of the notion that time reserve belongs to individual nodes. Some assume that a single entity, the activity, can have only a single value of time reserve. Path Time Reserve is not a property of an activity, but rather of the path or paths on which the activity lies. We gain nothing from having a single value of Path Time Reserve. If we measure it at both events of each node in the project, however, we can quickly identify stretching problems and act to solve them. We can also understand the dynamical mechanics of a path to a greater degree.

For these reasons, the date altered compensation for stretching (DACS) process cannot be considered an acceptable project modeling technique. It does not invalidate the model, but it assumes a decision that is probably best made by the user. It also confuses path analysis.

Time Reserve Nodes

Another common technique establishes time reserve in a model through the use of nodes. To attain a Path Time Reserve of 30 days, we could insert a node just prior to a finish objective with a span of 30 days. The project modeling processes would treat this node just like any other activity. Its span, and therefore the reserve, would affect the early dates of all predecessor tasks. Distributing many of these nodes throughout the model just prior to major events would place the project's time reserves where we wanted them.

Figure 8-7

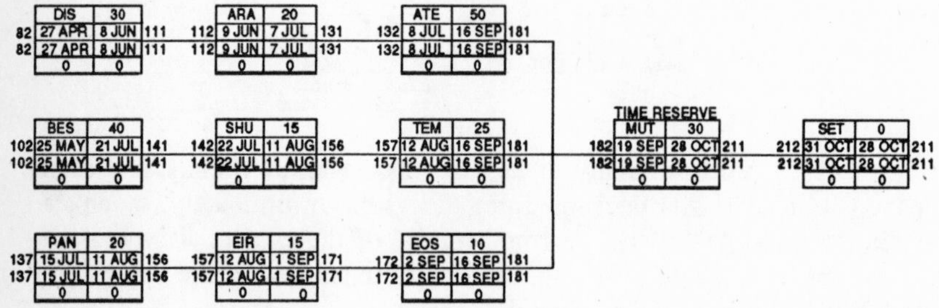

Figure 8-7 shows a project of three paths, all of which culminate at the events node SET. To establish 30 days of time reserve in this project, we insert a node, MUT, with a span of 30 days between the finish event and the project work. As we develop the project schedule from this model, we rework the plan until the Early Finish date of SET matches its required finish date. Because all of the work passes through the time reserve node MUT, it will all be scheduled to finish at least 30 days prior to the desired end date.

Because it builds time reserves into a schedule, this method does have some merits. Further, it makes this reserve very visible, since we need only look at the durations of the TIME RESERVE nodes. If any of these nodes' durations diminish, we will notice it quickly.

Every time a predecessor slips, however, it will push the early dates of both MUT and SET to the right. To compensate for this, we will have to reduce the duration of the time reserve node, MUT, by an amount equivalent to the slip in order to keep the end event's required schedule date as planned. Every time we do an impact analysis, we will have to adjust the durations of the time reserve nodes and reprocess the model. This accomplishes the same thing as proper implementation of the forward and backward passes of the CPM process (see Chapter 12), but at the cost of considerably more time and effort. For this reason, this technique is considered undesirable, although it violates no principle of time reserve management.

Negative Lag Values

Up to this point we have always expressed lag, like duration, as a positive value. A popular school of thought supports assigning negative values to lags on constraints, however, to model or simulate the overlap of two tasks. Figure 8-8 shows an example of negative lag. Node GROUCHO is constrained by the finish of node HARPOOO with a negative 10 days of lag, that is, node GROUCHO can start 10 days before HARPOOO finishes.

This may seem a reasonable means of modeling task overlap. If we follow the sequence of this path, however, we see a strange sequence in which activity HARPOO finishes, and then as a result, task GROUCHO starts 10 days earlier. The forward pass process would calculate an Early Start for GROUCHO after the Early Finish of HARPOO:

ES of GROUCHO = EF of HARPOO + lag + 1
= 159 + (−10) + 1 = day 150 (July 27)

To make this sequence feasible, time would have to flow in reverse, which has never happened except in science fiction. No work can be initiated as a result of something that occurs later, because work, like time, cannot flow in reverse.

Figure 8-8

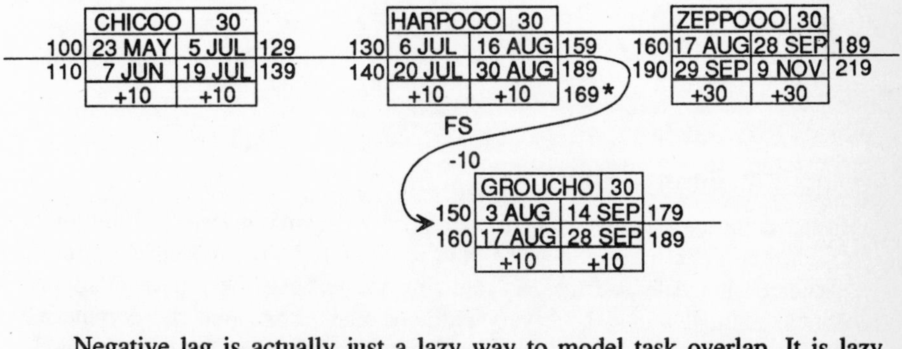

Negative lag is actually just a lazy way to model task overlap. It is lazy because it replaces two correct relationships with a single, impossible one. The interface of work must match the sequence in which time and effort flow, chronologically from left to right. If a relationship exists between the work accomplished in HARPOO and GROUCHO, then we must determine two things:

1. How much of the predecessor must be accomplished before the successor can start (giving a positive lag on a start-to-start constraint)

2. How much of the successor remains to be finished after the predecessor finishes (giving a positive lag on a finish-to-finish constraint)

An example of the resulting model appears in Figure 8-9. Here we model the interface of the two tasks with two constraints, a start-to-start and a finish-to-finish. The SS constraint allows node GROUCHO to start after completion of the first 20 days of HARPOO. This corresponds exactly to the relationship illustrated in Figure 8-8. However, if we model the interface of these two tasks with only this single constraint, then the finish of node HARPOO has no impact on GROUCHO.

Figure 8-9 includes the interface of the two node finishes, which did not show up in Figure 8-8. The model with the negative lag ignores this relationship entirely. If a model ignores such relationships, it sacrifices some ability to simulate impact correctly. For all of these reasons, this technique is not recommended.

Figure 8-9

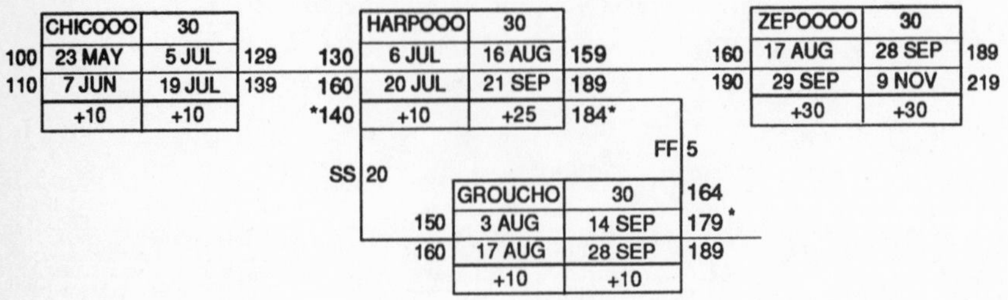

Zeroing Time Reserve Paths

In another common practice, some modelers zero the time reserve of the critical path such that all activities along the path report zero PTR. In a normal project the expected start date and the required finish date band the time available to the critical path. It would be a very unlikely coincidence that the cumulative span of the work of the critical path is identical in length to the time available to this path. The odds are enormous that the path would have either positive or negative time reserve.

Negative time reserve should motivate us to rework the plan to establish sufficient positive reserve. Some believe we should cease this process once we raise the path's time reserve to zero. In addition, they divide any positive reserve of the path among the tasks along it, increasing their spans accordingly. Either way, the length of the critical path becomes identical to the time available to the path such that we create the zero time reserve.

Distributing the time reserves of the path to the tasks in effect eliminates them as a reserve as it makes them a property of the tasks. This is equivalent to doling out the cost reserves of the project. In either case it delegates the responsibility for managing the project's reserves to the individual tasks.

Whether or not this is good technique depends on the project management's philosophy. If they wish to control the project's outcome, time and cost reserves would be invaluable tools for forward-thinking, proactive analysis. On the other hand, if management is content to let the project happen and respond to problems as they occur, then control of reserves might not be so important. In this latter case, relinquishing control of a project's time reserves is consistent with the management philosophy. However for those interested in proactive methods, the process of distributing time reserves among tasks is bad practice.

THE RIDICULOUS

Another category of modeling methods goes beyond bad techniques to become ridiculous. These techniques either invalidate the time reserve calculations or in other ways create severe problems in our modeling.

Imposed Dates

Dates imposed on the model calendar fall into six types. Chapters 4 and 5 discussed four of these date types: start-no-earlier-than (SNE), finish-no-earlier-than (FNE), finish-no-later-than (FNL), and start-no-later-than (SNL) dates. We will now explore the remaining two date types, start-on (SON) and finish-on (FON) dates, or *imposed dates*. These dates act on either the start end (start-on dates) or the finish end (finish-on dates) of a node with an identical effect. They are mandatory dates, so they become both the early date and the late date. Figure 8-10 shows an example of a finish on (FON) date. A start on (SON) date would have the identical effect on the Early Start and Late Start calculations.

Figure 8-10

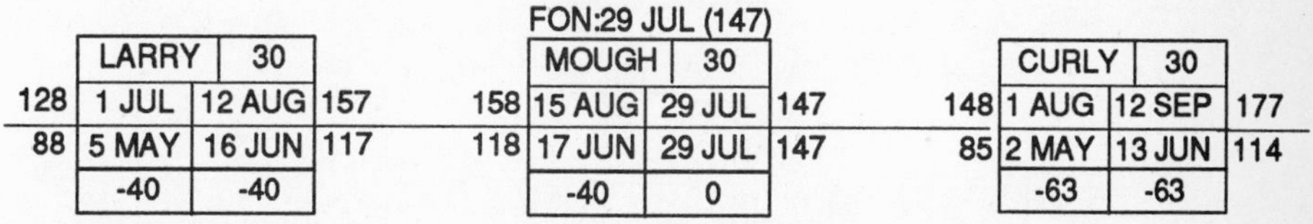

Node MOUGH's finish-on (FON) date requirement of July 29 becomes both the task's Early Finish and Late Finish dates, regardless of the model interfaces. The predecessor of MOUGH, task LARRY, finishes 10 days after MOUGH is supposed to finish. This does not delay the Early Finish date of MOUGH, however, which remains July 29. This is what a finish-on (FON) date does.

Likewise the successor of MOUGH, task CURLY, has a more stringent schedule requirement for MOUGH's finish (May 2), but this does not matter either. The Late Finish of MOUGH will still be July 29, as set by the FON date.

The imposed date ignores constraint relationships. The finish-on date will control the schedule of a task regardless of other requirements. If we happened to be modeling a solar eclipse, this date method might be of some use since no constraint that Homo Sapiens controls can have any effect on the event. In this case, any constraint would be invalid anyway.

Some event dates are critical and management will achieve them at virtually any cost. This does not make these dates finish-on or start-on dates in our model any more than declaring that an event will happen on a certain date makes it happen. This is precisely why we use modeling methods—to identify problems that threaten our critical project milestones. Instead of taking critical events as given, we must make them the focus of our time reserve management processes in order to maximize the odds of success. This strategy recognizes that all project events depend on other project efforts, and no event occurs until completion of its predecessors.

Figure 8-11 illustrates the example from Figure 8-10 modeled properly. We have specified July 29 as an FNL date on node MOUGH. This date is essential. The CPM processes show that the predecessors of the finish of MOUGH will not be completed in time to make our required date. This is a problem that we must try to resolve, but we have a more serious problem.

The FNL date is not MOUGH's LF date. This means that a constraint successor creates an even more stringent requirement on the finish of MOUGH to meet another downstream schedule objective. The project is in serious trouble, as indicated by the magnitude of the negative time reserve (–103 days).

Figure 8-11

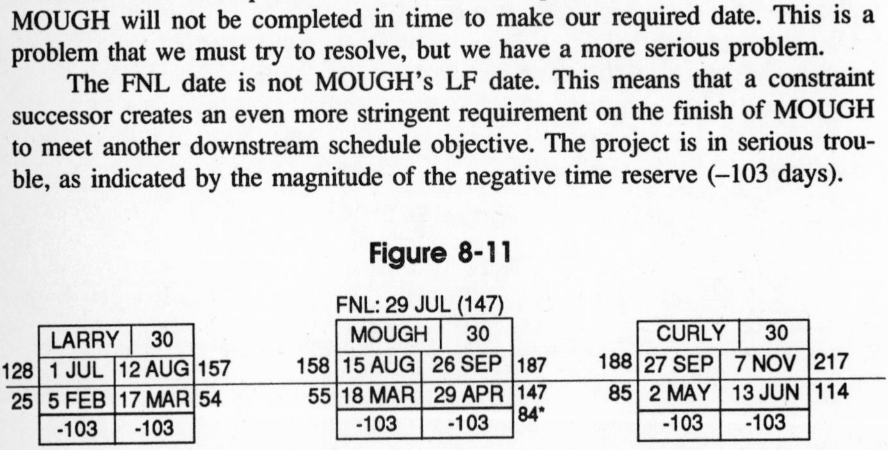

Figure 8-11 reveals the problems that we must correct in order to meet our schedule requirements. The finish date for MOUGH in Figure 8-10 only reveals the date that we need to make; it offers no understanding of the problems that threaten our project. Since the start-on (SON) and finish-on (FON) date mechanisms have no basis in reality, they contribute very little to the practical modeling process.

Hanging Events Nodes

As was previously discussed, the start date of an events node (a node with DU = 0) appears later than its finish date. This is simply a perception problem. Nevertheless, attempting to correct this apparent problem has spawned an entire family of invalid modeling techniques. The worst of these, the *hanging node* technique, is the culmination of a long chain of flawed logic.

Figure 8-12 illustrates an example of this problem. In a simple model fragment we need to report the start of node BH as an event, so we create an events node BS (the first mistake). We would logically insert this node between nodes LS and BH, but, bothered by the apparent date disparity of an events node (the second mistake), we give it a duration of one day (the third mistake). This makes the Early Start and Early Finish dates the same calendar date, but it shoves the early dates of the node's successors (nodes BH and BG) to the right by one day. We normally correct this problem in one of two ways (the fourth mistake): we either set a negative lag of one day on the task's successor (BH) or hang the events node outside the main path, as shown in Figure 8-12.

Figure 8-12

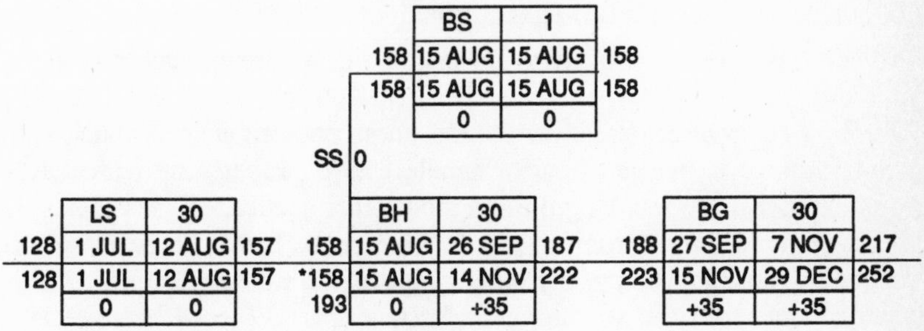

We discussed the first, negative-lag solution previously. In the second solution, we use a start-to-start constraint with no lag to move the events node out of the mainstream logic. This appears to do the trick as the start date (ES) of BH and the start (ES) and finish (EF) dates of BS are all the same date (August 15). Everything seems well and good.

What have we done to time reserve calculation, though? We have invalidated our model and created a good example of what can happen when we lose sight of the modeling objectives to focus on details. Node BS has no successor and is therefore a finish node. It also has no FNL date, so it is a hanging finish. The normal process defaults the Late Finish date of such a node to its Early Finish date, which makes the time reserve at this node zero (SPTR and FPTR = 0).

This is bad enough, but the backward pass process calculates the Late Start date of BH from the invalid Late Start date of BS. The constraint from BS requires an earlier date (August 15) than the Late Finish date of BH less the task's duration (October 18), so the Late Start date of BH is driven by the hanging finish. The situation deteriorates further as the process imposes this invalid value on the late date calculations of node BH's predecessors, corrupting all of their late dates, as well. The time reserve of this path should be +35, but the hanging finish events node corrupts the proper calculation of all of the predecessors of node BS.

When two or more paths merge, the path with the least time reserve prevails. Therefore any time we create a hanging finish events node, we zero out any positive time reserve to the left of the node. We report an event, but as a consequence we corrupt our time reserve analysis. To correct this, we have three options:

1. Default the Late Finishes of all hanging ends to some distant downstream date like the project end date plus six months
2. Establish an FNL date on any hanging finish events node that does not corrupt the model's late date calculations
3. Constrain the finish of node BS to some downstream successor node

Our long chain of errors leaves us with these poor choices. A much better and simpler choice is to learn how to handle events properly in a Precedence model, as explained at great length throughout other sections of this text.

The technique for reporting events in Precedence has fathered more erroneous modeling techniques than any other single aspect of the discipline. All this confusion is unnecessary. If we simply understand that in Precedence events are node ends and model them as such, we can avoid mistakes that invalidate our time reserve calculations.

Excessive Target Dates

Sometimes, scheduling concerns override the interests of project modeling. We can become so preoccupied with getting the model to report specific dates that we virtually ignore the analysis methods. This leads to *date rigging,* in which we contrive the model and its parameters to yield predetermined schedule dates with little or no regard for valid modeling processes.

Of the many and diverse date rigging methods, the most common is the misuse of date targets, specifically the date predecessors (SNE and FNE dates). The goal of our process in this case is to create a modelfile with activities and events in agreement with a schedule. Rather than deriving the dates through constraint logic, lag values, and task durations, we simply apply date targets to the nodes. We impose SNE and FNE dates on virtually every activity in the model, perpetuating the schedule data. This keeps the model parameters and our predetermined schedule in perfect agreement, but otherwise it has marginal value.

To maintain a project model as a valid analytical tool, it must generate its dates by constraint logic, lags, task durations, and the proper application of date targets. This establishes a valid model on which to base analytical processes. Without this foundation, the data in the modelfile at best replicates the input, adding nothing.

Some modelers even calculate spans for tasks from the date targets:

$$DU = FNE - SNE + 1$$

If constraints are added in such a way as to avoid putting a predecessor in conflict with the date predecessors, the forward pass dates of the model will also equate to the schedule dates. These durations are redundant, as the date targets (SNE and FNE dates) would act as no-earlier-than dates on each node end. As long as no predecessors violate these date predecessors, the calculated early dates will be identical to the schedule dates. Either way, the model is rigged.

There are many other means of rigging schedules through modeling. An enormous amount of time and effort have gone into this process. It is an excellent example of how misguided direction, coupled with ignorance, can lead to very complex and useless operations.

Unfortunately, when we use the process in this manner, it provides no value added (NVA) and the effort spent on it is all wasted. This kind of model can produce no valid analytical data; at best it outputs what is input. It is all just an overcomplicated means of duplicating a schedule, along with a very expensive graphics process.

SUMMARY

When choosing among the many ways to model various characteristics of a project, we must avoid any method that may invalidate the model as an analytical mechanism. Of the many variations of modeling techniques, some are good, some are bad, and some are ridiculous. We must evaluate any modeling technique in terms of how well it affects Path Time Reserve calculations, as well as other analytical and practical processes.

The Good

Date Targets Determined by Time Reserves. As an alternative technique to determine the expected (ES) date of a start node, we can subtract a start specified time reserve (SSTR) from the task's Late Start date. We can determine the Late Finish date of a finish node by adding the finish specified time reserve (FSTR) to the Node's Early Finish date.

Lag Values on Finish-to-Start Constraints. A finish-to-start constraint can carry a lag as long as it represents the accomplishment of project effort. This can occur in two situations:

1. A physical process like the curing of concrete
2. Work that must be taken into account, but needs no particular visibility in the model

Lags as a Percentage of Effort. Lag values on start-to-start (SS) and finish-to-finish constraints can be stated as a percentage of either the predecessor's span (for SS constraints) or that of the successor (for FF constraints). These percentages would be represented as additional parameters in the model. The CPM process calculates a current lag value using the most current durations every time it works through the model.

The Bad

Dates Altered to Compensate for Stretching. This process adjusts the Early Start date of an external predecessor stretched activity (EPSA) in the forward pass to maintain consistency with the task's duration (ES = EF − DU + 1). Likewise, it adjusts the Late Finish date of an external successor stretched activity (ESSA) in the backward pass to maintain consistency with the task's duration (LF = LS + DU − 1). This sets a predecessor date by its successor in the forward pass, and a successor date by its predecessor in the backward pass. This practice serves only to confuse path analysis.

Time Reserve Nodes. This technique defines time reserve as specific nodes labeled as such in the model. These normally are distributed throughout the model, often immediately preceding major milestones. As the project changes, the durations of these nodes must be adjusted. Although this practice does not invalidate the time reserve processes, it encumbers them and provides no advantage over late dates.

Negative Lag Values. This technique simulates task overlap through finish-to-start constraints with negative lags. It shows the start of a task as driven by an event in the future, which is illogical. This lazy method of modeling task overlap replaces two constraints with one.

Zeroing Critical Path Time Reserves. This technique distributes the time reserve of the critical path among the tasks along the path. This delegates the responsibility for time reserve management to each individual task. The project management affects only the initial dispersal.

The Ridiculous

Imposed Dates. This technique assigns specific dates called start-on (SON) and finish-on (FON) dates to activities. The start-on date fixes both Early Start and Late Start dates, and the finish-on date fixes both Early Finish and Late Finish dates. The technique ignores constraints or interfaces. Such a parameter would be useful only for modeling preordained events.

Hanging Events Nodes. Many methods have been devised to correct the apparent disparity between the start (ES/LS) and finish (EF/LF) dates of an events node (one with DU = 0). Many of these corrupt the analytical properties of the model. One common method establishes events nodes with a duration of one unit outside the main path. This usually corrupts the late date calculations because the hanging node sets any positive time reserve of its predecessors to zero.

Excessive Target Dates. Sometimes the modeling processes are used merely to portray already derived schedule data through a variety of methods collectively called *date rigging*. The most common method improperly utilizes date predecessors, imposing projected start (SNE) and finish (FNE) dates on every node that match its schedule. At best, this method outputs the data that is input, adding nothing and providing no analytical capability.

CHAPTER 9

Change and Status in the Modeling Process

In the time before work on a project actually begins, the plan of the project is developed, at least to some degree. The aim is to set forth a seemingly viable scheme that makes all of the project's goals appear attainable. The project management issues schedules that reflect this optimism, showing activities and events, if only on paper, to be performed in the time frames necessary to meet schedule objectives. In this blissful phase of the project, no potential problems can dull the naive optimism or the desire to get on with the job.

Shortly after the project starts, things begin to unravel. Some tasks are accomplished exactly as planned, some even finish early, but many are late. The late completion of tasks begins to delay other tasks.

On top of schedule slippage, requirements change, the design changes, material arrives late, outputs fail tests, things get lost or buried in bureaucracy, manufacturing gets backlogged. Reality twists, mangles, and distorts the project plan in an infinite combination of ways, proving once again that the most perfect principle of project management is Murphy's law:

Anything that can go wrong will go wrong!

This cynical prophecy seems to control the progress of the project completely. Yet in spite of all this, the project must succeed; schedule and cost objectives must still be met. We cannot neglect to plan due to uncertainty, complexity, and the inability to perform. Instead, these are all compelling reasons to plan throughout the life of a project. Only in this way can we have any hope of overcoming the adversity that can plague a project.

Once these deviations begin to affect a project (usually on the first day, and almost certainly by the second day) the original project model loses validity. Yet the need for the benefits of project modeling remains as strong, or becomes even stronger. In fact, in control over the on-going project, the modeling processes could yield their greatest benefits; this is where they could provide significant return on investment. This is also where they have historically failed to deliver. The key to success versus failure is the practical knowledge of how to harness the CPM processes to the on-going project management task. We will now begin that study.

We assess status and the effects of change in two phases:

1. Incorporate the current project forces (change and status) into the project model.
2. Apply the CPM process to the modified model to help diagnose and then correct any problems.

In this chapter, we will examine the first phase thoroughly and begin to explore the second. There is, however, considerable investigation that must precede our thorough consideration of the second phase.

CHANGE AND STATUS

Two forces influence an on-going project: change and status. *Change* is the directed adjustment of the project to accommodate new or altered requirements, which generally result from the knowledge gained through the progress or lack of progress of the project. The model must reflect this change in the project, including adding, deleting, or rearranging the sequence and interface between tasks. We must alter the model to accommodate these changes according to the same principles by which we built the model in the first place.

We change a model of an on-going project in three steps:

1. Identify the need for a change.
2. Evaluate the scope of the change.
3. Incorporate the changes into the project model, and ultimately the project.

The modeling process aids evaluation by modeling the change and then, through the CPM process, determining its potential impact on the balance of the project. Through the model, we can evaluate many change scenarios and determine the optimal solution, all before expending the actual resources to make the change. We can redevelop and then test our plan before we put it into actual practice. This process will be further explored in Chapter 12 in the section titled "What-If Analysis."

Project status (or schedule status), the second force that acts on projects, results from the accomplishment, projected accomplishment, or lack of accomplishment of the activities and events of a project. This is the primary subject of this chapter.

To assess status, we model the effects of the achievement of work. This can take more than one form. We can look at such an interim model like the farmer who describes his work schedule as "that what's been done and that what's ain't been done."

The point in time at which the work that has been accomplished meets the work yet to be accomplished is an event called the status date (SD). We measure work performance at this point in time. Anything to the left of (earlier than) this time should have been accomplished, while anything to the right of (later than) this time remains to be accomplished.

We set a common point for the entire project to prevent confusing the status of one task with that of another. Common choices include the end of the month, the end of the accounting month, the end of the week, or the end of the day. The status date should take the form of a date (e.g., January 31, February 16, etc.) and it should accompany any statement of status. It is also a date expressed from the finish perspective (end of unit).

Separating tasks that have been accomplished from tasks that remain to be accomplished is probably too simple a division. We actually group statused activities into three categories:

1. Those that have started and finished (completed tasks).
2. Those that have started, but not yet finished (on-going tasks).
3. Those that have neither started nor finished, but are due in the near term (soon-to-come-due tasks).

Each of these situations is handled differently, so we must examine each individually. The first type of task lies entirely to the left of the status date; the third type lies entirely to the right of the status date. The second type of task straddles the status date, starting to its left and finishing to its right.

To begin, let us analyze the basic principle of status, which defines the relationship between activity and event accomplishment and the specific model parameters that represent this data. We could perhaps model the status of tasks in the forward pass (in terms of early dates) or the backward pass (in terms of late dates). Which is correct? Remember that the actual project progresses from the chronological left to the right, just as time flows from the past to the present to the future. Impact of status also flows left to right. Naturally, then, this is the proper way to model status, that is in the forward pass, the left to right flowing process. Therefore, we simulate the progress or status of each task in terms of its early dates. The Early Start and Early Finish tell us when we can expect to accomplish our tasks, the basic message of status.

As the actual (past) or projected (future) start and finish dates of a task move out (right) or in (left), the Early Start and Early Finish of the task must track this movement. It is most important that this is done in a way that preserves the principles of the CPM process, mainly to allow the valid calculation of project time reserves, while still reflecting the correct schedule information.

COMPLETED ACTIVITIES

Actual Start Date

When a task has actually begun, we need to associate the date it started with its Early Start date. We do this by affixing a new attribute to the task called the actual start (AES) date. This becomes the Early Start date of the task, regardless of any constraint or target date. Examine the model fragment shown in Figure 9-1.

Figure 9-1

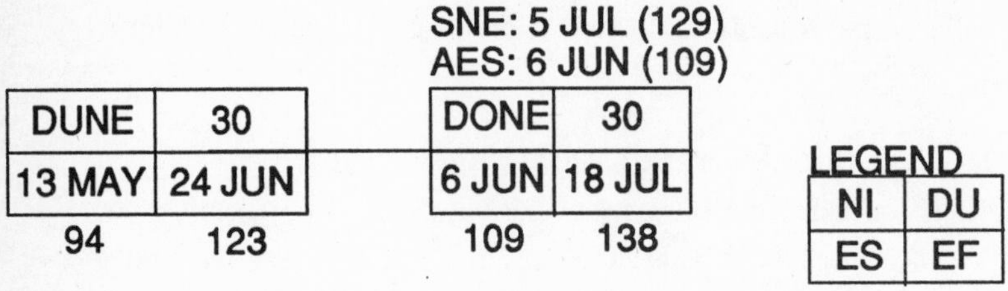

Task DONE has a finish-to-start (FS) predecessor, DUNE, and a date predecessor (SNE date). The constraint from DUNE would give an ES for DONE of June 27. The date predecessor would give an ES for DONE of July 5. DONE actually started on June 6, however, as the AES date indicates. This becomes the ES of DONE. When a task has an actual start date (AES), this becomes its Early Start date. The latest early date rule simply does not apply here.

The start of DUNE creates an obvious inconsistency in the logic of the model. How can task DONE depend on the finish of task DUNE when DUNE does not finish until June 24 and DONE actually started on June 6. Its start without the completion of DUNE violates the constraint (DUNE-DONE).

The date predecessor (SNE) of DONE also seems invalid. A date predecessor behaves like any other predecessor, but in this case the activity started in spite of it. These inconsistencies should be identified and corrected. When an actual date conflicts with other model parameters, we should note the inconsistency, then reconcile and correct the erroneous parameters.

To find the Early Finish (EF) of DONE, we add its duration to its ES date (June 6) and subtract one unit. The finish of the task is dependent upon the actual start date of the task and its duration. Figure 9-2 illustrates another situation involving actual start dates (AES). The start of task GONE has two predecessors, the finish of GOON (March 25) and its date predecessor (April 8), but neither of these dates drive the Early Start of GONE. It actually started on April 25, and that becomes its Early Start date. In this case, the start of the task has been delayed beyond when its predecessor relationships would have allowed it to start.

This situation does not constitute an inconsistency in the parameters, but as we might expect, the delay does affect the project through erosion of Path Time Reserve. This will be examined more thoroughly later in this chapter. To calculate the Early Finish of GONE add its duration to its Early Start date and subtract one unit (ES + DU – 1).

In summary, the rule states that whenever an actual start date is an attribute of a task, that date also becomes the Early Start date of the task, regardless of other model parameters. All subsequent early dates are calculated relative to the actual date.

Figure 9-2

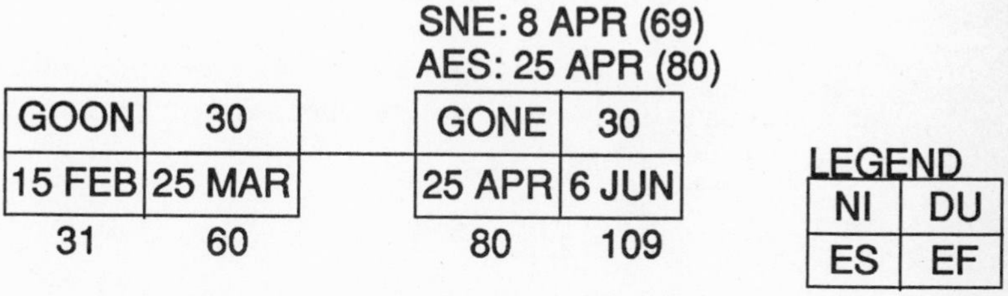

Actual Finish Date

When a task actually finishes, we need to associate the date with its Early Finish (EF) date. Again, we add an attribute, called the actual finish (AEF) date, to the node in the model. Any date value in this attribute becomes the Early Finish of the task. No other finish predecessor, including the task duration, drives the Early Finish date. The model fragment in Figure 9-3 illustrates this point.

The finish of task WENT has two predecessors, its own start and the finish of WUNT (September 1). Both of these date are later than the task's actual finish date, but again the latest early date rule does not apply. Any actual finish (AEF) date becomes the Early Finish date of the task. Again the predecessor relationship is in conflict with the actual status. This condition should be recognized, and then reconciled.

Any task that has an actual finish (AEF) date should also have an actual start (AES) date since no task can finish without ever starting. A node that shows an AEF date, but no AES date is simply missing information. This creates an inconsistency in our model's status that requires investigation and correction.

Figure 9-3

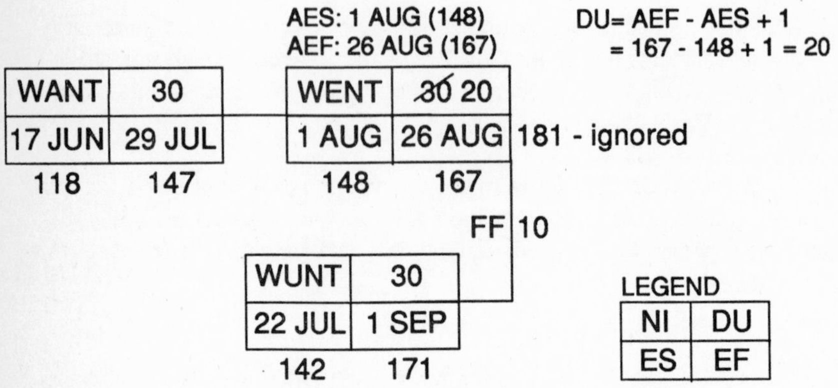

When both actual dates are present, the difference between them redefines the duration of the task:

$$DU = AEF - AES + 1$$

This value replaces any estimate of the task's span. We generally retain the original duration (ODU) as another task attribute. Comparing it with the actual duration aids impact analysis and forms a basis for future estimates of a similar activity's span.

To store the original duration, we must determine just how to define that term. This definition requires an understanding of status and control as well as baseline retention, which will be discussed later. At this point, we need to understand only two principles regarding tasks that have been completed (AEF and AES are both known). First, we will alter a completed task's duration based on its actual dates; second, we will retain a value for the original duration to aid analysis.

It is important to update the duration of a task, not only to keep it consistent with the actual dates, but also to allow an accurate backward pass. One prevalent school of thought ignores backward pass calculations for completed tasks. Some even extend this exclusion, ignoring tasks that have only started, as well. These practitioners either believe that these tasks have no late dates or set their late dates equal to their early dates, reasoning that a finished task has no time reserve. Remember, however, that time reserve is not the property of a task, but of the path or paths on which it lies. Completed tasks still lie on paths, and they still contribute to the lengths of these paths. We find that the forward and backward pass dates and resultant time reserve values for all of the completed tasks benefit path analysis significantly. However, to accomplish this requires the duration of tasks to be maintained consistently with their status.

To better understand all of this, we need to examine a more complex problem. Figure 9-4 illustrates a model before status assessment. Examination reveals that path along -MBA-PHD- has a negative PTR. It is the worst-case path in the example. The path along -DDS-MFA-LLD- has 30 days of PTR and the worst-case path through -MED- has only 25 days of PTR. We then assess status as of May 16 (SD).

Figure 9-4

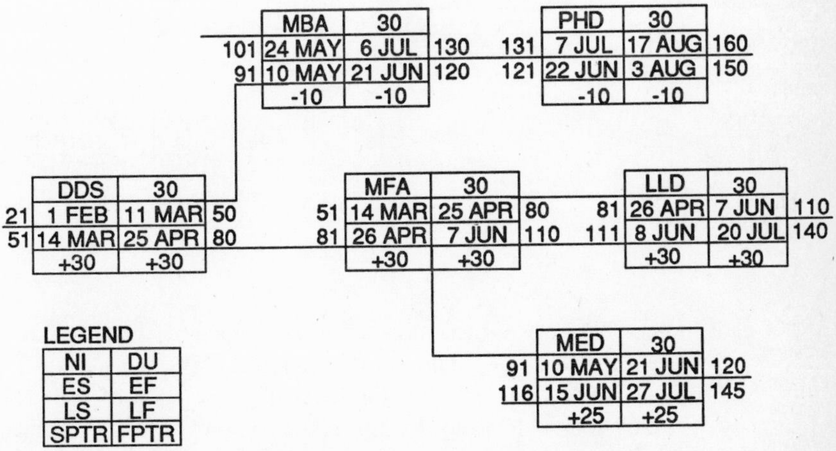

Task DDS actually started on February 1 and actually finished on March 25. Task MFA actually started on March 21 and actually finished on May 16. Let us update the model with this information, calculate the CPM parameters, and analyze the data.

First, we append the actual dates as attributes to these two tasks, and then calculate new durations for these tasks. After running normal forward and backward passes, we calculate current PTR values. Remember that we set Early Start and Early Finish dates to actual dates on any node for which we have these values.

Figure 9-5 illustrates the results of these processes and we can quickly identify the impacts of the status. Nothing has changed, however, in the path along -MBA-PHD-. It still has a negative 10 days of reserve and none of the calculated dates have been affected. It remains the worst-case path in the (visible) project and it continues to threaten some schedule objective to the right. At least the current status did not exacerbate this situation.

Figure 9-5

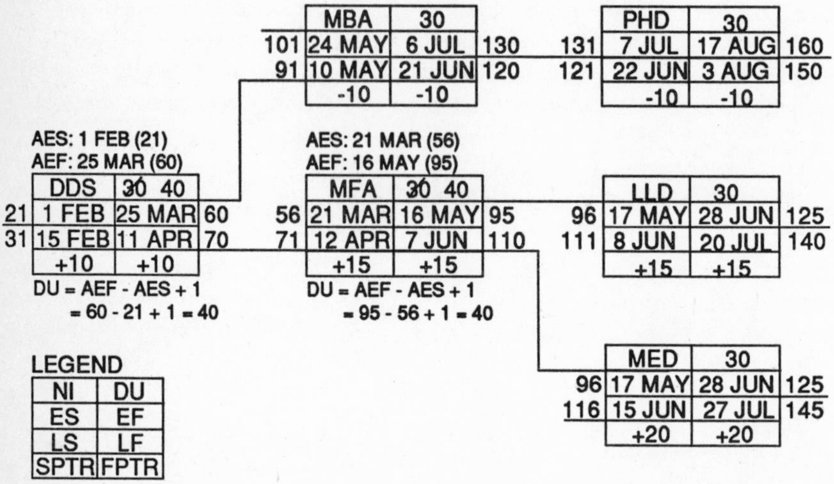

The path along -DDS-MFA-LLD- lost 15 days of its time reserve (from +30 to +15 days) due to the 10-day delay in the finish of DDS and the 10-day delay in the finish of MFA. How can we add 20 days to the span of the tasks along the path yet only reduce PTR by 15 days? The reason for this is because the actual start of MFA preceded the actual finish of DDS by five days, despite the finish-to-start constraint between them.

While the overlap of the last 5 days of task DDS and the first 5 days of task MFA has limited the net increase in the path length to only 15 days, it has also created an inconsistency between actual dates and the model's logic. The constraint between DDS and MFA is no longer consistent with the project's status; it actually should be a start-to-start with a 35-day lag, which we can determine from the PTR values of the two nodes. Node DDS has only 10 days, which reflects the full 20 days of erosion, while MFA reflects the correct 15 days of erosion. Whenever the value of PTR changes across a constraint, something else is influencing the successor. In this case, this is the inconsistency in status.

Notice, further, that the slip of the finish of DDS has no impact on MBA. The slip of the finish of MFA does, however, affect the starts of both LLD and MED, though in different ways. Task LLD has no other predecessor, so it is affected directly, its start and finish slipping 15 days. MED does have another predecessor of its start, the activity that determined its ES in the first example. However, the delayed finish of MFA overrides the other MED predecessor and pushes its start and finish date forward five days. PTR in this path shows a like decrease (+25 to +20).

This simple example shows clearly how the calculation of late dates on completed tasks can aid analysis. These dates and the PTR values they produce help us not only to assess project status, but also to understand the influences of that status. The result is a more thorough analysis. In order to derive correct results, we must correlate the task durations with their actual start (AES) and completion (AEF) dates.

Each slip affects the paths of the model in a different way, which we can determine through effects to successors early dates and all path PTRs. We can assess the amount of PTR erosion (15 days) as well as the rate of erosion (50%). Chapter 12 will more thoroughly explain how to use this information to identify problems and to mitigate their adverse impacts.

Summary: Actual Start and Finish Dates

A task's actual start (AES) date is an attribute of the task and becomes the Early Start date of the task. A task's actual finish (AEF) date is an attribute of the task and becomes the Early Finish date of the task. We recalculate the duration of a completed task as follows:

$$\text{Actual DU} = \text{AEF} - \text{AES} + 1$$

When recalculating the backward pass through finished activities, we substitute in these actual durations.

The Path Time Reserve for completed tasks resulting from these calculations contributes valuable information for assessing how and to what extent actual status affects remaining paths. When the actual duration of a task exceeds its original duration, the task increases the length of the path or paths on which it lies by the difference. The slip can decrease Path Time Reserve of these paths by up to the same amount.

ON-GOING TASKS

Tasks that have started but not yet finished are called *on-going tasks*. In effect, they straddle the status date; part of the task is done before the date, and part will be completed after it. Since such a task has been initiated, its actual start (AES) date becomes its Early Start (ES) date in the model. As the task has not actually finished, its Early Finish (EF) date is determined by the forward pass process.

We can expect most on-going tasks to finish in the near term, although this also depends on their durations and any other finish predecessors. To estimate the date of this near-term finish when assessing status, we update a new parameter called the *projected finish (PEF) date* by one of four methods. Two of these methods calculate the projected finish date based on an estimate of how much work on the task has been completed and the other two methods calculate the date based on an estimate of how much work remains. Two of these methods also require a reference point external to the task, the status date (SD).

Percentage Complete Method

Percentage complete (PC) is an integer value between zero and 100 that reflects an estimate of how much of a task has been completed as of the status date (SD). This percentage is measured relative to the task's duration. The method calculates the remaining duration (RDU) of the task as the task duration less the completed portion of the task. Adding the remaining duration to the status date gives the projected finish (PEF) date of the task.

Let us find the date on which a 30-day task that is one-third, or 33 percent, completed as of November 4 (day 216) can be expected to finish. First, we calculate the remaining duration of the task:

$$\begin{aligned} RDU &= DU - (DU \times PC / 100) \\ &= 30 - (30 \times 33 / 100) \\ &= 30 - 10 = 20 \text{ days} \end{aligned}$$

We then add the remaining duration of the task to the status date to find the PEF date:

$$PEF = SD + RDU$$
$$= \text{day } 216 \text{ (Nov. 4)} + 20 \text{ days} = \text{day } 236$$

The percentage complete method of assessing on-going task status tells us that the 236th workday, December 6, is the projected finish of the task based on the given conditions. This projected finish date functions like a finish-no-earlier-than date. It becomes the Early Finish (EF) date of the task provided it is the latest early date, that is, no other predecessors of the task's finish (FF constraints) force a later date.

After calculating a task's projected finish date (PEF), we refigure the task's duration as the difference between this date and the task's actual start date (AES):

$$DU = PEF - AES + 1$$

This yields an Early Finish date that is identical to the PEF date because the AES date is also the task's ES date. We will also use this revised duration to calculate a valid Late Start date in the backward pass.

Let us further explore this process through another example. Figure 9-6 adds status information to a previously used model fragment. Task ALPHA started and finished on the dates shown. Task DELTA is on-going. It started on March 14 and is 60 percent completed, as of the 15th of April (SD).

Figure 9-6

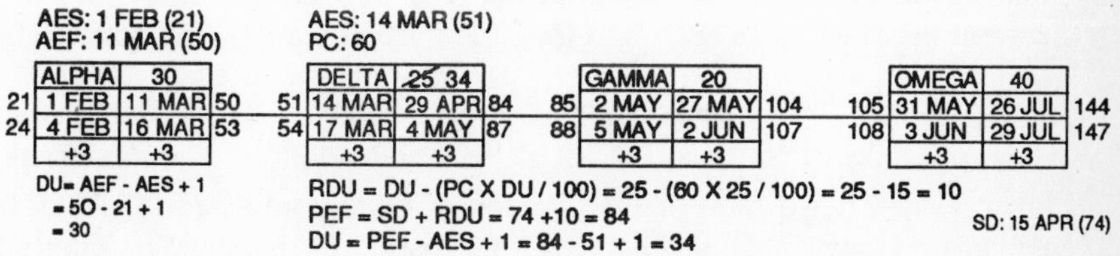

The Early Start date of task DELTA is its AES date (March 14). We can then determine the task's Early Finish date using the percentage complete method by calculating the remaining duration of the on-going task from its duration and the percentage complete value (RDU = 10). We can then calculate the projected finish date of the task by adding the remaining duration to the status date (PEF = 84/29 Apr). Finally, we calculate the duration of the task from its projected finish and actual start dates (34). Rerunning the forward and backward passes reveals that DELTA's PEF is its latest early date, and therefore its EF.

Having calculated this data, we must analyze it to assess the impact of current status on the balance of the project. Task ALPHA started and completed on schedule, leaving the path unchanged. Task DELTA started as planned, but its finish is estimated to slip nine days (duration increases from 25 to 34 days). In a simple linear model, this erodes Path Time Reserve that same nine days (+12 to +3). This is clear without the CPM process, but we can continue our analysis to compare the rate of PTR erosion to the rate of progress of the project. We have lost 75 percent of the Path Time Reserve (9 of the 12 days) before completing half of the project (54 ÷ 124 = .44). This indicates a serious problem to which the project management must react immediately before it gets worse.

The percentage complete method of calculating a task's finish date takes a very simple approach to a fairly complex concept. It would assume, for example, that a half-completed task with a duration of 30 days would take another 15 days to complete. The estimate of how much of the task is completed, though subjective, is generally fairly reliable. The problem is that it has no causal relationship to the time necessary to finish the remainder of the task. The nature of the work or other circumstances could drag the first half of a 30-day task out for six months, and it may actually only take 10 days to finish the last half. The reverse of this could be equally true. This percentage complete method ignores such potential deviations. The correlation it assumes between the completed portion of a task, its duration, and the amount of work that remains to be done is nebulous at best.

An alternative percent complete method calculates the projected finish (PEF) date of a task based on the portion of the work completed. It gives a figure for the duration of the work completed (DUWC) as the number of work days between the status date (SD) and the task's actual start (AES) date:

$$DUWC = SD - AES + 1$$

It then establishes a ratio of the duration of work completed (DUWC) to the percentage complete (PC) and equates it to a ratio of the remaining duration (RDU) to the percentage of work remaining (PWR), which equals 100 – PC. We can then solve for the remaining duration of the task.

$$DUWC / PC = RDU / PWR$$
$$RDU = DUWC \times PWR / PC$$

Adding the remaining duration to the status date then gives the PEF date of the task. Let us revisit a previous example to see how this method works.

TaskA, with a duration of 30 days, actually started on February 1. As of March 4 (using the five-day work week) we have estimated its percentage complete (PC) as 33 percent. To calculate the remaining duration (RDU) of TaskA based on this data, we must first calculate the duration of the work completed:

$$\text{DUWC} = \text{SD} - \text{AES} + 1$$
$$= \text{day } 45 \text{ (March 4)} - \text{day } 21 \text{ (Feb. 1)} + 1 = 25 \text{ days}$$

Thus, 25 workdays have passed since TaskA began, and in this time one-third of the task has been completed. We can now calculate the remaining duration of the work:

$$\text{RDU} = \text{DUWC} \times \text{PWR} / \text{PC}$$
$$= 25 \text{ days} \times 67 / 33 = 50 \text{ days}$$

We add this value to the status date to determine the task's projected finish date:

$$\text{PEF} = \text{SD} + \text{RDU}$$
$$= \text{day } 45 \text{ (March 4)} + 50 \text{ days} = \text{day } 95 \text{ (May 16)}$$

Completing the first third of the task took 25 days, so we can expect to take another 50 days to finish the other two-thirds of the task. Then we recalculate the duration of the task as the difference between its AES and PEF dates (75 days) or the sum of RDU and DUWC (DU = RDU + DUWC = 50 + 25 = 75). In the forward pass, we treat the projected finish (PEF) date as a finish-no-earlier-than date to calculate the task's EF date. In this case, there is no other later predecessor of the task's finish (AS + DU – 1 = PEF), so the PEF is also the EF date.

Another example will show how this percent complete process works in more detail. In Figure 9-7, we will calculate the projected finish of DELTA from Figure 9-4 based on this alternative method. In this method, we first calculate the span of the completed work as the difference between the actual start date and the status date (24 days). We then calculate the percentage of work remaining by subtracting the percentage complete from 100—if 60 percent of a task has been completed, 40 percent remains to be done. We then figure the remaining duration of the task (16 days) from the ratio of the duration of the work completed to the percentage complete and the duration of the work remaining to the percentage of work remaining. We then add this remaining duration to the status date to find the projected finish of the on-going task (workday 90 or May 9). Finally, we recalculate the duration of the task from this projected finish date and its actual start date (DU = 40) or the sum of RDU and

DUWC (DU = RDU + DUWC = 16 + 24 = 40). The normal forward and backward passes can then resume.

Figure 9-7

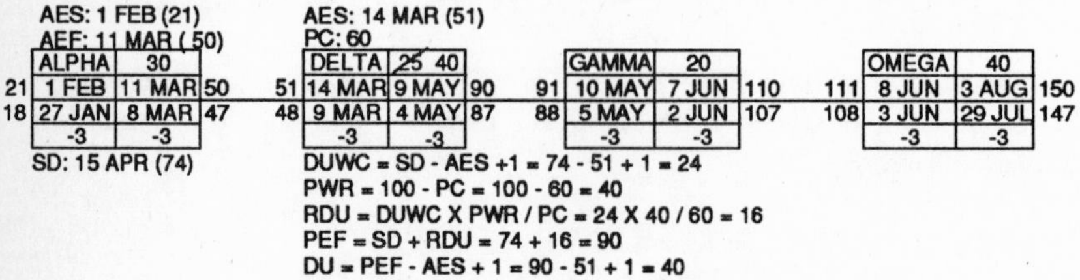

From the same status data, we have now estimated that the finish of task DELTA has slipped 15 days (DU increases from 25 to 40 days) instead of 9 days, as we found with the first percentage complete method. This alternative percentage complete yields entirely different results because it is very different from the first method. It uses some additional data and is based on a different premise.

Though this second method appears on the surface to be more valid, it still tries to quantify a very complex relationship with equations. It assumes a definite relationship between the length of time necessary to finish one portion of a task and that necessary for any other portion of a task. If it takes 10 days to accomplish 10 percent of a task, this method assumes it will take 20 days to finish 20 percent of the task, and 100 days to finish all of the task. Real-world tasks are rarely accomplished in such purely linear progressions. Basically this method only indirectly answers the fundamental questions: How much work remains and how long will it take? A process that gives a direct answer to these questions may be much better.

Remaining Duration

Remaining duration estimates the number of time units (days) necessary to finish a task as a numeric value. Rather than calculating this value as a function of the work accomplished, we estimate it directly. This reduces the source of error to our ability to estimate this value, avoiding the compound error of estimating something else and then deriving the final value based on suspect assumptions. As in the percentage complete method, we must take the status date (SD) as a point of reference.

As an example, let us determine the projected finish (PEF) date of a task with a planned duration of 30 days and an estimated remaining duration (RDU) of 40 days relative to a status date of September 2 (workday 172). Notice in this case that the duration of the remaining work (40) exceeds the original duration for the entire task (30 days). Unfortunately, significantly underestimating the task's scope makes this quite possible.

We can expect the task to finish on:

$$\text{PEF} = \text{SD} + \text{RDU}$$
$$= \text{day } 172 \text{ (Sept. 2)} + 40 \text{ days} = \text{day } 212 \text{ (Oct. 31)}$$

We then recalculate its span using the equation:

$$\text{DU} = \text{PEF} - \text{AES} + 1$$

Another example (Figure 9-8) will show more clearly how this process works. As before, task ALPHA started and finished on time and task DELTA is on-going. In this case, we estimate that it will take an additional 15 days beyond our status date (April 15) to finish the task.

Figure 9-8

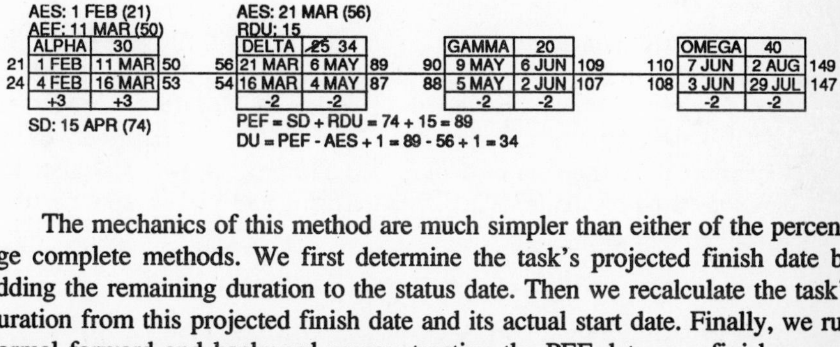

The mechanics of this method are much simpler than either of the percentage complete methods. We first determine the task's projected finish date by adding the remaining duration to the status date. Then we recalculate the task's duration from this projected finish date and its actual start date. Finally, we run normal forward and backward passes, treating the PEF date as a finish-no-earlier-than date. The recalculated duration would also yield a finish date identical to the PEF date, and allow correct date calculations in the backward pass.

This calculation shows that the finish of task DELTA slipped nine days (DU increased from 25 to 34 days), causing a nine-day erosion of PTR. We know that the path suffered a 14-day PTR erosion (from +12 to –2 days). What happened to the other five days?

Closer examination reveals another slip in the project besides the finish of DELTA. Task DELTA's start was also delayed, and by 5 days (56 – 50 – 1 = 5). This slip and the 9-day slip together account for all of the 14 days of PTR erosion. The analysis of completed activities reveals this second impact. By examining the change in the PTR across the constraint ALPHA-DELTA (+3 to

–2), we knew we had experienced a 5-day delay. Since DELTA has no other predecessors, the delay had to be caused by a slip in the form of either an SNE or an AES date.

In the RDU method, the original duration of the task and its actual start (AES) date do not affect the calculation of its projected finish date. The only parameter, an estimate of how much time any remaining work will require, addresses the point much more directly. This method gives us a sharply improved picture of the status of an on-going task. Another method incorporates this technique with a more practical twist, though.

Earned Value

Earned value (EV) effectively measures task accomplishment by weighting lengths of time (duration increments) to reflect expected amounts of accomplishment. It gains its effectiveness at portraying status, however, at the cost of much more complex information requirements. This technique is an element of a much more extensive cost-schedule integration process, which will be covered briefly in Chapter 16.

Estimated or Projected Finish Date

This method directly estimates the date on which a task will finish, which is its projected finish (PEF). For example, TaskA has a duration of 30 days, an Actual Start Date of February 1, and a projected finish date of May 16.

The technique establishes this projected finish (PEF) date regardless of the original duration. In fact, before beginning the CPM process, we recalculate the task's duration based on the projected finish. This would yield a new duration in our example of 75 days:

$$\begin{aligned} DU &= PEF - AES + 1 \\ &= \text{day 95 (May 16)} - \text{day 21 (February 1)} + 1 \\ &= \text{75 days} \end{aligned}$$

This method of assessing on-going task status assumes an accurate estimate of the amount of work remaining and its duration. In this way, the method is virtually identical to the remaining duration method except that:

1. It needs no status date (SD).
2. It requires a definitive commitment to a specific date. A statement of a remaining duration is not as specific, because it requires a point of reference. This is easily muddled.

These differences give this method an advantage over the remaining duration method. It has a disadvantage in that it assumes that the date comes from an honest estimate of the duration of the remaining work, instead of thin air. A sloppy estimate will give an invalid PEF date, tainting any subsequent analysis. With this stipulation, however, the projected finish date is the preferred method of gauging the status of on-going tasks, and it will be our principle method for reflecting projected status.

Before we proceed, however, we will review another example to ensure that we understand both the mechanics of this process and the associated analysis.

Figure 9-9

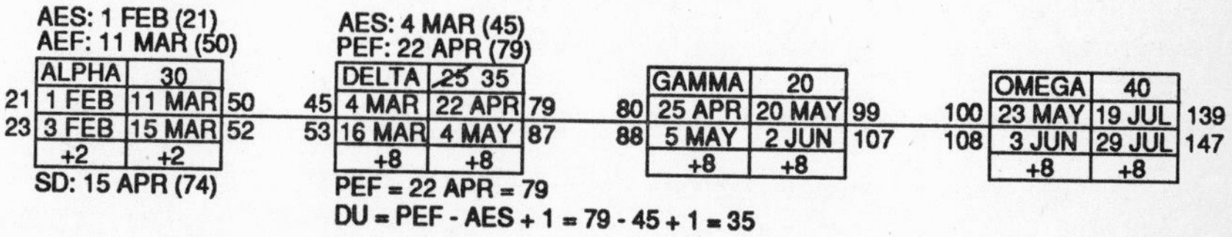

In Figure 9-9, Task ALPHA started on February 1 and finished on March 11. Task DELTA started on March 4 and is projected to finish on April 22. To assess this status in terms of the parameters of Task DELTA, we begin by recalculating the duration of this node as the difference between the actual start date and the projected finish date. Normal forward and backward passes can then proceed. The projected finish (PEF) date of DELTA becomes its Early Finish date because it is the latest early date. This gives the same date as that derived from the recalculated duration added to the task's Early Start date.

Again, we can see a change in the value of PTR across the constraint ALPHA-DELTA, but in this case the value increases (+2 to +8). Remember, any time PTR changes across a constraint, something else is influencing the path. DELTA started on March 4, even though its predecessor did not finish until March 11. ALPHA and DELTA are partially overlapped yet interfaced with an FS constraint. We must resolve this obvious inconsistency between the current status and the model logic. In essence, the status reflects a start-to-start constraint relationship with a 24-day lag.

Overall, PTR diminished four days (+12 to +8 days). By comparing the duration of DELTA before the status check to that after the check, we can see that the finish of this task slipped 10 days, which should be identical to the reduction of the PTR. The 6-day overlap between ALPHA and DELTA, however, recovers 6 days of PTR leaving a net impact of 4 days. The PTR measured at ALPHA, which precedes the advanced start of DELTA, illustrates the total impact of the slip of DELTA based on the original constraints (10 days of PTR erosion: +12 to +2 days). The PTR of the remainder of the path beyond the overlap reflects the net condition, a 4-day erosion.

The Relationship of On-Going Status Parameters

Selected parameters of one method of assessing status of on-going tasks can often supplement calculations from another method. For instance, the remaining duration (RDU) of a task and its projected finish (PEF) date can both be calculated from the percentage complete (PC), and vice versa. As a rule, we check the status of a task through either the remaining duration (RDU) or projected finish (PEF) date and then calculate its percentage complete (PC) as a measurement parameter. In other words, we derive it from other values, it does not drive those values.

Other Predecessor Impacts

Any of the methods of measuring on-going task status may bring a finish predecessor (FF constraint) of a task into conflict with the status. This predecessor would make the Early Finish of the task occur later than the projected finish, creating a conflict between the model logic and the status. The CPM process must flag such a condition so that we can resolve the contradiction.

This is why we always assess the projected finish date of a task and then impose this value as a finish-no-earlier-than date in the forward pass. Otherwise, when we determined the task's projected finish, we could overlook the impact of a predecessor of the task's finish, which would not be noticed until it was actually needed. By calculating the task's Early Finish date in the forward pass we detect any external precedessor (FF constraint) that could drive the Early Finish of the task in the form of a difference between the EF date and the PEF date.

Summary—On-Going Tasks

All of this creates a complex, interconnected list of processes that accomplish the same thing—finding the Early Finish date of an on-going task. In the interest of simplicity and consistency, we will use only two methods. One will be discussed in Chapter 16 (earned value). The other, the most direct means of reaching the end result, consists of projecting the Early Finish of the node directly.

We can do this most simply through a projected finish (PEF) date. We will relegate all the other methods so far discussed (PC and RDU) to the role of adjunct parameters. We will maintain them as utilities, but we will not use or maintain them in the mainstream of the modeling processes. We will now summarize the principles of modeling on-going tasks, that is, tasks that have started but not finished.

The date of a task's actual start (its AES date) becomes its Early Start date. We can project the Early Finish of a task by two means:

1. Estimate the projected finish (PEF) date directly.
2. Calculate the PEF by the amount of earned value, as discussed in Chapter 16.

After determining the PEF date of an on-going task, we recalculate its duration as follows:

$$DU = PEF - AES + 1$$

The PEF date acts as a finish-no-earlier-than (FNE) date in the EF calculation. Since an actual start (AES) date replaces the Early Start date for all on-going tasks, the EF date calculated by adding the new duration to the ES date will always equal the PEF date.

An external predecessor with an FF constraint can drive the task's EF date to a later date. The backward pass processes data for on-going tasks based on revised durations, keeping the tasks' Late Start dates consistent with current conditions.

These calculations give Path Time Reserve values for on-going tasks, which help identify and quantify the effects of current status on remaining paths. Any difference between the revised duration of a task and its original duration increases or decreases the length of the path or paths by the difference. Time reserves will reflect net effects.

NEAR-TERM TASK STATUS

Tasks that have not yet started but are expected to begin in the near term, perhaps within a month, are called *near-term* or *soon-to-come-due tasks*. These tasks lie completely to the right of the status date, but not very far to the right. By reassessing and updating their status rather than waiting for actual status, we get an earlier indication of impending trouble.

Projected Start Dates

As the start of a task approaches, we can better understand its scope, requirements, resource needs and availability, and ultimately our ability to accomplish it according to plan. From this information, we update our projection of the task's expected start and finish. This requires that we first assess the status of its predecessors, as they strongly influence its potential start date. Other forces, especially resource availability, could also affect its start date. To satisfy all these considerations, we determine another activity attribute: the projected start (PES) date of the task.

Since the activity has not actually started, the predecessors of the task's start can still affect its ES. Therefore the projected start date functions like a start-no-earlier-than (SNE) date, stating that the task can start no earlier than the projected start date, although predecessor constraints could delay its start further.

Comparing the task's Early Start date to its PES date can yield valuable information. If these two dates are equal, the projected start date accurately represents the task's status. If the ES date is later than the PES date (it could not be earlier), then either a predecessor threatens the task's projected start, or the activity interface logic is in error. Either way, we should determine the source of the threat and attempt to reconcile it before it has any actual impact.

Projected Finish Dates

When updating the status of the finish of a near-term task, we might logically try to employ the same methods we used to assess the status of on-going tasks. Only one of the four methods for reviewing on-going tasks works for near-term tasks. Three of the methods (PC, RDU, and EV) determine the projected finish date of a task based on its partial completion, precluding their use for tasks that have not started. We can, however, project a finish date for a near-term task directly by the projected finish (PEF) date method, which can then replace the Early Finish date.

Again, other predecessors of the node's finish could threaten this date. Therefore the PEF functions like a finish-no-earlier-than date. We can recognize delays caused by predecessors by comparing the PEF date of the task to its Early Finish date. Identical dates indicate that the projected schedule does not contradict the constraint logic. Any difference between these dates indicates that the constraint logic of the model conflicts with the schedule of the task. We should identify and resolve this contradiction.

When a task has both PES and PEF dates, we need to recalculate its duration. Two projected dates dictate a task's span just as two actual dates do. This revised duration can be calculated from the equation:

$$DU = PEF - PES + 1$$

This projected value should be used in both the forward and backward pass calculations.

Other Near-Term Status Scenarios

A task that does not have both PES and PEF dates must have one of three other possible arrangements. For the first two, a task has either a projected start (PES) date or a projected finish (PEF) date, but not both. Of the several ways to handle this situation, each with its advantages and disadvantages, we choose the best compromise. We will treat the present projected date value as a no-earlier-than date, but we will not recalculate the task's duration.

A task with a PES date, but no PEF date, has an Early Start date determined by the latest of the predecessors of its start, including the PES date. The EF of this task is then the ES plus the duration or a later date as specified by any other successor of its finish.

A task with a PEF date, but no PES date, has an Early Start date determined by the latest of the predecessors of its start. The EF of this task depends on the latest predecessor of its finish, including its ES plus its duration, the PEF date, and any other FF predecessor. Since we do not recalculate the duration, the task's Early Start date plus the duration will probably not equal its projected finish date.

For the third possible combination of projected dates, a task could have neither a projected start date nor a projected finish date. Such a task has no status information, and we handle it just like any other task in the forward and backward passes. Its Early Start date comes from the latest of the predecessors of its start; its Early Finish date comes from the latest of the predecessors of its finish.

In a final method of assessing near-term status we change only the Early Finish of a task. We revise the task's duration to respond to indications that the task will take more or less time than previously estimated. The Early Finish date of the task will change accordingly in the forward pass equation:

$$EF = ES + DU - 1$$

The resulting value would become the Early Finish of the task unless another predecessor of its finish made it finish later.

Examples can better illustrate the understanding of status of near-term tasks. In the model fragment in Figure 9-10, the first task, ALPHA, is completed. The second and third tasks (DELTA and GAMMA) have both projected start and finish dates, and the last task, OMEGA has only a projected finish date. Let us resolve the status of this series of tasks.

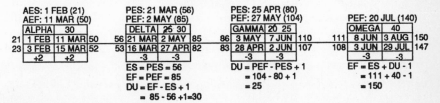

Figure 9-10

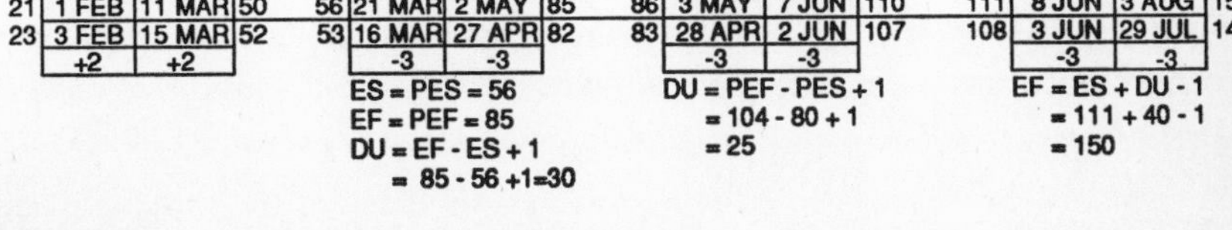

We set the Early Start (ES) and Early Finish (EF) of ALPHA equal to its actual start (AES) and actual finish (AEF) dates. These actual dates still give a duration of 30 days.

We compare the projected start (PES) date of Task DELTA with its other predecessors (the EF of ALPHA), selecting the latest early date as its projected start date (56 > 51). The difference between the projected dates of DELTA gives the task a new duration of 30 days. Task DELTA's Early Finish date, the projected finish (PEF) date, equals its Early Start plus the new duration minus one day. This task's finish has no other predecessors, so its status and the model relationships are not in conflict.

The projected start (PES) date of task GAMMA, however, is not the latest early date (86 > 80). Another predecessor overrides its Early Start: the constraint DELTA-GAMMA. We still recalculate the task's duration as the difference between PEF and PES dates, or 25 days. We then add this value to the task's ES date (day 86) for calculation of the task's Early Finish date. Since this date is later than the PEF date, it becomes the Early Finish date of the task. Comparing GAMMA's projected dates to its early dates reveals that the task's status conflicts with the constraint logic (80 to 86, 104 to 110). Another interface appears to prevent us from meeting our projected status. Before we proceed with our analysis, we should resolve all status conflicts of this type.

Task OMEGA has no projected start date, but it does have a projected finish date. Therefore its Early Start date depends on its predecessor, the finish-to-start constraint GAMMA-OMEGA. Since the task has only one projected date, we do not recalculate its duration. Adding its original duration to its Early Start date gives a later start date (day 150) than its projected finish date (day 140) so the sum becomes the Early Finish date of the task. Again, comparing the task's EF date to its PEF date shows a conflict (140 to 150). This means that a predecessor threatens its expected completion.

After we have modeled the status, we can then analyze the resultant information. The last three tasks together reflect a PTR of –3 days, but Task ALPHA reports +2 days. The project started with 12 days of PTR, so it has lost 15 days. Where did this time reserve go? Slips added 5 days each to the spans of tasks DELTA and GAMMA. This accounts for 10 days of the erosion. The other 5 days of erosion came from a delay in the initiation of Task DELTA. These three factors (two span increases and one delay) collectively result in the 15 days of Path Time Reserve erosion.

The PTR measured at Task ALPHA reflects only the 10-day impact of the increased path length (from +12 to +2 days). Across the constraint ALPHA-DELTA, the time reserve of the path changes again (from +2 to –3 days), and we cannot trace this erosion to the confluence of any other path. Examination reveals that DELTA's projected start date is delayed by 5 days. This delay affects all successor tasks as well.

Any actual dates become the nodes' early dates. The projected dates function only as additional parameters in the forward pass. These values become the nodes' early dates only if they are the latest dates. Projected dates become the early dates of Task DELTA, but other predecessors drive the early dates of GAMMA and OMEGA.

The reason for this is that we are still contending with projected or forecasted information. Until the events occur, our analysis must assess and allow for factors that can adversely affect our schedule. We still intend to achieve projected dates, but other variables may affect them. Our status process is designed to help us identify these threats, so we can confront and perhaps resolve them.

A second example (Figure 9-11) illustrates a different set of conditions. As in the first example, Task ALPHA is completed. Task DELTA has the same projected start and finish dates as before, but GAMMA has a new set of dates. Task OMEGA has only a projected start date.

Figure 9-11

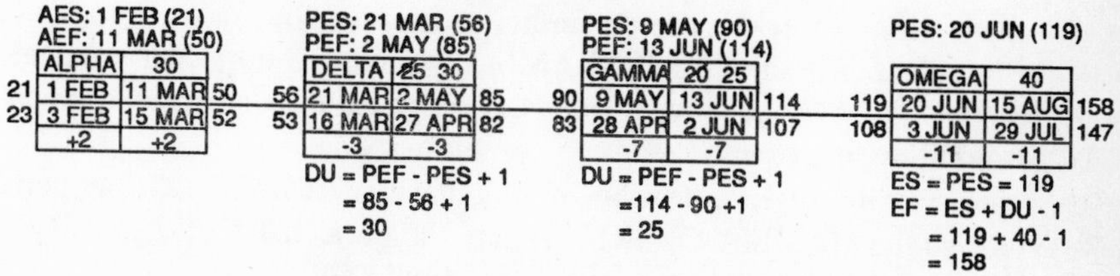

As ALPHA and DELTA contain the same information as before, we need not reconsider them. The status information for GAMMA and OMEGA is new, though. The projected start date of GAMMA (day 90) is the latest early date of all predecessors, making it the Early Start date of the node.

With both a projected finish date and a projected start date on this node, we will recalculate the duration. Adding this value (25 days) to the Early Start date results in a date identical to the projected finish date (day 114). Notice that this is always true when the status date at the start end of the node (the AES or PES) becomes the node's Early Start date because:

$$PEF = ES + DU - 1 = EF$$

This projected finish date (day 114) becomes the node's Early Finish date if it is the latest predecessor. Since no other predecessors constrain the finish of GAMMA, this is true.

Two values affect the calculation of the Early Start of OMEGA: the Early Finish of GAMMA (plus the lag plus one day = day 115), and OMEGA's projected start date (day 119). The latest date, the projected start date, becomes the Early Start date of OMEGA. Since the node does not have a projected finish date, we do not recalculate the duration. OMEGA's Early Finish date is simply its Early Start date plus its span minus one day (EF = 119 + 40 – 1 = day 158). We run the backward pass using the newly recalculated durations, and then finally we compute Path Time Reserves. We can now examine the results of this completed CPM process.

The PTR changes from +2 to –3 days across the constraint ALPHA-DELTA, as in the first example. The +2 days on ALPHA represents a PTR erosion of 10 days (from +12 to +2 days) from the project's previous condition. Since neither the start date of this task nor the required finish date of the path changed, we must look for changes in the path length. Calculated durations for all four tasks show that the spans of DELTA and GAMMA have, in fact, increased by 5 days each, giving a total path length increase of 10 days. This amount of erosion is reflected in the PTR values of Task ALPHA. The erosion of PTR across the ALPHA-DELTA constraint comes from the delay (5 days) on the start of DELTA caused by its projected start date.

The Path Time Reserve suffered further erosion of four days from –3 to –7 days across the constraint DELTA-GAMMA. Again, the projected start date pushed the start of GAMMA forward four days past the date on which its constraints would allow it to start.

Still further time reserve erosion of four days (–7 to –11 days) appears across the constraint GAMMA-OMEGA. Again, the projected start date delayed the start (ES) beyond the date the predecessors would allow.

Clearly, when a projected start (PES) date is a task's latest predecessor, becoming its Early Start date, it delays the task's start date. This erodes Path Time Reserve by an amount equal to the length of the delay. Unless a projected date matches the previous status information, it will have an impact. Anytime we inject a projected start or finish into the model, there is an impact that must be identified. It either makes current status inconsistent with the model, creating a conflict with other predecessors, or it causes erosion of Path Time Reserve. As an important part of our analysis, we must identify tasks affected by projected dates and assess their impact on the project.

Summary: Near-Term Task Status

To model the status of a near-term or soon-to-come-due task, that is, one that has neither started nor completed, we estimate its start date in the form of a projected start (PES) date and its finish in the form of a projected finish (PEF) date or a change to the duration of the task. Projected start and finish dates act as no-earlier-than date predecessors on a task.

We recalculate the duration of a task that has both PES and PEF dates through the equation:

$$DU = PEF - PES + 1$$

We then process the forward and backward passes based on this revised duration. We do not recalculate the duration of a task that has only a projected start or a projected finish date, but not both.

The Path Time Reserve impacts of these calculations come from increases or decreases in duration of the near-term task and from projected delays in the task's start and finish dates.

SUMMARY

Two distinct objectives guided the material in this chapter:

1. Demonstrate how to simulate change and status data in the project model and the relationship of this data to the forward pass calculations.
2. Demonstrate how to assess and analyze impacts on a project of status and change.

Change is modeled in the same manner in which we built the original model. We add or delete tasks and constraints and/or we modify existing tasks and constraints.

The principles of modeling task status are simple, but the details can get fairly complex. Keeping the fundamental principle of status in mind, however, should help us keep the process in proper perspective:

> We assess task status based on either actual or projected early dates simulated in the forward pass process.

We model the status of a task through its forward pass parameters: SNE dates, FNE dates, ES, EF, DU, etc. We must do so in a way that preserves the proper calculation of the project's time reserves. This allows us to monitor the impact on projects of task status as changes in Path Time Reserves (PTR). This impact takes two forms:

1. Changes (increases or decreases) in the cumulative lengths of paths through duration changes for individual tasks.

2. Delays to tasks' start or finish dates.

Each of these affects Path Time Reserve differently, so noting the effects, we can determine their sources. Increases (or decreases) in task durations typically show up as PTR erosion (or increases) through the entire path, since these changes affect the cumulative length of the path. The impact of a delay (or a pull-in) in a start or finish date affects a specific point or event in the path, so it affects PTR only downstream (to the right) of this point.

Clearly, status data enriches our analytical capability enormously. We can maintain a model as it changes through time and continually reevaluate the impacts on the remainder of the project. This keeps us looking forward, managing proactively throughout the project's progress.

The many available methods of modeling task status can become very involved in combination, so we will summarize them once again. Completed tasks have both AES and AEF dates. We set such a task's ES date equal to its AES date, and its EF date equal to its AEF date. We then recalculate its duration as follows:

$$DU = AEF - AES + 1$$

If a completed task shows an AEF date, but no AES date, an error has occurred in the status assessment process. We set the task's ES date by the normal rules of the forward pass and set its EF date equal to its AEF date. We do not recalculate duration. Since the old value for duration threatens the accuracy of the backward pass data, we must identify and correct the erroneous status condition.

On-going tasks show AES dates, because they have begun, but lack AEF dates, because they have not yet completed. We assign projected finish (PEF) dates to such tasks based on specific status information, then recalculate their durations by the equation:

$$DU = PEF - AES + 1$$

As we initiate the forward pass, we set the Early Start of an on-going task equal to its actual start (AES) date. The recalculated duration will yield an Early Finish (EF) date for the task that matches its projected finish (PEF) date. This date becomes the Early Finish date of the task, unless some other predecessor of the task's finish (an FF constraint) results in a later date. This would make the Early Finish date later than the projected finish (PEF) date, creating a conflict with the task's status that would require reconciliation.

For a task that has an actual start (AES) date, but no other status information, we find its Early Finish date from the equation:

$$EF = ES + DU - 1 = AES + DU - 1$$

By an adjunct utility process, we can find the projected finish (PEF) dates of tasks based on other status parameters: percentage complete (PC) and remaining duration (RDU).

In the first of two percentage complete methods, we determine how much of a task has been accomplished as a percentage of its duration and express this PC value as a whole number (i.e., 20 percent is *20*, 55 percent is *55*, etc.). We calculate the remaining duration of the task based on the percentage complete and the original duration:

$$RDU = DU - (DU \times PC)/100$$

We then calculate the task's projected finish (PEF) date by adding the remaining duration (RDU) to the status date (SD):

$$PEF = SD + RDU$$

In the second percentage complete method, we again express the percentage of the total task that has been accomplished as a whole number (the PC value). We then calculate the duration of the work completed (DUWC) as the difference between the task's status date and its actual start date:

$$DUWC = SD - AES + 1$$

We find the percentage of work remaining (PWR) by subtracting the percentage complete (PC) from 100:

$$PWR = 100 - PC$$

We then determine the task's remaining duration (RDU) from the duration of work completed (DUWC), the percentage complete (PC), and the percentage of work remaining (PWR):

$$RDU = (DUWC \times PWR)/PC$$

Adding the task's remaining duration to the status date gives the projected finish date:

$$PEF = SD + RDU$$

In the remaining duration method, we add a direct estimate of the remaining duration (RDU) to the status date to find the projected finish date:

$$PEF = SD + RDU$$

Having determined the projected finish date, we can also determine the other status parameters:

$$RDU = PEF - SD$$
$$DUWC = SD - AES + 1$$
$$PC = DUWC/(DUWC + RDU)$$
$$PWR = RDU/(DUWC + RDU)$$

This method gives calculated values rather than estimates for these parameters.

Near-term tasks have neither started nor completed. Projected status parameters are fairly reliable for these tasks because only a short interval separates the status date and their initiation.

For a near-term task for which we have projected start and projected finish dates, we recalculate the duration:

$$DU = PEF - PES + 1$$

We treat these two dates (PES and PEF) as no-earlier-than dates in the forward pass calculations. We use the recalculated duration in the forward and backward pass processes.

For a near-term task with a projected start date, but no projected finish date, we do not recalculate the duration. We treat the PES date as a start-no-earlier-than (SNE) date, and determine the Early Finish of the task by the normal rules of the forward pass.

For a near-term task with a projected finish date, but no projected start date, we do not recalculate the duration. We determine the Early Start of the task by the normal rules of the forward pass, and treat its PEF date as a finish-no-earlier-than (FNE) date.

We treat a near-term task without projected start or projected finish dates like any other task in the forward and backward pass processes.

Status information for completed tasks, on-going tasks, or near-term tasks, all takes the form of actual dates (AES and/or AEF) and/or projected dates (PES and/or PEF). This provides us a consistent methodology by which to assess status for all nodes in a model.

CHAPTER 10

The Role of the Computer

WHY THE COMPUTER?

It does not take long working with the CPM process manually to arrive at some rather obvious conclusions:

1. The process is extremely slow and tedious.
2. It is easy to make mistakes, and each error can affect the entire balance of the process.

These problems impose severe limitations on the management of a project. In fact, manual calculations become impractical for any project larger than about 50 activities. Modeling most projects within this restriction yields only cursory results with too little detail to reveal anything of substance.

To really exploit the possibilities of CPM, we need to perform a large number of tedious mathematical calculations routinely, very quickly, and without error. A computer can perform mathematical calculations flawlessly and follow the rules of the CPM process without deviation, all in a fraction of the time a human needs to do this work. Therefore, to apply the CPM process practically, we must enter into a partnership with modern electronics. However, this assistance is not without its cost. We must master the difficulties of a computer system and its software on top of the already complex and difficult processes of the Critical Path Method. This creates a dual complexity for the user.

As in any partnership, we must carefully define the roles of the partners. The computer has a distinct clerical role, but the rational, human half of the partnership makes the relationship work.

As its primary role, the computer performs the calculations required by the CPM processes. At a minimum, the computer:

1. Performs the process of the forward pass according to its rules to determine the early dates (ES and EF) of each node
2. Performs the process of the backward pass according to its rules to determine the late dates (LF and LS) of each node
3. Calculates the Path Time Reserves of the project at each node (SPTR and FPTR)

The human role begins with collecting and loading the proper model information into the computer and continues through analysis of the output. The computer operates the mechanics of the Critical Path Method from information we provide. The computer outputs the early and late dates, the time reserves, and other pertinent model parameters for our analysis. The computer takes on the tedious, time-consuming, error-prone chores, freeing us to find the information and analyze and act on the processed data.

To gain the benefits of this partnership, we need to communicate with the computer's application software. To do this, we enter the project model information into a computer database from which the computer gets its input and to which it outputs the calculated CPM information. Thus, to understand our relationship with the application software, we must first understand the project model database.

THE PROJECT MODEL DATABASE STRUCTURE

Before we can begin our examination of the project model database, we must agree on some basic terminology. Unfortunately, terminology can vary considerably among different software packages. The meaning of even a general term like *database* can change. We will define each term in our own context and accept that these definitions will not be universal.

Database, for our application, means the total structure of all the data or information regarding a particular project. Each database will organize the information pertinent to the project in several structures.

One of these data structures contains the parameters that define the calendars of the project. We will call this the *calendar datafile* or simply the *calendar.*

Other data structures contain information about project structures, the organizational hierarchy, for example. We call these the *project structure datafiles.*

Still other data structures define the formats for graphic and tabular output products for the project. These *product profiles* are discussed in Chapter 15.

The final structure contains the specific activity, event, and constraint information that defines the project. We call this the *project modelfile.* To build this file, we break the project down into its elements and define each in a data structure called a *record.* Every activity in the project corresponds to an individual *activity* or *node record.* Every constraint in the project corresponds to an individual *constraint* or *node-to-node record.*

The modelfile stores event information in two forms. Events that merely define one end of a node are stored as part of the pertinent activity or node record. Some events are discrete individual elements of the project and are not the result of or initiation of one activity. They have many predecessors and/or successors. These events nodes are activity records with no (0) duration, and contain the information that is pertinent to the event, but use the same record structure as an activity. Therefore, we can define the project modelfile as the sum total of the activity (node) and constraint (node-to-node) records.

Specific parameters of activities, events, or constraints are stored within the records in structures called *fields.* We can think of fields as individual compartments within each record. As an activity, event, or constraint is defined by many parameters, each record contains many fields, each storing a specific piece of information regarding the record. These fields have names for identification.

Every field also has certain limitations imposed to specify the kind of information it can store, of which there are four *field types.*

Alphanumeric Fields

The first field type, the *alphanumeric field,* allows short, continuous (without spaces) strings of characters of a fixed length. These characters are limited to the letters of the alphabet (A to Z) and the numerals (0 to 9). The node identification parameter, the constraint type, and the calendar designation are all alphanumeric fields. The principal functions of alphanumeric fields include:

1. Unique record identifiers (e.g., node identifiers, record sequence identifiers, etc.).
2. Specific, predetermined information (e.g., a calendar designation, a constraint type, data flags, etc.).
3. Sorting, selecting, and summarizing data, as later chapters will explain.

Date Fields

The second field type, the *date field,* contains dates. These fields are quite common in a project modelfile record. Examples include the activity's Early Start, Late Start, Early Finish, Late Finish, start-no-earlier-than, finish-no-later-than, actual start, actual finish, projected start, and projected finish dates. Date fields represent date values in some specific format like the American standard date format in which numerical values specify the month, day, and year in that order (e.g., December 23, 1989 is 12/23/89). We will abbreviate the American standard format as "mm/dd/yy" (month/day/year). Two characters represent each value in the abbreviation.

The military date format is also common. Numerical values represent the year and day, and the month is abbreviated (e.g., December 23, 1989 is 23/Dec/89). We will abbreviate the military date format "dd/mmm/yy" (day/month/year). In this case, the three characters that represent the month indicate an abbreviation (e.g., Jan, Feb, Mar, Apr, etc.).

All formats specify dates in continuous strings of characters, separating the day, month, and year by delimiter characters, usually slashes (/) or dashes (-). This prevents the numbers from all running together and eliminates any need for leading zeros on single-digit days or months. Typically, American standard dates include slashes for delimiters, whereas military dates include dashes. The abbreviations of the date formats indicate appropriate delimiters. For example, "mm-dd-yy" indicates the American standard date format with dashes for delimiters, while "mm/dd/yy" indicates the same format with slashes for delimiters.

If the project model defines its base unit as something less than a day (e.g., a shift or an hour), then we must append some value to the date to specify the period within the day. To accommodate hourly units, we could add the time in 24-hour increments. The notation for 3 p.m. on February 6, 1989 would look like "6/Feb/89/15" in military format. The numeral *15* represents 3 p.m. (12 noon + 3 hours). For 6 a.m. we would indicate *6*, for 1 p.m. we would indicate *13*, and for 6 p.m. we would indicate *18*.

If the base unit were shifts, we could represent the graveyard shift as *1*, since it is the first of the new day, first shift as *2*, and second shift as *3*. For second shift on February 6, 1989, we would write "6/Feb/89/3." We can append these sub-day values to either date format.

We have discussed just two ways to represent time units of less than one day. Other systems of notation could be devised to accommodate partial day units, but we will work only in whole day units, so we simply introduce the concept.

For each project, we must choose one of the optional formats to represent dates, and specify all dates in a particular database the same way. Clarity requires one consistent time unit, one type of delimiter, and one format (e.g., days, dashes, and military). We may choose to represent date data in alternate forms in output reports (see Chapter 15), but the date format within a project's database should remain consistent.

All date fields require one other characteristic, which is to specify their relationship to the start or finish of the time unit (e.g., morning or evening). In this way we can define ES, LS, SNE, AES and PES dates as start or morning dates. EF, LF, FNL, AEF and PEF dates are finish or evening dates. Every other date type should be defined relative to either the start or finish of the time unit. This information is useful in the proper portrayal of dates in both product and process.

Numeric Fields

The third type of field, the *numeric field,* contains numeric values. Databases often define two separate types of numeric fields for integers and decimals. Integer fields would contain whole numbers and decimal fields would contain decimal values. Numeric fields store parameters or values that become input to or output from mathematical equations. Activity durations, constraint lags, and time reserves are all stored in numeric fields.

Some numeric fields, like Path Time Reserve, must allow for plus or minus signs. Any numeral with no sign represents a positive number. In the project modelfile, all of the numeric values introduced so far are whole or integer numbers. Thus, almost all numeric fields in the project modelfile are integer fields. The need for a few decimal fields will arise in connection with resource modeling and analysis.

Text Fields

The last field type, the *text field,* stores text, like the task's description. Text fields are relatively long, usually 200 characters or more, to accommodate thorough task descriptions and other explanatory information. Besides letters and numerals, they may contain any keyboard character, including: ! @ # $ % & *) - > < ? “ : ;.

Only text fields allow spaces; fields of all other types must contain continuous strings of characters. Spaces are obviously necessary to make text material understandable. Itisawkwardtodescribesomethingwithallofthewordsruntogether. Text fields can store any information contained in any of the other four field types.

Record Modelfiles

Each field in a record has a name which, for convenience, we can abbreviate to two, three, or four characters. For instance the task's duration is called the *DU*, and the node identifier is called the *NI*. In previous chapters we defined an abbreviation along with each activity and constraint parameter. These will now become the field names.

Particular fields correspond to the two kinds of records: activities and constraints. The characteristic fields determine the formats of activity records and constraint records. Event information is defined in fields contained in each activity record. Figure 10-1 shows some of these record types, and Appendix G displays them all. The partial list in Figure 10-1 illustrates fields that have been discussed so far. We will amend this list as the need arises.

We define the project in a way that the software can understand by entering the appropriate information in the fields of each activity and constraint record. Every record makes all its normal fields available, but we need not enter a value into every field on every record. A null value is pertinent information in the case of target dates, actual dates, projected dates, etc. We enter a value into a field only when it is necessary or for the fields that are flagged in Figure 10-1 with an asterisk (*), the *essential parameters* or *fields*. These must contain values.

Figure 10-2 illustrates a sample project model, for which we will generate a database to illustrate the conversion of a model into a modelfile. The modelfile records for this project model, with some of the record field values, appear in Figure 10-3.

The calendar designation *A1* is a coded reference to the five-day week calendar with eight holidays (1). The calendar designation B1 is a coded reference to the six-day week calendar with eight holidays (2). These will be explained later in this chapter.

When translating a model diagram into a project modelfile, each node generates one activity record and each line between two nodes generates one constraint record. This simplifies validation of input data. Each record must correspond to a node or line, and we need a record for every line and every node.

From the listed data, a computer CPM processor calculates the dates and time reserves shown in the project model diagram (Figure 10-2). The computer stores these ES, EF, LS, and LF values in the appropriate date fields and the SPTR and FPTR values in the appropriate numeric fields of each activity record after completing the calculations. These calculated fields for each activity record are shown in Figure 10-4.

This illustrates the software application's maintenance of both input and calculated data. Later we will learn how to portray this information in graphic and tabular form.

DATABASE FUNCTIONS

The user controls and manages the project model through the database. This requires the ability to enter data and then change it as necessary. We call this data manipulation the *data processes*. During the data processes, we communi-

Figure 10-1
Modelfile Record Fields

Field Name	*Field Type*	*Purpose*
Activity Record		
NI*	Alphanumeric	Node identifier
DU*	Integer	Activity duration
ODU	Integer	Original activity duration
SNE#	Date	Start-no-earlier-than date
FNL	Date	Finish-no-later-than date
AD	Text	Activity description
CAL*	Alphanumeric	Calendar
PES	Date	Projected start date
PEF	Date	Projected finish date
AES	Date	Actual start date
AEF	Date	Actual finish date
PC#	Integer	Percentage complete
RDU#	Integer	Remaining duration
ES$	Date	Early Start date
EF$	Date	Early Finish date
LS$	Date	Late Start date
LF$	Date	Late Finish date
SPTR$	Integer	Start-end Path Time Reserve
FPTR$	Integer	Finish-end Path Time Reserve
Additional Event Flags		
SEF	Alphanumeric	Start event flag
FEF	Alphanumeric	Finish event flag
SED	Text	Start event description
FED	Text	Finish event description
Constraint Record		
PNI*	Alphanumeric	Predecessor node identifier
SNI*	Alphanumeric	Successor node identifier
LAG*	Integer	Constraint lag
TYP*	Alphanumeric	Constraint type: FS, FF, SS, SF
CAL*	Alphanumeric	Calendar

Asterisk (*) indicates essential information.
Pound sign (#) indicates fields that will be deleted in practical application.
Dollar sign ($) indicates fields that are calculated by the CPM process.

Figure 10-2

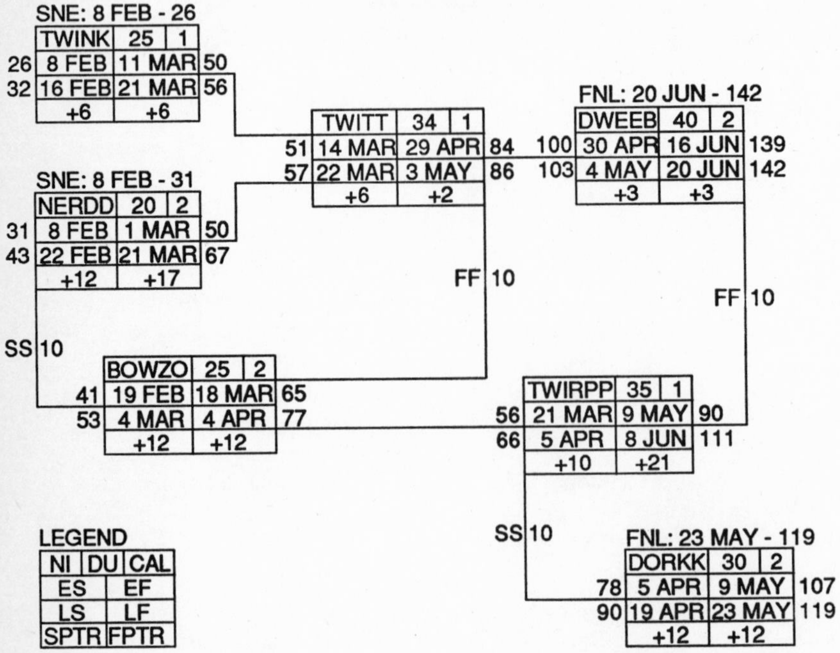

cate with the computer primarily through a keyboard or mouse. This limits the number of ways that we can accomplish the data processes. The first of these data processes is called *data entry* or *data load.*

Data Entry or Load

In this operation, we load or enter new activity and constraint records into the database in one of a variety of ways. The simplest, but slowest is the ambiguously named user-friendly (UF) method. At the other extreme is the full screen edit (FSE). The user-friendly method meticuously leads the user through data entry for each field of each record. Figure 10-5 illustrates an example of a method of user friendly data entry.

Our example divides the computer screen into a series of compartments, each of which contains a specific field value. Every compartment is labeled with its appropriate field name, and other text on the screen usually explains features and fields. The cursor keys, tab and backspace keys, or mouse move us into the first position of the desired field where we type the necessary value, and then move on to the next field. We can easily identify the fields where we are enter-

Figure 10-3

Activities

NI	DU	CAL	SNE	FNL
TWINK	25	A1	8-Feb-88	
NERDD	20	B1	8-Feb-88	
BOWZO	25	B1	11-Mar-88	
TWITT	30	A1		
DWEEB	40	B1	20-Jun-88	
TWIRP	35	A1		
DORKK	30	B1	23-May-88	

Constraints

PNI	SNI	LAG	TYP	CAL
TWINK	TWITT		FS	A1
NERDD	TWITT		FS	A1
NERDD	BOWZO	10	SS	B1
BOWZO	TWITT	10	FF	A1
BOWZO	TWIRP		FS	A1
TWITT	DWEEB		FS	B1
TWIRP	DWEEB	10	FF	B1
TWIRP	DORKK	10	SS	A1

Note: No activity descriptions, events, or status information have yet been defined for this model.

Figure 10-4

Activities

NI	ES	EF	LF	LS	SPTR	FPTR
TWINK	8-Feb-88	11-Mar-88	21-Mar-88	16-Feb-88	+6	+6
NERDD	8-Feb-88	1-Mar-88	21-Mar-88	22-Feb-88	+12	+17
BOWZO	19-Feb-88	18-Mar-88	4-Apr-88	4-Mar-88	+12	+12
TWITT	14-Mar-88	29-Apr-88	3-May-88	22-Mar-88	+6	+2
DWEEB	30-Apr-88	16-Jun-88	20-Jun-88	4-May-88	+3	+3
TWIRP	21-Mar-88	9-May-88	8-Jun-88	5-Apr-88	+10	+21
DORKK	5-Apr-88	9-May-88	23-May-88	19-Apr-88	+12	+12

Note: The figure omits activities' durations, calendar designations, and date target fields because they would force the table beyond the page width.

ing data. The box length on the screen corresponds to the field length, showing how much of each field we have filled.

Figure 10-5

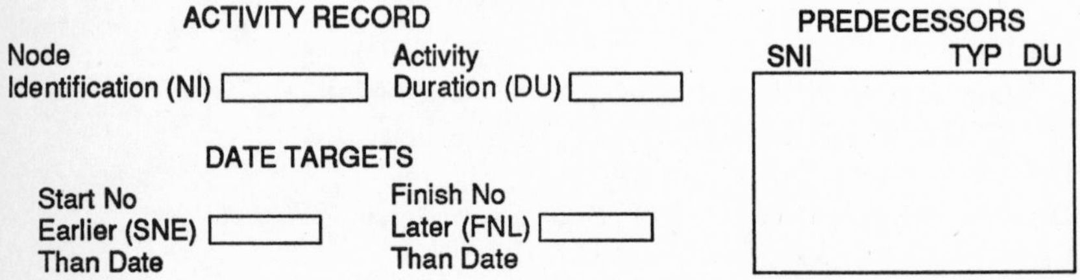

If we notice an error in entered data, we can return to the appropriate field box and reenter the correct value. When we are satisfied with the information we have loaded, pressing the Enter key loads the record into the project modelfile. We then get a blank data-entry screen on which to enter the next record. With the user-friendly method, we can enter constraints on a separate screen form or as predecessors or successors of activities. The example in Figure 10-5 employs the latter technique.

This method is called user-friendly because it leads the user through the data-entry process step-by-step and makes it as simple as possible. We need only provide the values for the fields of each record. The simple and explicit interface to the computer allows a virtual novice to load data into the modelfile. Someone who is not familiar with the computer might find this assistance invaluable.

Users learn data entry quickly, however, most becoming proficient by the time they enter 50 or so records. At this point the user-friendly method becomes an obstruction rather than a help. The user's need for help from the computer interface gives way to a need for maximum efficiency and speed.

A full screen edit method provides the fastest means of data entry. This scheme shows the available fields in columnns and the records as horizontal rows. The cursor keys or the tab and backspace keys move us from field to field, where we enter values as necessary. Review Figure 10-3 to see an example of an FSE data entry screen. It looks exactly like a listing of the record data.

When evaluating a computer CPM system, we should evaluate the data-entry function. A good system will provide both a user-friendly interface and a full screen edit capability for faster work. Some provide other methods that lie somewhere between these two extremes. It is important for our computer application to accommodate a wide range of user ability.

The large number of fields necessary to define activity or constraint records usually will not all fit on one screen in either the user-friendly or the full screen edit method. To overcome this, UF methods provide additional data entry screens. The FSE method holds the balance of the fields in off-screen columns. Certain keys then scroll the display left and right. They usually list the column for the first field (NI) at the left of all screens to properly align the data. Likewise, certain keys scroll the display down through the data below the first screen.

Record Change or Modification

Having entered data into a modelfile, we then need to access individual records to correct, change, or update the information. This is called the *record change process*.

Normally, we do not change all records in a modelfile at one time. For large projects, we probably never change more than 10 percent of the records in any one session. This raises the need for a means of selecting the records to change. We can do this by identifying a unique feature of the desired record. Usually, we specify the unique node identifier (NI) value for an activity or the unique pair of node identifier codes (PNI and SNI) for a constraint in a query to our application software. The computer then recalls the records with the specified node identifiers (for activities) or pairs of node identifiers (for constraints). Some pairs of nodes are bound by two constraints (an SS and an FF). If we identify such a constraint pair in a query, the computer recalls both constraints for review, and we simply ignore the constraint we are not changing.

When we need to change several records, we can save time by making a list of the activity record node identifiers (NIs) and the constraint record node identifiers (PNIs and SNIs). Our application software would then present each record in turn.

We can change the data in several ways, including, as for data entry, a simple (user-friendly) method and an expeditious (full screen edit) method. A user-friendly technique would display the same screen we used for data entry. It would show one record at a time, and we would simply key through the record, stopping at each field we need to change and overtyping the old value with the new value. Pressing the enter key then returns the updated record to the modelfile and allows us to move on to the next record.

The record change process in the full screen edit method resembles the FSE data-entry method. We call up all of the records to be changed and they appear on the screen just as they looked in data entry. We move the cursor to the fields to be changed and overtype the old value with the new. When we have finished altering the records in the display, we hit the enter key and load the revised data back into the modelfile. Changing even as few as 10 records proceeds much more quickly in the FSE method than in the UF method. Still, the user should be able to choose between speed and simplicity.

The need sometimes arises to change all records that share some criterion, for instance all activities with SNE dates, all activities with durations greater than 50, or all activities with ESs between one date and another. To do this, we identify the conditions and then ask the software to retrieve the appropriate records.

Sometimes we need to reset a value in a field to zero, that is, to delete a field value. We do this in the same way we make any other changes to a record. Most commonly, we either enter a null value (zero) in the field or delete the data with the space bar. To remove a start-no-earlier-than date from an activity record, we would call up the record through the change process and overtype the existing date value with zero or in some other way delete the date. The activity then has no date predecessor.

We may also need to set a specified field on several records with the same value. This need would not arise often for the fields that have been introduced to this point, but we might, for example, wish to change the calendar for all activities that lie on negative time reserve paths to to a six-day (or seven-day) work week. This requires that we first identify all the records on which we wish to change a field to a common value, then we select the field and the desired value. The computer then seeks out the appropriate records and enters the desired value into the proper field of each. When we work with structure fields, the ability to set a field to a specified value on many records will be invaluable.

Deleting Records

Often we must eliminate records from the modelfile in a process called *record deletion*. We must first identify the record, or records, we want to delete. Again, we identify the record by some unique value in one of its fields. The delete facility of the software then deletes the record from the modelfile.

Anyone who deletes records will eventually delete one in error. The realization of this error often occurs the instant after the record disappears. Therefore, the software should have an *oops facility* that stores the last record or set of records deleted. This protects us for at least some period of time, since we can recall the record or records that were last deleted and reenter them into the modelfile. This deleted-data retention process should have some controls and conditions, as it would be impractical to store all deleted data indefinitely. Storing only the most recently deleted data is one way of imposing such control.

Section Summary

As an introduction to the project modeling database, we defined the project modelfile, which contains the elements of the project's model. A project modelfile contains a record for every activity and constraint in the project. It stores event information as part of, or in additional fields of, certain activity records.

Each record contains many fields which store specific attributes of each activity, event, or constraint. Field types define the kind of information they can contain: alphanumeric, integer, decimal, date, and text values.

We enter and control the data in the project modelfile through three *data processes*: record entry, change, and delete. By these processes, we maintain the correct information in the modelfile. Software packages provide a variety of means to accomplish each data process. User-friendly (UF) methods provide a very simple, step-by-step process while full screen edit (FSE) methods assume some proficiency, but speed up the work.

THE CALENDAR DATAFILE

Another important project datafile contains information on the project's calendar. It is called the *calendar datafile*. This file must define three parameters: the time unit, the work-day patterns, and the holiday tables.

Time Units

As we have seen, we calculate durations and lags in terms of the time unit. With this unit, hours, shifts, days, or weeks, we measure the length of work. Specifying the basic unit defines the duration or lag values: DU = 60, 60 hours versus 60 shifts, 60 days, or 60 weeks. The calendar datafile application provides several ways to specify the time unit. For instance, we can express the number of units per day. In this way, the workday would be stated as one unit. Shifts would be expressed as three, and hourly units would be expressed as 24 for a full day or 16 for two shifts.

This method cannot accommodate work units longer than one day because no field type accepts fractions (e.g., 1/5 or 1/7) except in decimal form. Units longer than a day require an alternative form of definition, typically in days per unit. For instance, we would express a week-long time unit as the number of workdays per week (e.g., seven or five). We would express a unit of one month as 30 days per unit.

For our purposes, the basic time unit will always be one unit per day. This is the most common and certainly the most practical unit for project modeling.

The Workday Pattern

The workday pattern distinguishes normal periods of work from normal non-work periods. The standard five-day week excludes Saturday and Sunday as workdays. We need to be able to exclude these days universally instead of identifying every Saturday and Sunday. Other defined calendars might well include these days as workdays and exclude, for example, Wednesday and Thursday. In each case, normally one or two specific days of the week are nonwork days, and we need to specify these days over the entire calendar.

We could define the work week either in terms of the days it includes or those it excludes. For example, we could define a five-day week with a normal weekend either as every day but Saturday and Sunday, or as the days from Monday through Friday.

For periods of less than a day, the definition of nonworkday patterns becomes more involved. When shifts are the work unit, we must express the weekend as including everything from first shift on Saturday through third shift on Sunday night, or third shift on Friday night through second shift on Sunday, as the case may be. If the unit is hours, the workday definition becomes even more specific.

Since we must often work weekends to relieve schedule problems, we need more than one work-week pattern in our calendar. If, for instance, the unit is workdays and Saturday and Sunday are considered nonworkdays, we need at least three work patterns. The first is the normal five-day week, with Saturday and Sunday off. The second is the six-day week that excludes only Sunday. The third work pattern is a seven-day week with no off days. These three patterns would enable us to define any combination of overtime days. For simplicity, we designate each of these work patterns with a code. We might call the five-day work-week pattern *A*, the six-day work week pattern *B*, and the seven-day work week pattern *C*. If the need arose for a work-week pattern with Friday and Saturday as the nonworkdays, we could define it and label it work pattern *D*.

Holidays

Holidays are specific nonworkdays in every year. Christmas, Thanksgiving, the Fourth of July, Labor Day, and Memorial Day are examples. Unfortunately, it is difficult to specify holidays universally. We could express December 25 as a holiday for all years, but Thanksgiving, Memorial Day, and Labor Day fall on different calendar days every year. Some must also exclude the Friday after Thanksgiving as a holiday. When some holidays fall on a weekend, the Monday after or the Friday before also becomes a nonworkday. In some work schemes, if the Fourth of July falls on a Thursday, then Friday is also a holiday. The rules governing holidays differ widely between companies, countries, and the government. The British, for instance, do not celebrate the Fourth of July. These rules may also vary from year to year within a company. These problems impose two requirements on the database:

1. We must establish a separate holiday table for each organization, including any government agencies, involved in the project.

2. These tables must specifiy a date for each holiday for every year in the duration of the project.

To meet these requirements, we must list all holidays, year by year, in specific holiday tables, and label each table for easy identification. For instance, Table 0 might allow for no holidays, Table 1 might list holidays for company number 1, Table 2 might list holidays for company number 2, Table 3 might list holidays for the federal government, and so on.

We must also develop specific holiday tables for six- and seven-day week calendars along with those for five-day week calendars. Often, a six- or seven-day week calendar must exclude the weekend following a holiday so the holiday list must include these dates, as well. Combining these tables with the five-day week calendar excludes these Saturday holidays automatically.

Let us examine an example of a holiday table for the two calendars defined in Chapter 3 (Figures 3-4 and 3-5):

1-2 Jan	4 Jul	24-26 Dec
1-2 Apr	5 Sep	
30 May	24-26 Nov	

This table covers only a single year. An actual table would list holidays over several years. January 2, April 2, November 26, and December 24 are Saturdays but they are included in the table because they would be considered holidays in the six-day week calendar.

The Node—The Constraint and the Calendar

Individual activities in a project may require different calendars and different holiday tables, so we need to specify both parameters on each activity. A two-digit field called the *calendar field* or *CAL* stores this information. The first character of this field specifies the calendar (e.g., *A* for a five-day work week, *B* for a six-day work week, etc.). The second character of the field specifies the holiday table (e.g., *0* for no holidays, *1* for the holiday table for company number 1, etc.). In Figure 10-3, we specified each activity to work on calendar A1 or B1, which indicated the five-day work week (A) with eight holidays (1) or the six-day work week (B) with eight holidays (1).

Whenever a constraint contains a lag value, it must relate to a specific calendar as well. The easiest course is to allow the user a field to define the appropriate calendar. It is also correct to automatically reflect the lag on an SS constraint relative to the predecessor activity and the lag on an FF constraint relative to the successor activity. Any lag on an FS constraint is independent.

Calendar Utilities

Besides helping to define the calendars, the calendar datafile should provide other very useful features called *calendar utilities.*

Overtime Reserve Calculations. To calculate available overtime or calendar reserve requires the identification of work time included on one calendar that another calendar excludes. Thus, the calendar datafile should provide a utility to make this calculation. After we specify a pair of dates between which to calculate available overtime, the utility begins its process.

First, it calculates the number of holidays between the two dates for each of the defined holiday tables. Next, it determines the number of normal off days as the difference between the two dates on each calendar. The smallest value comes from the base work week; other values come from longer work-week patterns which include Saturdays and/or Sundays as workdays. The utility subtracts the smallest value from any other values, and the difference is the number of normal nonworkdays. In our example, the difference between the two specified dates on calendars B and A would be the number of Saturdays available; the difference between calendars C and A would be the number of Saturdays and Sundays. These become part of the available overtime value.

To determine the final component of overtime reserve, the utility calculates available offshifts. This figure depends on the base work unit of the calendar. If the base unit is workdays, then the utility assumes one shift. It then sums second and third shifts on every workday, or it may add half of a second shift for every workday, depending on how much of this time we choose to include in overtime. The number of normal workdays between the two dates is also the number of available extended shifts.

Let us calculate the number of nonworkdays between two dates using our six-day and five-day calendars to see how this works. This example requires one holiday table, Table 1, in addition to the two calendars, A for the five-day week and B for the six-day week. To determine how much overtime reserve is available between January 11 and July 21, we first determine the number of holidays between the two dates. Arranging the holiday table in chronological order shows four holidays between the two dates: April 1 and 2 (April 2 being a Saturday), May 30, and July 4.

Second, we determine the number of normal nonworkdays, in this case the number of available Saturdays. Calendar A (the five-day week) includes 124 workdays between the two dates:

$$\begin{aligned} DU &= FIN - ST + 1 \\ &= \text{day } 141 \text{ (July 21)} - \text{day } 18 \text{ (Jan. 11)} + 1 \\ &= 124 \text{ days} \end{aligned}$$

Calendar B (the six-day week) includes 151 workdays between the two dates:

$$\begin{aligned} DU &= FIN - ST + 1 \\ &= \text{day } 168 \text{ (July 21)} - \text{day } 18 \text{ (Jan. 11)} + 1 \\ &= 151 \text{ days} \end{aligned}$$

The number of Saturdays between these two dates is the difference between these two numbers, or 27 (151 – 124 = 27). Notice that we do not count Saturday, April 2 as it is considered a holiday.

Finally, we determine the number of offshifts available for overtime. With workdays as the basic time unit, we assume a single shift per day. This leaves one second shift and one third shift per day available for work. Between January 11 and July 21 we counted 124 normal workdays, so we have 124 second and 124 third shifts of available overtime. In total, between the two specified dates, we counted 4 holidays, 27 Saturdays, 124 second shifts, and 124 third shifts of available overtime. If we worked only Saturdays and half of a second shift (4 hours) each workday, we would add 87 equivalent workdays.

Date and Duration Calculations. A calendar should also provide a utility to calculate the duration between two dates, the start date of a task based on its finish date and duration, or the finish date of a task based on its start date and duration. This utility can calculate any one of these values from the other two using the basic date math equation:

$$FIN = ST + DU - 1$$

Figure 10-6

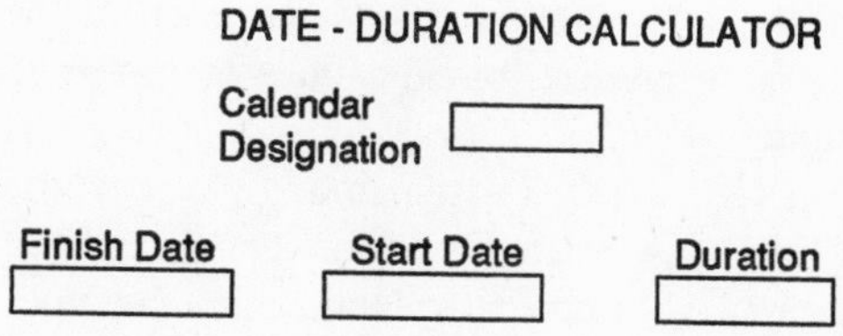

Besides the values for the equation, the processor needs us to indicate the appropriate calendar and holiday tables, since work patterns can affect these calculations. Figure 10-6 shows a simple way to accommodate this need. We enter the calendar designation (e.g., A1) and two of the values (e.g., FIN — 26 Aug, and ST — 1 Jun), then press the Enter key or a programmed function key to direct the utility to calculate the missing value (in this case, 62), which is either a date or duration. This utility vastly simplifies the establishment and maintenance of project models.

SUMMARY

As its primary role in the CPM process, the computer performs calculations based on project model data. Specifically, it:

1. Performs the forward and backward passes, calculating the early and late dates of each activity
2. Calculates the project's Path Time Reserves (SPTR and FPTR)

The user inputs activity and constraint parameters essential for the computer's operation. For activities, these include the node identifier code (NI), duration (DU), calendar designation (CAL), date targets (FNL and SNE), and status information (AES, AEF, PES, and PEF). For constraints, these include the predecessor node identifier (PNI), successor node identifier (SNI), constraint type (TYP), lag value (LAG), and calendar designation (CAL).

The computer and the user communicate through the project database. Part of this database, the project modelfile, contains the information on the activities, events, and constraints of the project. Each activity and constraint corresponds to a separate record in the project modelfile. Event information is stored as additional parameters of existing activity records or in additional activity records, called events node which have no (0) duration. Each record contains several fields, each of which stores a specific parameter.

There are four types of fields. Alphanumeric fields store continuous, fixed-length strings of alphabetic and numeric characters. Each of these fields stores either a unique parameter, a predetermined value, or information for sorting, selecting, and summarizing data.

Date fields store specific dates. They maintain day, month, and year information in a consistent format, separated by delimiter characters.

Numeric fields accept only numeric values, which become input to or output from mathematical calculations. Numeric fields are sometimes divided into two categories: integer and decimal.

Text fields store long descriptive or explanatory information. They accept any keyboard character, including blank spaces. Figure 10-1 gives a partial list of necessary activity, event, and constraint fields and Appendix G lists them in full.

To control the project modelfile, we enter (load) records, change existing records, and delete records through the software interface. User-friendly interfaces do this through a simple, thoroughly definitive, and slow process. Full screen edit interfaces provide the fastest means to do this, but require some proficiency from the user.

The calendar datafile stores all calendar information. It identifies each specific workday pattern and holiday table with an alphanumeric designation.

To define a calendar, we must specify a time unit, workday pattern, and holiday table. The time unit defines the basic unit in terms of hours, shifts, days, weeks, etc., in which all duration and lag values will be measured. The workday pattern specifies the normal workdays and nonworkdays. A single project may employ more than one workday pattern, and the datafile designates each with a unique character or characters. The holiday table specifies all holidays for each year and for all workday patterns.

Two additional calendar utilities, an overtime calculator and a date/duration calculator, perform calculations based on calendar information. The overtime calculator determines the amount of available overtime periods, including holidays, nonworkdays, and off-shifts, between two user-specified dates. The date/duration calculator determines any start date, finish date, or duration from given values for the other two and a specified calendar.

CHAPTER 11

Additional Roles of the Computer

In addition to the basic CPM and calendar processes, we can look to the computer application to accomplish certain other functions. These adjunct processes will be discussed over the next several chapters. The initial discussion focuses on:

1. Setting modelfile flags
2. Detecting model errors
3. Data sort and selection.

MODELFILE FLAGS

The computer can facilitate the user's analysis of the CPM project model data output by flagging certain conditions in the project modelfile, including:

1. The critical or least-time-reserve path of the project.
2. The location, amount, and cause of any external constraint stretching.
3. Start and finish nodes. It can also indicate whether these nodes have proper initiation targets.

To determine these flags, the software must scan the data, and find and mark the records that fit the criteria. The flags appear as computer-calculated values in fields called flag fields established for this purpose within each record.

Critical Path Flag

The necessary critical path flag (CPF) is a two-character alphanumeric field in activity records or a single-character alphanumeric field in constraint records that identifies the activities and constraints that lie on the critical path. As stated, we define the critical path as the path in the project with the least Path Time Reserve (PTR).

In an activity record, a flag value of *SF* in the CPF field indicates that the critical path goes completely through the node. A flag value of *F* in the CPF field indicates that the critical path passes through only the final portion of the node, that is, the amount of the node's duration defined by the lag on a finish-to-finish constraint from a predecessor. A flag value of *S* indicates that the critical path passes only through the initial portion of the task, the amount defined by the lag on a start-to-start constraint to a successor.

A value in the CPF field of a constraint record indicates a particular interface that attaches two nodes along the critical path. Such a flag will appear as a single value, *X* for example, the presence or absence of which indicates that an interface does or does not lie on the critical path. Collectively these flags identify all nodes and constraints on the critical path of the project.

The computer sets these flags immediately after it completes the calculations for the CPM processes (the forward and backward passes and time reserve calculations) because it needs these values. The process begins with determination of the terminus node in one of three ways. The first method identifies the true critical path of the project by locating the node in the project with the least time reserve. When several nodes share the same least value, this method selects the one with the latest EF date (the latest in the schedule). If the model has date successors (finish-no-later-than dates) internal to the model, then this node may not be a finish node. An internal node could reflect the least path time reserve as the result of an internal FNL date. We can find the critical or least time reserve path to this node.

The second method for determining the terminus node limits the search for the node with the least time reserve to finish nodes. This corresponds to an alternative definition of the critical path that requires it to terminate at a project finish. We can find the critical or least time reserve path to this node.

The third method recognizes the fact that a least-time-reserve path exists to any node in the project. It therefore allows the user to specify any node in the project as the terminus node in order to find its critical path.

The three methods differ in their definitions of the terminus of the critical path. Once this has been defined, the process of identifying the least time reserve path to the terminus node is identical.

Step 1. The process compares the value of Start-end Path Time Reserve (SPTR) of the node to its Finish-end Path Time Reserve (FPTR). If they are equal, then the critical path goes completely through the node. The computer sets the value *SF* in the critical path flag (CPF) field of this node's activity record and proceeds to step 2. If FPTR is less than SPTR, indicating forward pass constraint stretching in the node, it jumps to step 4.

Step 2. The process examines the predecessors of the terminus node's start. For finish-to-start predecessors, it looks for the least value of FPTR. For start-to-start predecessors, it looks for the lowest SPTR. It flags the CPF field of the constraint (X) with the lowest SPTR or FPTR value, and then proceeds to step 3 for the SS constraint and step 1 for the FS constraint.

Step 3. If a predecessor with a start-to-start constraint has the least PTR value of all the terminus node's predecessors, the computer flags this predecessor's CPF field with an *S*. The critical path goes only through the portion of this task equal to the lag on the start-to-start constraint. If a predecessor with a finish-to-start constraint has the least PTR value, the computer moves to this node and proceeds back to step 2.

Step 4. If FPTR is less than SPTR, the critical path goes only through the end portion of the node, so the computer marks the CPF field in the node's activity record with the value *F*. It then finds the predecessor of the node's finish with the least value of FPTR (assuming the model includes no SF constraints). It flags the CPF field of the record of the constraint between these two nodes (X), moves to the new node, and proceeds back to step 1.

This process continues, tracking a path all the way back to a start node. The path may pass through some completed or on-going activities. These remain part of the critical path, even though we can do little or nothing to influence their outcomes. They still contribute their sequence and spans to the cumulative length of the path.

Besides flagging the critical path, the software application can also derive certain parameters relative to this path. Summing the durations of the nodes that lie entirely on the critical path (CPF = SF) and the lag values of constraints that lie on the critical path (CPF = X) determines its cumulative length.

$$\text{Length of the work along a path} = (\Sigma\ \text{Dus} + \Sigma\ \text{Lags})\ \text{along path}$$

This value is the cumulative length of any path but it does not, however, take into account delays caused by other date or constraint predecessors, which diminish time reserves. The difference between the SPTR value of the first node of the path and the FPTR value of the terminus node is the total amount of delay along the path.

Length of delay along a path =
(SPTR at start of path) – (FPTR at finish of the path)

In addition, subtracting the Early Start date of the first node from the Late Finish date of the terminus node, and adding a day for the start-finish convention, gives the time available to the path.

Time available to a path = (LF of Path) – (ES of path) + 1

These additional parameters of the project and the critical path (or any path) will be very useful in our analysis.

This entire process is necessary to track, flag, and parameterize the least-time-reserve path through a model. Later chapters will explain how to use and graphically portray this information.

Stretching Flags

We need flags to indicate external constraint stretching so that, at least for forward pass stretching, we can understand where we need to make decisions regarding the affected schedule. Such flags also indicate the effect of one path on the middle of another. After the computer software accomplishes the CPM calculations, we need it to flag any constraint stretching, as well.

Two flag fields on each activity node identify the location and amount of any stretching. The first flag identifies stretched nodes in a three-character alphanumeric field called the *activity stretching flag* (ASF) field. When stretching occurs in the forward pass (when SPTR exceeds FPTR) the software sets this flag to the value *FPS* (forward pass stretching). When constraint stretching occurs in the backward pass (when SPTR is less than FPTR) the software will set the value *BPS* (backward pass stretching) in the ASF field. Nodes that are not stretched (those with identical values of SPTR and FPTR) have a null value in the ASF field.

The second stretching flag on each activity record indicates the amount of stretching. Since this is determined mathematically, it is stored in a numeric (integer) field. This *amount of stretching* (AOS) field equals the absolute value of the difference between Start-end Path Time Reserve and Finish-end Path Time Reserve ($|AOS| = SPTR - FPTR$). We need only the absolute value of this calculation, as the sign (+ or –) indicates only the pass on which the stretching occurred, which is indicated already in the ASF field.

To summarize, any activity with a value in the ASF field has been stretched by a constraint in either the forward pass (indicated by an FPS flag) or the backward pass (indicated by a BPS flag) by the number of work units in the AOS field.

Besides this information, we need to determine which predecessor causes forward pass stretching or which successor causes backward pass stretching. Constraint records provide this information in a stretching flag field that indicates the stretched node. The software determines which constraint causes any stretching, then places the appropriate node identifier code (the stretched node's NI) in this field in the constraint record. This alphanumeric field is the same length as the node identifier code. It is called the *stretched node identifier* (SNID).

By scanning the modelfile activity records, we can recognize a node that is stretched in the forward pass by the value *FPS* in the ASF field. We then look for the stretched node's identifier (NI) in the SNID field of any constraints. The predecessor of this constraint caused the stretching. Conversely, any node with the flag *BPS* in the ASF field is stretched in the backward pass. We then look for the stretched node's identifier in the SNID field of any constraints. The successor of this constraint caused the stretching.

A condition sometimes arises that can inhibit this identification. It is possible, though very unusual, for a node to be stretched in the forward pass by one constraint and in the backward pass by another. Although this condition is rare, it remains a possibility, so we must learn how to cope with it.

If the stretching in the forward pass does not equal that in the backward pass, the flags will still alert us to the stretching, but the AOS value will reflect the difference between the forward pass stretching and the backward pass stretching. The field does not give the correct amount in either pass, but rather a net value. In analyzing the stretching, we should detect this disparity.

Unfortunately, it is possible, though doubly remote, that the amount of stretching (AOS) in the forward pass could equal the amount in the backward pass. In such a case, the method would fail to detect stretching in either pass. The SPTR and FPTR values of such a node would both diminish by the same amount, leaving no difference between them for the process to detect. We are saved from such an error only by the extreme rarity of the occurrence.

With these rare exceptions, this process will identify almost all stretching by type (FPS or BPS), amount (AOS) and source (SNID). We can then assess the situation and decide how to respond to the condition.

Start and Finish Node Flags

As another essential function, the software can indicate the project's logical start and finish nodes and determine whether or not they have proper date targets.

Remember, we must impose an external date predecessor (either an SNE, an AES date, or SSTR) on every start node to establish its Early Start date, and we must assign an external date successor (an FNL date) or FSTR on every finish node to establish its Late Finish date. Since any node can be both a start and a finish, we need two flags, a *start node flag* (SNF) and a *finish node flag* (FNF). These are both two-character alphanumeric fields.

A properly targeted logical start node is a start node with a means of establishing its ES date. The proper techniques for accommodating this was discussed in Chapter 4. Properly targeted start nodes would show the value *ST* for "start," in the SNF field. If we have not properly set the Early Start date of a logical start node in a project model (i.e., one that has no predecessors of its start), then its SNF flag field would show the value *HS* for "hanging start."

Similarly, a properly targeted logical finish node is a finish node with a means of establishing its LF date (Chapter 5). Properly established finish nodes would show the value *FN*, for "finish," in the FNF field. If we have not properly set the Late Finish date of any logical finish node (i.e., one that has no successor of its finish), then the FNF field would show the value *HF* for "hanging finish."

To accomplish the forward and backward passes with hanging starts or finishes in the project model, we must presume some value for the missing dates. The computer will set the Early Start or Late Finish dates of nodes flagged HS or HF to default values. Typically for the ES date of a hanging start, it will arbitrarily insert a global project start date. It will also set the LF date of a hanging finish to the node's EF date. Since all subsequent calculations will be based on these dates, any hanging node's default dates will probably introduce errors into the project model. To avoid this, we should rectify the problem by locating hanging nodes and supplying any missing dates.

Nodes flagged ST or FN are start or finish nodes, respectively, with proper targets established for generation of their Early Start or Late Finish dates. The forward and backward passes can begin accurately with these values. This also means, however, that the starts or finishes of these nodes depend on something external to the project. Early dates for start nodes are independent of predecessors and late dates for finish nodes are independent of successors. To keep the model in agreement with the project, we should ensure that we have imposed accurate dates. Further, we must verify that any identified start or finish node is in fact a valid start or finish in the first place.

Database Flag Example

We can use the sample model in Figure 11-1 to illustrate flags that the computer processing sets.

Figure 11-1

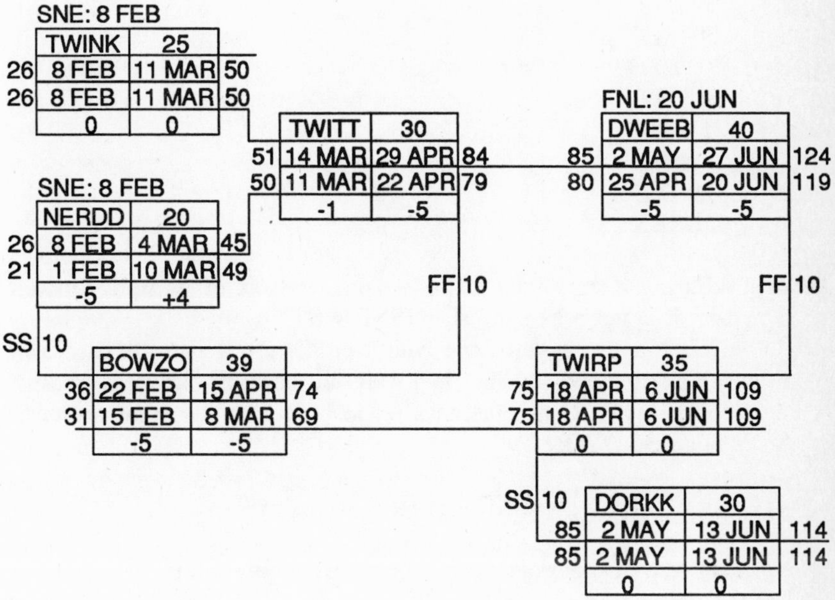

Activities

NI	SNF	FNF	CPF	ASF	AOS
TWINK	ST				
NERDD	ST		S	BPS	9
BOWZO			SF		
TWITT			F	FPS	4
TWIRP					
DWEEB			SF		
DORKK		HF			

Constraints

PNI	SNI	TYP	LAG	CPF	SNID
TWINK	TWITT	FS			
NERDD	TWITT	FS			
NERDD	BOWZO	SS	10	X	NERDD
BOWZO	TWITT	FF	10	X	TWITT
BOWZO	TWIRP	FS			
TWITT	DWEEB	FS		X	
TWIRP	DWEEB	FF	10		
TWIRP	DORKK	SS	10		

Node TWINK is a start node; it has no predecessors of its start. Since it has an SNE date, it is not a hanging start (SNF = ST).

Node NERDD is also a start node with a proper target date (SNF = ST). It is also constraint stretched (SPTR ≠ FPTR) in the backward pass (ASF = BPS because SPTR < FPTR) by nine days (AOS = 9). The successor responsible for this is node BOWZO:

$$|AOS| = SPTR - FPTR = -5 - 4 = 9$$

The critical path of the project (–5) passes through the initial portion (first 10 days) of the task as indicated by the CPF = S.

Node BOWZO is on the path with the least time reserve (–5) which makes it part of the critical path. The path goes through the entire node (CPF = SF because SPTR = FPTR).

Node TWITT is stretched (SPTR ≠ FPTR) in the forward pass (ASF = FPS because SPTR > FPTR) by four days (AOS = 4). The predecessor BOWZO is responsible for this:

$$|AOS| = SPTR - FPTR = -1 - (-5) = -1 + 5 = 4$$

The critical path of the project (–5) passes through the final portion (last 10 days) of the task as indicated by the CPF = F.

Node DWEEB is also on the path with the least time reserve (–5), and therefore part of the critical path. The path goes through the entire node (CPF = SF because SPTR = FPTR). This path continues through an unseen successor.

Node DORKK is a finish node; it has no successors of its finish. It is also lacking a proper date successor (FNL), so it is a hanging finish (FNF = HF). The LF of the task *defaults* to the node's EF date.

Node TWIRP has no flags. It does lie on a zero time reserve path (0) and even though it is not on the critical path, it is on a path with concern for at least one schedule objective.

The SS constraint NERDD-BOWZO causes the backward pass stretching of NERDD, as indicated by the value in the SNID field.

The FF constraint BOWZO-TWITT causes the forward pass stretching of TWITT, as indicated by the value in the SNID field.

The FS constraint TWITT-DWEEB lies along the least-time-reserve path, so it shows a flag (X) in the CPF field. Constraints NERDD-BOWZO (SS), and BOWZO-TWITT (FF) also lie on the critical path and are so flagged (X) in the CPF field.

These flags help us identify problems or certain special conditions in our project. Such information enhances the utility of our software.

PROJECT MODELFILE ERRORS

When constructing project model datafiles, entering inaccurate or inappropriate data can cause errors in the CPM process, some of them significant enough to totally invalidate it. We have discussed one such type of error, hanging starts and finishes. We need to examine others, as well, and we need our software to help identify them, if at all possible.

Loops

Models form loops when constraints tie a series of nodes together so that a node becomes a predecessor or successor of itself. Figure 11-2 shows an example project model segment in which nodes A, B, and C are tied together by constraints A-B, B-C, and C-A.

Figure 11-2

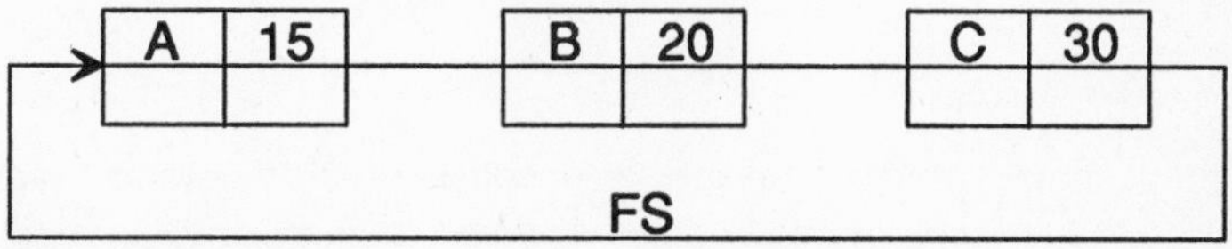

The EF of C depends on the ES of C, which depends on the EF of B, which depends on the ES of B, which depends on the EF of A, which depends on the ES of A, which depends on the EF of C. Therefore, the EF of C depends on the EF of C, forming a loop. Every calculated date of the nodes in a loop depends on itself, so the CPM process cannot calculate a date without having that date as input.

The loop in Figure 11-2 is obvious. In an actual modelfile, loops can be very long and complicated, passing through many activities and constraints. This can make them very difficult to find.

Fortunately, a computerized process can automatically identify any and all loops. It then outputs a list of the nodes and constraints involved in the loop. This is only the first step, however, because this list only narrows the search. The user must still determine which constraint causes the loop (A-B, B-C, or C-A).

The most effective way to isolate the problem is to sketch a project model fragment including the nodes and constraints in the loop. We must keep this diagram in agreement with the left-right convention. The looping constraint will generally appear in the diagram as the one that violates the convention and flows backward from right to left.

Often, loops result, not from logic errors, but instead from data input errors. Entering the wrong node identifier as either a predecessor (PNI) or successor (SNI) in a constraint record can create a loop. The simple example in Figure 11-3 illustrates how this can happen.

This example contains no actual loop, but the modelfile does contain an error. An *A* entered instead of a 1 in the third position of the successor node identifier field in the second constraint caused the loop. The software identifies the three nodes and two constraints as a loop. Diagramming this list and comparing the results to the correct model should reveal the error.

Figure 11-3

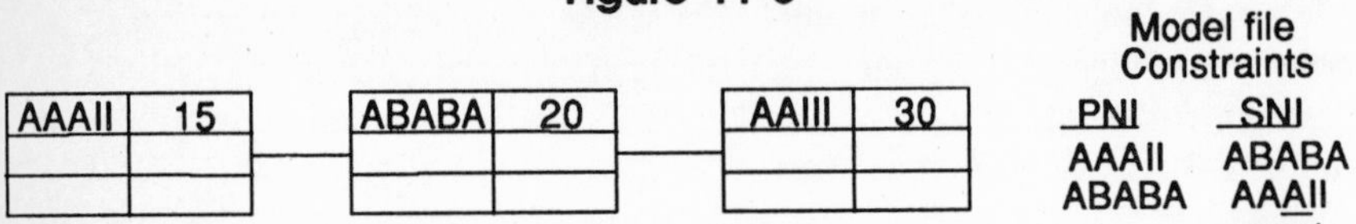

Unloops

Unfortunately, some software packages erroneously identify certain situations as loops. They test for loops by methods other than the basic criteria; a calculated date cannot be dependent upon itself. Our first example (Figure 11-4) constrains a predecessor to a successor at both the start (SS) and finish (FF) ends. One node is the predecessor in both constraints and the other node is the successor in both constraints. Some software packages would report this as a loop.

This is not a loop, however, because no calculated date depends on itself. Let us examine the early date calculations:

The ES of A is determined by its predecessors.

Figure 11-4

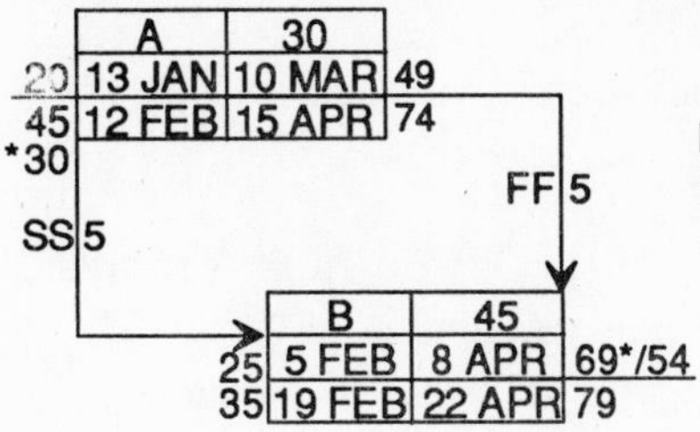

NOT A LOOP

Node Ident	Node DU	PNI	SNI	Typ	Lag
		Constraints			
A	30	A	B	SS	5
B	45	A	B	FF	5

LEGEND

NI	DU
ES	EF
LS	LF

The EF of A is determined by the ES of A plus its duration (30).

The ES of B is determined by the ES of A plus the lag (5) on the constraint.

The EF of B is determined by the later of the ES of B plus its duration (45), or the EF of A plus the lag (5) on the constraint.

All of these dates can be calculated, and the results appear in Figure 11-4. Likewise, the late dates can be calculated by the rules of the backward pass. No calculated date requires itself as input, so this cannot be a loop.

The second example (Figure 11-5) resembles Figure 11-4, except that one node is constrained to another at the start end and then reverses this relationship at the finish end. One node is the predecessor in a start-to-start relationship and the successor in a finish-to-finish relationship. The other node is the opposite, the successor in the start-to-start constraint and the predecessor in the finish-to-

Figure 11-5

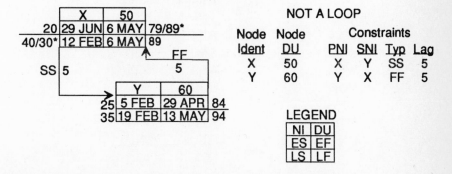

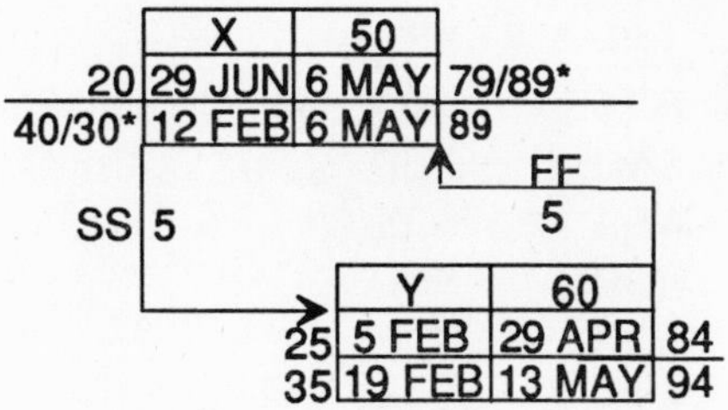

NOT A LOOP

Node Ident	Node DU	PNI	SNI	Typ	Lag
		Constraints			
X	50	X	Y	SS	5
Y	60	Y	X	FF	5

LEGEND

NI	DU
ES	EF
LS	LF

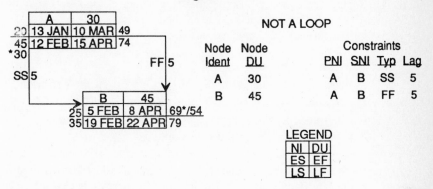

finish constraint. Software packages very commonly identify such relationships as loops, but again, we can demonstrate that this is not a loop by checking whether any of the calculated dates depend on themselves.

Examining the forward pass calculations, we can see that:

The ES of X is determined by its predecessors.
The ES of Y is determined by the ES of X plus the lag (5) on the constraint.
The EF of Y is determined by the ES of Y plus its duration (60).
The EF of X is determined by the later of the ES of X plus its duration (50) or the EF of Y plus the lag (5) on the constraint.

Each of these dates can be calculated, with the results illustrated in Figure 11-5. Likewise the late dates can be calculated by the normal backward pass process. This situation does not qualify as a loop, either.

Figures 11-4 and 11-5 demonstrate the two most common situations erroneously identified as loops. The software must identify any loops in the model because they disrupt the proper CPM processes. It must, however, identify only true loops. False reports limit our ability to model the project adequately.

Constraint Node Incompatibility

When we enter records for a project model, we must ensure that every node identified in a constraint record as a predecessor or a successor also appears as a separate node record. When a nonexistent node identifier (NI) appears in a constraint either as a predecessor (PNI) or successor (SNI), we have an error of *constraint node incompatibility* (CNI). To better illustrate this, Figure 11-6 lists nodes and constraints with examples of this error.

Figure 11-6

	Constraint Identifiers	
Node Identifiers	PNI	SNI
FOX	FOX	SAC
SAC	SAC	WED
WEA	WEA	HOH
HOH	HOH	SIA
SIA	HOH	UTE
UTE	ATE	ONA
ONA	SIP	ONE
	FOX	HOH
	SAC	UTE

Constraint SAC-WED has a valid predecessor (SAC), but the successor (WED) does not appear in the list of node identifiers. This constraint is invalid due to CNI. The relationship ATE-ONA has a valid successor (ONA), but the predecessor (ATE) does not exist, making this an invalid CNI relationship. In constraint SIP-ONE, neither the predecessor (SIP) nor the successor (ONE) are defined as nodes, so this constraint is also invalid. The predecessors and successors of the other six relationships (FOX-SAC, WEA-HOH, HOH-SIA, HOH-UTE, FOX-HOH, and SAC-UTE) have been identified as nodes, so these are valid constraints.

The software must identify this problem whenever it is present, and it should do so before executing the CPM process. It can make this check immediately prior to beginning the CPM process as part of a data validation routine, or it can check the records as the user enters them. The latter method catches the error as it is made, but it also requires that both nodes in a constraint be entered into the modelfile before the constraint can be entered. This makes careful planning of the sequence of data entry necessary. In many cases, this imposes more grief than it adds value, so we prefer that software identify any constraint node incompatibilities in separate data validation checks just prior to CPM processing.

Duplicate Node Identifiers and Duplicate Constraint Pairs

The primary requirement for node identifier codes is that each code value be unique within a modelfile. In loading data, the user could easily enter the same node identifier on more than one activity record. Accidental duplication of an entered node record would create such an error leaving two identical records in the modelfile. A similar error could result from accidental duplication of just the node code, leaving two different records with the same identifier in the modelfile.

Either way, this error is fatal. Any constraint to which this duplicated node identifier is either a predecessor or a successor would reference two nodes in the project model, confusing the interface.

The software must identify this condition either in a data validation check just prior to CPM processing or at the time the node records are loaded. In this case, the user would suffer no adverse impact from the problem being identified at the time of the data entry, so this is the best time to do it.

Duplicating constraints creates a different, and less troublesome, problem than duplicating only node identifiers. Some duplicates can coexist and some cannot. We saw in discussion of unloops that two nodes can, and frequently do, bear both start-to-start and finish-to-finish relationships to one another. Such a pair of constraints between the same two nodes could be valid. Two nodes cannot, however, bear both finish-to-start and finish-to-finish relationships to one another. This is inconsistent. Likewise, two finish-to-finish constraints with different lag or calendar values could not bind the same two nodes together. These are examples of invalid pairs of constraints between two nodes.

It is best to have the software identify any constraint records in which the same two node identifiers appear. This check should be made just prior to CPM processing during data validation. The user can then separate erroneous relationships from valid ones. Such an error would not prove fatal, in any case.

Out-of-Sequence Updating

Entering dates into the modelfile, either as targets or status values, creates two new opportunities to invalidate the model. We could enter status dates that contradict model constraint logic or we could enter a holiday or a nonworkday by mistake. The first of these problems we call *updating out of sequence* (UOS), and we call the second *calendar incompatibility* (CI). We will examine each of these in turn.

If we apply an actual status date (projected status will be reconciled through the forward pass) that conflicts with the logic of the model to an activity or event, we create an inconsistency, and we should correct it. These update out of sequence (UOS) problems can take four basic forms:

1. Applying an actual start date on a task before all predecessors of its start reach completion.
2. Applying an actual finish date on a task before all predecessors of its finish reach completion.
3. Applying an actual start date that precedes the dates of the predecessors of the task's start.
4. Applying an actual finish date that precedes the dates of the predecessors of the task's finish.

The software should identify any node that shows a UOS error, and classify the error by the form (1-4) it violates.

Examples may show these conditions most clearly. In Figure 11-7, the actual start of node B conflicts with the predecessor of its start, node A, since that predecessor has not actually finished. This violates the constraint relationship between the start of B and the finish of A, creating a UOS error of the first type. The actual start date of node C also contradicts the constraint from its predecessor, node B, because that node also lacks an actual finish date. The actual start date of node C (June 6) also precedes the projected finish date of a predecessor (June 10), so node C exhibits both the first and third types of out-of-sequence updating error.

Figure 11-7

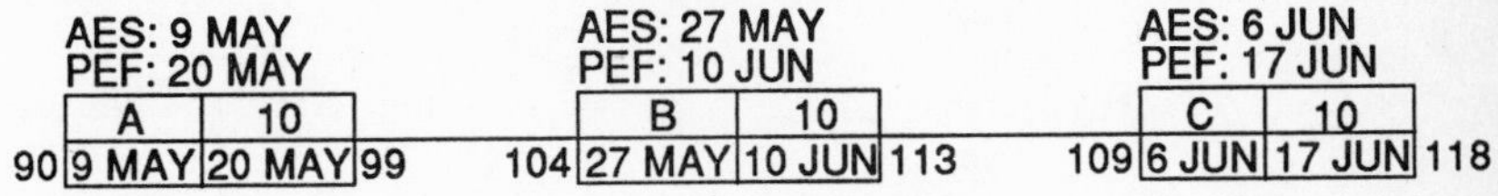

The software's data validation process should identify node B as containing a UOS error of the first type. It should also identify node C as containing UOS errors of the first and third types.

Figure 11-8

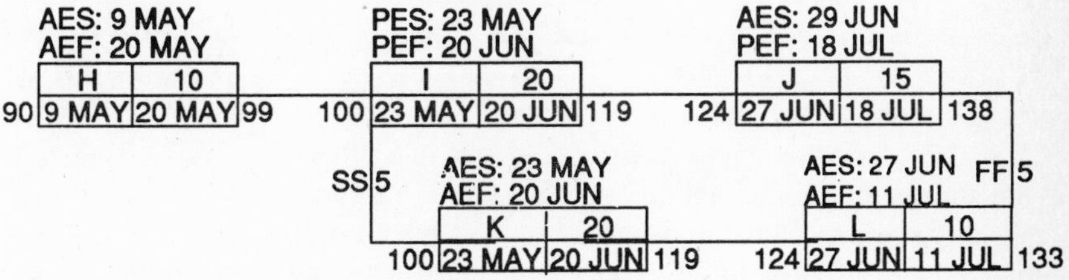

Figure 11-8 illustrates other examples of updating out of sequence violations. The actual start of node J is updated out of sequence because its predecessor (node I) has not finished, in fact it has not started, creating a type 1 error. Node K has started and finished, but the predecessor of its start (node I) has not actually started. Node K also contains a type 1 updating out of sequence error. It also contains a type 3 error because the actual start date (May 23) is earlier than its predecessor's projected start date plus the lag (May 23 + 5 days).

The actual finish date on node L is out of sequence because one of the predecessors of its finish (the finish of node J) has not reached completion, a type 2 UOS error. Node L's actual finish date (July 11) is also earlier than the projected finish date of one of its predecessors plus the lag (July 18 + 5 days), creating a type 4 error.

In total, the software's data validation function should register several UOS problems. It should identify node J as containing a type 1 UOS error. It should identify node K as containing type 1 and 3 UOS errors. Finally, it should identify node L as containing type 2 and 4 UOS errors.

A final example (Figure 11-9) illustrates several nodes with status dates that violate the model sequence. The actual start date of node N (May 23) is earlier than the finish of its predecessor (May 27), a type 3 violation. The actual finish date of node Q (June 17) is earlier than its predecessor date, its own actual start date (June 27). This is a type 4 violation.

Figure 11-9

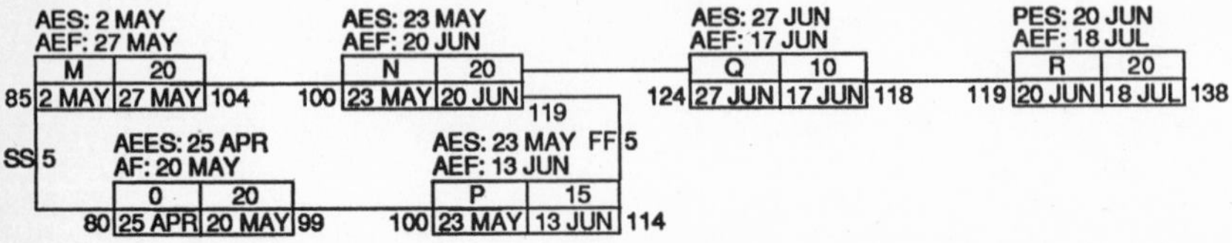

Node R has an actual finish date, but it shows only a projected date for its own start—it has not yet begun. This is a type 2 violation. The actual start date of node O (April 25) is earlier than the start of its predecessor (M) plus the lag (May 2 + 5 days), another type 3 violation. Finally, the actual finish date of node P (June 13) is earlier than the finish date of one of its predecessors, the finish of N, plus the lag (June 20 + 5 days). This is a type 4 violation. Five of the six nodes in this example have update out of sequence (UOS) errors. The software should flag each, stating specific types of errors.

These examples illustrate several ways in which erroneous entry of status data can cause conflict with the model logic. The errors all fall into one of four categories. If the software identifies both the nodes that we have updated out of sequence and the type of each error, it can simplify our job of finding and fixing the problem.

Updating out of sequence does not cause a fatal error, as a loop does. The rules of the forward pass still allow the calculation of dates, as all three examples illustrate. However, the inconsistency of status with the model compromises the resulting data. For this reason, we must identify and correct such mistakes.

Calendar Incompatibility

Of the two types of mistakes that may occur in entering status dates, updating out of sequence is the first. The second type arises when the user inputs any dates that conflict with the task's calendar. Inputting a Saturday as an actual start date for a task with a normal five-day work week is one example. We could also input a holiday as an FNL date. The software data validation function should detect and flag such inconsistencies so we can correct them prior to CPM processing. We will also, however, create a set of rules to prevent this condition from disrupting the CPM process.

When a user-input date falls on a nonworkday as established by the task's calendar, the software will move the date for calculations. If we make this mistake with a start date (SNE, AES, PES), the computer process will move it forward to the next valid workday. If, for example, we place an actual start date on a Saturday, the software will move the date to the following Monday. If we misplace a finish date (FNL, AEF, PEF), the software will move it backward to the previous valid workday. If we give a Saturday as an actual finish date, the computer process will move it to the previous Friday.

These rules allow the forward and backward passes to proceed, but we must also check during data validation for nodes that have dates that conflict with the task's calendar. This double check is necessary because the rules make certain assumptions that may not be true. The incorrect calendar may be listed for the task, or the input date may be wrong. Ultimately, the user must resolve the inconsistency after the software flags it.

Section Summary

In addition to setting special flags to highlight specific conditions within a model, the CPM software should flag certain error conditions in the modelfile. It must also supply additional information for some errors. We need to be informed of at least the following errors:

1. Hanging starts and finishes
2. Loops
3. Constraint node incompatibilities (CNI)
4. Duplicate node identifiers
5. Duplicate constraint identifiers
6. Updating out of sequence (UOS)
7. Calendar incompatibilities (CI)

DATA SORT AND SELECTION

Once we have built a project model, then we have to control the information. This means dealing with hundreds, perhaps thousands, of data records. The software must provide ways to organize and control our interface with the database. Two of the ways in which this is accomplished is to allow the user to extract a specific portion of the data, and to control the sequence or order in which the data is portrayed. We call these processes *data selection (filtering)* and *data sort (order)*.

The diagram of a project model in Figure 11-10 will help illustrate data selection and sorting. Instead of showing the usual model information, we have grouped the seven nodes under color labels. Nodes NERDD, TWITT, and DWEEB are labeled RED; nodes BOWZO and TWIRP are labeled BLUE; nodes TWINK and DORKK are labeled GREEN.

The software's data selection function would segregate from the modelfile a list of the activities in one or more of these color groups from all the others.

Since smaller groupings are easier to understand, they help us concentrate our examination. We can manage a large database by controlling a series of subgroupings of the data in this way, and avoid the difficulties in handling the data in its entirety.

Figure 11-10

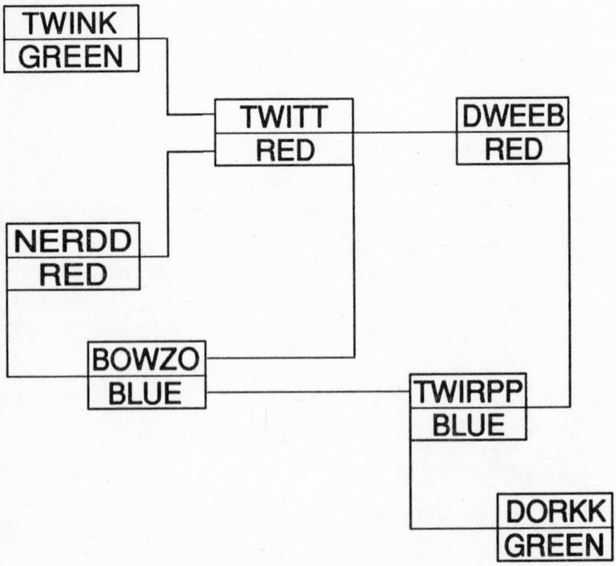

The software's data sorting function allows us to order the data by grouping activities together and then displaying the groups in some controlled sequence. This sequence is determined by the criteria by which we order the data. For instance, we can order dates chronologically and numbers sequentially. We can arrange word-type data (color specifications for example) in alphabetic order, frequently shortened to *alpha order*.

A list of activities grouped or sorted by arbitrarily assigned colors probably would not give a meaningful organization to our data. This is a simple example of a more appropriate sort and select data, like the organizations that perform the tasks or the locations at which they are performed. Proper criteria vary among specific projects. For instance, tasks performed under government contracts usually must conform to a work breakdown structure (WBS), but commercial projects might not even know what a WBS is. Some projects are completed in phases. Different contractors or agencies may perform aspects of the work. We may need to organize data in relation to all of these factors.

We call these collective groupings that result the *project structures.* Almost all projects require the creation of some kind of project structures. Generally at least two, frequently three, and sometimes many criteria overlay our definition of each project task.

The basic modelfile provides no information about project structures. To provide this, we must add fields to the records to store sort and selection criteria. These are called the *sort/selection fields* or the *project structure fields.*

These structure fields are almost always alphanumeric because some structure data is alphabetic, some is numeric, and some combines alphabetic and numeric information. Sorting the values in fields requires a way to distinguish between the sequence of alphabetic values and that of numeric values. Unfortunately different software packages sort these fields differently. Some sort the alphabetic characters ahead of the numeric characters, and others reverse this order. We will define the sequence as numeric first, then alphabetic. The value 1 would come before 9, which would precede A, which would precede Q, etc. We call this *alphanumeric sequence* or *alphanumeric order.*

Typically, the software selects and sorts data simultaneously; that is, these processes are complementary. We select a section of the modelfile to examine and then order the selected data.

Examine Figure 11-11 to see how this works. We have identified five activities where the records for each have two structure fields, one for the location of the work and one for the department accomplishing the work. This work is done in three locales (Tripoli, Timbuktu, and Thule) by three engineering departments (AENG, BENG, and CENG) and two fabrication departments (AFAB and BFAB). To select out all activities performed in Timbuktu, we select by the location field for the value TIMBUKTU. The software generates list 1. To order all the data by location, we sort the entire modelfile by the location field. The software produces list 3 (with THULE first, TIMBUKTU second, then finally TRIPOLI). Since all values begin with T, the sort mechanism uses the second character, then the third, etc., if necessary.

We can also combine the two functions of sort and selection. To select the work done by the engineering departments and then order this data by location, we first select by the department field for the value "_ENG." The value we state as a selection criterion gives the second, third, and fourth characters of the department field. This ignores the first character, so it captures the data on all three engineering departments. Such a partial definition is commonly called a *subfield.* We order the selected data by the location field (i.e., THULE first, then TIMBUKTU, and finally TRIPOLI just as for list 3, but for fewer records). The software outputs list 2.

Figure 11-11

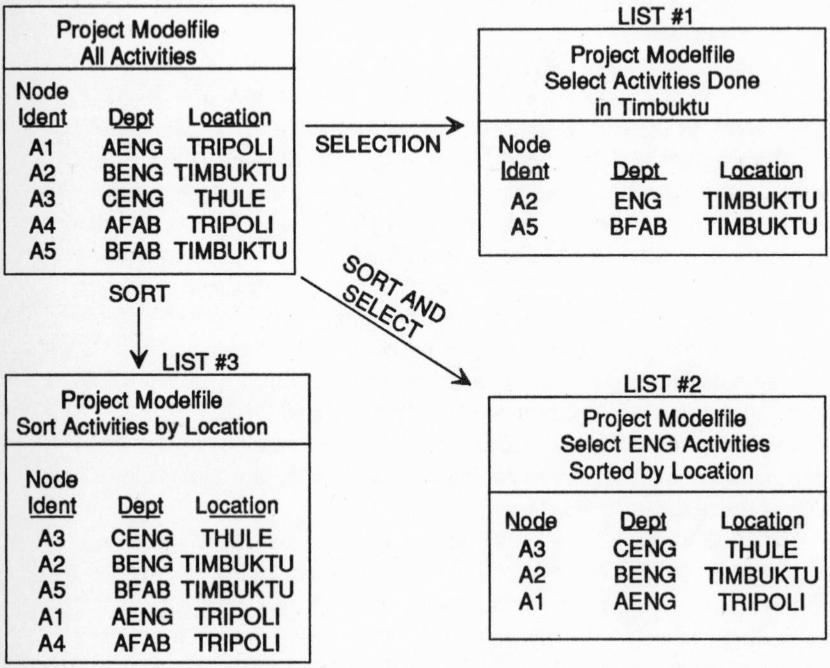

Field Coding Principles

We call the values used in the fields for selection and sorting *field coding*. It can become a powerful tool if we follow certain guidelines or disciplines, especially as we get into output products. First, however, we must learn the principles of structure field coding.

The first rule requires that we *define and document a standard coding structure for each project*. We must define structure fields we will use and state the specific data parameter that each field will contain. The fields must be loaded with predetermined and consistently stated values. The sort and selection processes do not actually select, for example, any work done in Timbuktu. Rather, they select and/or order on the specific value "TIMBUKTU" in a certain field. If we have entered that exact value in that specific field for the activities accomplished in Timbuktu, the software will return the desired results.

If some records store the location information in one field while others store it in another, we have a problem. If we specify one field as the selection field, then we will miss the records that store location values in another field. Even if all the records have locations coded in the same field, all must code the information in exactly the same way. If some of the location fields are coded "TIMBUKTU" while others are coded "TEMBUCTOO" and still others "TYMBUCTWO," selecting the activities performed in "TIMBUKTU" will return only a portion of the desired data.

To prevent these problems, we must establish a coding structure by defining a code for the parameter to be stored in each structure field and then specifying the acceptable values for that code. We must maintain this discipline as we load data and then double check our work.

The second rule requires that the *defined coding structure encompass all of the project's requirements for sorting and selecting data.* We must build the sort and selection structure along with the project database. We should list every factor that has any bearing on the project (location, functions, organizations, phases, etc.), since these become the criteria for sort and selection. We should also think about what criteria could arise in the future, and change and maintain our structure as necessary.

The project structure and field code values must be as dynamic as the database itself. They must accommodate new growth. If they do not, we may find ourselves unable to find or control information in our own database without considerable recoding. If we anticipate growth from the beginning, however, we can usually prevent this dilemma. Then when new criteria arise, we simply add new fields to the coding structure. If a need arises for new values in an existing field (e.g., a new location or function is added to the project) we simply add them to the appropriate parameter list. This allows the structure to grow and change with the project.

The third rule requires that *codes within each field have a consistent length.* We must avoid fields with some values three characters long, some four, and some five. All values should be one length to prevent problems with left and right justification. When a field value occupies only a portion of a field (i.e., a four-character value in an eight-character field), the software places it either all the way to the left (left justification) or all the way to the right (right justification). It fills the balance of the field with nulls or zeros. Since this character is the first value in an alphanumeric sort, the sequence of a list of values can vary depending upon where the zeros fall. Figure 11-12 lists codes with varying character lengths for country names loaded into a six-character alphanumeric field. The resulting sequences with left and right justification show the problems this can create.

Figure 11-12

Data List	Left-Justified Order	Right-Justified Order
USA	CAN000	0000IT
USSR	CHILE0	0000NZ
GDR	GDR000	0000UK
UK	IT0000	000CAN
NZ	MEX000	000GDR
IT	NZ0000	000MEX
MEX	PERU00	000USA
PERU	SPAIN0	00PERU
CHILE	UK0000	00USSR
SPAIN	USA000	0CHILE
CAN	USSR00	0SPAIN

The figure shows leading or trailing zeros in place of nulls to illustrate why the sequence of the left-justified list differs from that of the right-justified list. If however, the values all had the same number of characters, then the order of the list would be identical whether left- or right-justified.

The fourth rule requires that *we accommodate any desired data sequence in our parameter values*. The earlier color criteria had the alphabetic sequence BLUE, GREEN, and RED. Suppose, however, that the sequence RED, BLUE, and GREEN better suited our information needs. To achieve this sequence, we could dedicate the first character in the field to defining sequence. To follow both rules three and four, we would define the codes as: BBLUE, CGRNN, and AREDD. The sorting process will place these values in the preferred sequence.

Comparing this list of field values to the original list shows that the values are becoming cryptic. As we define values to conform with still more requirements, we will come to realize that designing code values requires a trade-off between succinct, cryptic values and longer, more easily understandable values. We can load short, cryptic codes faster with fewer chances for keystroke errors, but they also are harder to read and become even less recognizable when errors are made. To maximize the benefits and minimize the drawbacks, *we need to design concise, yet communicative code values*. This is the fifth rule.

Many records will contain the same field values in the structure fields, since multiple activities share the same location or function, among other characteristics. To order data, we must define the sequence for records with identical values. The sixth rule governs this, requiring that *we designate as many fields for data order as necessary to order all records*. The sorting function orders records with identical first (primary) field values by their second (secondary) field values. It orders records with identical primary and secondary values by their third (tertiary) field values, and so on.

As an example, we might designate the sequence as follows: the department field (primary), then the EF of the activity (secondary), and then its ES (tertiary). The sort function would order those activities within the same department by the chronological sequence of their expected finish dates. It would order activities within the same department and scheduled to finish on the same date chronologically by their expected start dates.

Normally, a software package maintains a default criterion by which to sort records with identical sort criteria values. As a default criterion, we will order the records as they occur in the modelfile or in the sequence in which we entered the data. Thus, the sort function would order activities within the same department that finished and started on the same date in the same sequence in which they were originally entered into the modelfile. We could, of course, establish a fourth order criterion, but three criteria generally provide adequate sequence guidelines for most records.

These six rules are important. Following them automatically instills much of the necessary discipline for generating, loading, and maintaining project structure information in the modelfile records.

Coding Various Data Types

We will need to load many different kinds of data into structure fields. Certain principles must guide the design of codes for each case. These principles turn the six rules of data structure and coding into a procedure for designing code values.

1. List all known values for each structure field.
2. Do not change values that are part of an existing hierarchical structure (e.g., WBS, SOW, etc.).
3. Enter numeric values unchanged into alphanumeric fields.
4. For textual values:
   Abbreviate the values to 5 characters or less.
   Assign all code values the same number of characters.
   Designate one character of the code to define the preferred sequence.
   Use standard acronyms, if possible.
5. Provide space for enough characters to facilitate some visual recognition of field values.

Let us look at some examples to demonstrate how these rules work. Numeric values, like codes that represent departments of a company, are rather cryptic. Still, they should be entered directly as they are without changes, but in an alphanumeric structure field, not a numeric field. The purpose of these values is sort and selection, not mathematical processes.

To structure a list of values with varying lengths, like locations, we abbreviate each to a common length and designate one character to indicate preferred sequence. The left-hand column of Figure 11-13 shows a list of locations in the preferred sequence. The other columns give examples of appropriate codes in one-, two-, three-, or four-character fields. Notice that, as the field length gets shorter, the values become more cryptic. The three- or four-character field values provide better visual recognition; the single- and two-character fields are simply too cryptic.

Figure 11-13

Values	One Character	Two Characters	Three Characters	Four Characters
Los Angeles	A	1L	1LA	1LAN
Denver	B	2N	2DN	2DEN
Dallas	C	3D	3DL	3DAL
Detroit	D	4T	4DT	4DET
San Diego	E	5S	5SD	5SND
San Francisco	F	6F	6SF	6SNF

Some values infer some kind of hierarchy, such as various major functions and subfunctions. To reflect this in the structure, we create code fields within code fields. We call them *subfields*. The list below is an example of organizations or functions which have a structural relationship. The first character of the field indicates preferred sequence, the second character indicates the subfunction, and the third, fourth, and fifth characters indicate the function.

Figure 11-14

Value	Code
Electrical Engineering	2EENG
Mechanical Engineering	1MENG
Structural Engineering	3SENG
Software Specifications	1SSFW
Software Coding	2CSFW
Software Verification	3VSFW

Examine the Code column of Figure 11-14. The value *E* in the second subfield refers to the electrical branch of the Engineering department; the value *M* represents the mechanical branch. The value *S*, however, represents both the structural branch of the Engineering department and the specifications branch of the Software department. This creates no conflict, because the code value is not the second subfield all by itself, but rather the combination of the last two subfields. Structural Engineering is *SENG* and Software Specifications is *SSFW*. By combining two or more subfields to create another subfield, we can form an enormous number of combinations, providing us tremendous flexibility in our sorting and selecting.

To select all Engineering department activities, we designate the value *ENG* in the third subfield (characters 3, 4, and 5). To select all Electrical Engineering branch activities, we designate the value *2EENG*, the entire field. We could do the same with the Software activities (designating *SFW*) or a subfunction within the Software department (for Coding, designating *2CSFW*). This creates a hierarchical coding structure that enhances functionality. Within one coding structure, we can designate both functions and subfunctions.

Applying these principles for structure design ensures proper coding, but this is not the only way to create adequate code field values. Many acceptable variations would work for all of the cases we have examined. The bottom line is to develop codes that do the job properly. One word is key to all structure field coding: discipline. Valid principles applied with discipline create a powerful tool to manage and control a project database. Without discipline, database anarchy reigns and disaster quickly follows.

The Node Identifier as a Code Field

The node identifier code is an alphanumeric field, so we can use it for sort and selection criterion. Up to this point, we have simply assigned any alphanumeric values as *node identifier codes*. The character string has had to meet only one requirement, that it be unique within the project to provide a simple, exact means of identifying the pair of activities involved in each constraint. There is no reason, however, why these values cannot also carry some coding information. In fact, we will find that augmenting node identifier codes with some project structure information can provide a better visual relationship with the project constraints. We can actually identify predecessors and successors through these code values without having to refer to a listing of all the activities, which will benefit our problem analyses.

However, additional rules govern the use of node identifiers as code fields. We cannot use the NI field in exactly the same manner that we use any other structural field. For instance, other code fields can have identical code values for many activities. Node identifiers must have a unique value for every activity;

this is the first rule governing their use. This does not mean that we must use unique criteria for each node identifier code; instead, we dedicate a subfield within the total NI code for a random value. Thus, in an eight-character field, the first six characters can provide field coding while the last two characters provide uniqueness. Even if the first six characters are identical in many NI values, the last two characters allow us to make each total NI code unique.

Other code fields require no universal structure. A designated code field may pertain only to a portion of the project activities. The activities of a department like Engineering may require values in a code field that have no significance to another department like Procurement. The Procurement activities would either have null values in this field or a different structure altogether. All activities must have a node identifier code, though. Therefore, the second rule for node identifier codes requires that they have a format with universal application; the structure must have relevance to each and every activity.

The third rule requires a practical reaction to change. Field code criteria and associated values often change as the project evolves, and we must keep up and maintain these dynamics. To update any other code field when a value changes, we simply change the field value on that one record. If, however, the code criteria for a node identifier code change, we must not only change that activity record, but every other occurrence of the node in the constraint records. Instead of a single value, we could end up changing two, three, five, or many more values, depending on how often the activity interfaces with others in the project. The average would be three or four. The third rule for node identifier codes, therefore, requires that we minimize the changes to them. We should encode only general and very stable criteria in node identifiers.

These three rules, along with the previously discussed guidelines, provide a basis for structuring node identifiers to provide valuable information. As long as we keep to these principles, there is no reason not to include sort and selection criteria in the coding of the node identifiers. In fact, the practice provides many advantages. We mentioned earlier that this can help us better visualize the constraint relationships. Further, by encoding some sort and selection information into this field, we free up another field for other criteria, or we may have to code one less field in the record, which reduces keystrokes and opportunities to make entry errors. To gain these benefits, we need only extend our criteria and coding disciplines to the node identifier codes.

SUMMARY

In addition to performing the basic calculations for the CPM process, we also look to our computer to accomplish other tasks. First, it must set values in certain flag fields on each record either prior to, during, or after performing the CPM calculations, as appropriate:

1. It must flag the least-time-reserve path of the project by setting values in the CPF fields of activities and constraints on that path after the user specifies the terminus node.
2. It must flag any constraint stretching by setting values in the ASF field. It must specify the amount of stretching in the AOS field, and identify the predecessor (FPS) or successor (BPS) that causes the stretching in the SNID flag field of appropriate constraints.
3. It must flag all logical start nodes (in the SNF flag field) and finish nodes (in the FNF flag field) and designate those start and finish nodes that are properly date targeted (ST or FN) or improperly date targeted (HS or HF).

Certain data conditions cause serious or damaging errors in the CPM process. The software should identify each of these either prior to, during, or after the CPM process, as appropriate:

1. In a loop, the calculation of a node's early and/or late dates requires its own dates as input. A loop is a fatal error, and the software must identify any loop, listing the activities and constraints involved.
2. In constraint node incompatibility (CNI flag field), a node identified in a constraint record does not exist as an activity record. Any such inconsistency is a fatal error and the software must identify it.
3. Two or more activity records having the same node identifier code creates a fatal error that requires identification.
4. The same pair of node identifier codes in two or more constraints may or may not be an error. The software must identify all cases of duplicated node pairs for the user to sort out. This is not a fatal error.
5. The presence or absence of actual status dates and the date values themselves may create inconsistencies with the model's logic. Although these are not fatal errors, each inconsistency must be specifically identified (in the UOS field) so the user can resolve the contradiction.
6. Any date entered into a record may conflict with the designated calendar of the record. This is not a fatal error, but it is an inconsistency that the software must identify (in the CI flag field) and the user must correct.

Computers give the user great power to sort and select data by certain criteria. We can use this feature to great advantage in project modeling:

1. The software's data sorting function allows the user to control the sequence in which the records appear on the monitor or in reports.
2. The software's data selection function allows the user to designate or screen certain records for review either on the monitor or in reports.

To the normal model parameters in the records, we add structure code fields to store values for sorting and selection. The design and use of structure code fields must follow certain disciplines:

1. The entire project must share a standard coding structure. We must define the fields to be used, what parameter each contains, and a listing of acceptable values for each parameter.
2. The project coding structure must be developed in the beginning of the project to encompass all data sort and selection requirements. It must accommodate growth and change over the project's life cycle.
3. Each code field should allow values of one length (in number of characters).
4. Designated code values must accommodate the preferred sequence of data.
5. As we develop the specific code values, we must consider the trade-off between longer more communicative values against shorter cryptic values.
6. When we designate the preferred order of data, we must list levels of criteria sufficient to sequence all necessary records. Usually we need to designate primary, secondary, and tertiary fields to accomplish this. The default record sequence is the order in which the records were entered into the database.

Additional rules govern sort and selection criteria in node identifier codes:

1. We must always reserve a portion of the field (at least two characters) for random characters to ensure that all node identifier codes remain unique.
2. The code criteria must apply to all activities in the project.
3. We must code only stable criteria in the node identifier to minimize changes.

CHAPTER 12

Modeling Techniques in a Practical Project Environment

Up to this point, we have introduced many individual project modeling processes: the forward and backward passes, time reserve calculations, various record flags, data entry, changing, and deleting, etc. By themselves, these are merely phenomena; no single calculation is inherently useful in the project management process. The user must combine these processes in a practical scheme to bring them out of the esoteric world of mere theory and into value added application. We must develop a comprehensive methodology that encompasses all of these processes and unifies them into a single project modeling tool.

The key to this overall methodology lies in an understanding of the relationship between the project model and the project schedule. Earlier chapters alluded to this relationship, but did not explore it in any depth because the many facets of the CPM process must first be understood independently. Only now can we mold these pieces into a complete project modeling scheme that will satisfy the requirements of more cost effective project management.

As we will see, though, the result is not a single scheme. The relationship between the schedule and the model with which we begin the project differs from the relationship by which we control the project. As the project evolves, so does the schedule-model relationship, and so does the project modeling methodology. We begin with an exploration of the preproject relationship between the schedule and the project model and the methodology that is necessary to accommodate this relationship.

THE PREPROJECT SCHEDULE-MODEL RELATIONSHIP

Before any project begins, its managers should establish a plan that maps out its anticipated accomplishments. This requires the collection of many elements: principal objectives and the work necessary to accomplish them, estimates of how long the work will take, and the necessary labor and other resources. From these elements, we build a desired sequence of work, defining the relationships between the discrete tasks. Once this information is collected, developed at the working level and translated into model elements, it becomes the project model.

Bottom-Up versus Top-Down Modeling

In the initial development of a working project plan, we encounter our first obstacle, the age-old decision between top-down planning and bottom-up planning. We must resolve this fundamental conflict of methodology to know where to begin a project plan. A discussion of the advantages and disadvantages of each method will lead us to an answer.

Top-down planning starts with the project as a whole, visualized as a single activity that spans the time from startup to completion. We then determine all of the delivery dates and any other significant milestones that mark progress toward the successful completion of the project. We can then begin to decompose the work of the project into separate efforts necessary to accomplish each of the major milestones. Now, instead of one activity representing the project, we have one for each major project event.

Further breaking these tasks down to meet product, function, organization, and location considerations provides new levels of schedule definition, each of which expands the amount of detail by factors that range from 2-to-1 to 10-to-1. The initial project breaks down into from 2 to 10 tasks. Each of these tasks in turn breaks down into from 2 to 10 tasks. This process can continue downward to define work as detailed as daily, or even hourly, tasks.

Every detail task at every level of decomposition is part of a task at the next higher level. Each task rolls up into a higher-level task in what is called a *parent-child relationship.* Successive rounds of decomposition thus form an entire hierarchy of parent-child relationships that structure the work of the project from the summary to the detail.

This is the primary advantage of top-down planning. Developing the detail plan by breaking the total project down in successive layers to the working level helps ensure that all of the project's objectives influence the lower-level plans, and that these plans mesh together to satisfy the project schedule objectives. In effect, top-down planning ensures that the sum of the activities at any schedule level always equals the project whole. This prevents overlooking any portion of the project. Each facet of the schedule is developed to satisfy a particular project objective, such that all are achieved in this plan. Besides these important benefits of this methodology, the hierarchical structure it derives provides the basis for an excellent project management tool called summarization, as explained in Chapter 13.

The principal, and fatal, disadvantage of this technique is that the spans derived for the detail tasks depend fundamentally on the time available to the total project. All task durations come strictly from apportionment of the available time. The same task could be allotted a duration of six days, six weeks, or six months, depending on the overall span of the project and the negotiation process used to define each facet of the project. With too tight a time frame, we simply "shoe horn" the tasks to fit. With too loose a time frame, we stretch the effort to consume the available time. The time required by the scope of the work is not the primary driver of the task spans in the plan, whereas the real task spans cannot avoid this constraint. By itself, top-down planning cannot arrive at a workable task-level project plan. Yet, it can provide many benefits to our planning and management efforts if we can but avoid its fatal pitfall.

The *bottom-up approach* avoids this weakness by beginning with the scope of individual tasks. It defines the work of the project as relatively detailed tasks. Then working-level managers balance their assessments of the scope of these tasks against the necessary and available resources to arrive at the span of each detail task. Mapping these activities together, while preserving the preferred sequence of the project and the interrelationship of the tasks, yields a model of the project. This model represents all of the work, including its time spans and the interrelationship of its parts, in a single structure. A simple forward pass would then yield a project schedule at the working level of detail.

The principal problem with this method is fairly obvious. It typically schedules the project delivery dates several years later than they are required. The bottom-up approach ignores the time constraints on the objective, so it will get done when it is convenient, not when the organization needs it. This method has another disadvantage in that, starting at a low level of detail, it lacks the structure to ensure that the entire scope of the project is considered. The method could overlook some portion of the project.

Both the top-down and bottom-up approaches have significant advantages, and, alone, each has significant disadvantages. The advantages of one seem, however, to offset the disadvantages of the other. The correct choice, perhaps not surprisingly, is to utilize both techniques. This will, we hope, let us maximize the advantages of each while avoiding the pitfalls of both. Beginning with the top-down method, we only decompose the work down to a level just above the working level. We then develop a detailed schedule for the working level with a bottom-up method. Unfortunately, meshing these two plans together never gives a smooth fit; in fact, the disparity could be enormous.

SCHEDULE-MODEL RECONCILIATION

The real preproject planning process begins when we start to reconcile these two plans. This requires reasonable, intelligent adjustments to both sets of data until they merge into a single, viable scheme. We must develop a balance between resource, task span, the interface of work, and project objectives. This iterative process takes considerable coordination between the working levels and the management levels of the project, and project modeling techniques can contribute much to it.

The forward pass process simulates bottom-up project planning through the model. Date successors on the appropriate events in the model simulate top-down event requirements. Modeled in this way, any sequence of work that does not support the timely accomplishment of the project milestones will show up as negative Path Time Reserve (negative SPTR and/or FPTR values). The forward and backward passes aid us in establishing a viable plan that sets reasonable goals, yet meets the time objectives of the project. At this point, the schedule of the project and the early dates of the model are the same. This is the preproject relationship.

To resolve bottom-up and top-down schemes into a single plan, we must correct any negative Path Time Reserve problems through the Path Time Reserve equation:

$$PTR = TA - CSP$$

where:

PTR = Path Time Reserve
TA = the time available to the path
CSP = the cumulative sequence of the path

We must either increase the time available to the paths in trouble or decrease the cumulative length of these paths, or combine both tactics. We can increase the time available to a path by three methods.

1. We might attempt to start some paths early. This might require the resolution of some external interfaces to bring some resources in early.
2. We could also supplement some paths with overtime or multiple shifts. For instance, we could work on a series of tasks six days a week rather than five. Alternatively, we could add a second shift to work on a series of tasks. Either of these or other, similar methods would add work time to a sequence of tasks. Starting a program on overtime might not, how-

ever, seem an optimum solution. We should try to reserve this tactic for downstream problems.

3. As a third option, we could accomplish some of the milestones late. At first this might seem a poor choice, but it may not be totally unacceptable if we miss only internal schedule objectives and continue to meet the primary goals of the project.

Individually or in combination, these solutions add working time to a path to solve a time reserve problem. We can also add work time to the other half of the equation, increasing a path's time reserve, by decreasing its cumulative sequence. Basically, we can do this in two ways: decrease the spans of tasks or alter the sequence of the work.

We can decrease the span of a task if we find a way to do the work in less time. We can do this most directly by adding resources on the theory that two people can accomplish a task faster than one. The key to this process is keeping changes to a task's duration reasonable. We can easily change a duration value on an activity record; it is an entirely different matter to accomplish the activity within the reduced time span. Remember, the objective is not to get the model to come out on schedule, but to get the project to come out on schedule.

We could also reduce the duration of a task by deferring or deleting some of its scope. Reducing the amount of work reduces the time required to complete it. However, this does not achieve the schedule objectives, but merely delays them.

The second general method of decreasing cumulative path length involves altering the sequence of the path. Rearranging the interfaces of the work can often reduce its overall time requirements. This method usually overlaps tasks that were initially related by finish-to-start constraints, giving them start-to-start and/or finish-to-finish constraints with lags. This shortens the span of the overall path by the total of all overlaps.

Any overlapping of tasks has two other potential effects, though. First, it increases the total resource requirements of the project during the period of overlap. Can the project provide these additional resources? If not, one task cannot begin until another ends, freeing resources. If the project can provide the additional resources, however, overlapping tasks proves a very effective means of reducing the total sequence of work. Early identification of problems that require overlapping greatly aids this resource planning effort.

The second effect of overlapping concerns risk. Initiating a task with only a portion of its predecessors completed increases technical risks to benefit the schedule. We must weigh these risks against the potential positive schedule improvements. Some risks may be too great to take the chance while others appear acceptable. We must evaluate our options carefully here.

No doubt, this initial reconciliation of the project's requirements with its work effort can become very complex. We must gather extensive information concerning the work effort of the project. We then organize this data into a model through which to resolve any conflicts and problems. This requires us to make decisions based on our best estimates for the project. Undoubtedly, we will have to combine and adapt all our problem-solving techniques, and we will have to accept some risks. We must do all this in a very short span of time. Eventually, however, we will match anticipated problems with hopeful resolutions, addressing each time reserve problem in turn until we create a viable plan. This allows us to initiate the project with a roadmap and a time table to guide us toward our objectives.

But neither the complexity of the decisions nor the uncertainties on which we must base them justifies a failure to plan. What project could be completed, or even rationally initiated, without a detailed gameplan of how it intends to accomplish its objectives? It may be easier to initiate a project with only vague notions of the route to success, but this course almost always results in degradation of some or all of the project's objectives. Only organizations that have the discipline and intelligence to thoroughly plan have a chance of achieving all of their project goals. Ultimately, only these organizations will survive over the long term in the increasingly competitive project management environment.

Establishing Positive Time Reserve

Managers commonly make the mistake of continuing the reconciliation process between the top-down and bottom-up plans only until they increase negative time reserve values to zero. This makes every schedule date on these paths sacred. It leaves no margin for error. This is analogous to doling out all of the cost reserves before the project begins. Eliminating negative time reserves removed deficits. Now we must plan in profits—time profits.

It is critical that we continue our reconciliation process until we establish reasonable amounts of time reserve along every path before we initiate a project. This will provide us some flexibility to accommodate unexpected needs and deviations from the plan. It gives us the additional option of spending some of our time reserves in managing the project to completion.

How much time reserve is enough, though? Unfortunately, enough is what it takes to meet all our schedule objectives, and we won't know this for certain until we finish each objective. We can, however, try to establish time reserves based on intelligent estimates of how much each path needs.

Two factors affect these estimates. The first factor, the overall duration of the path, governs reserves for general problems of unexpected needs and activities exceeding their planned scopes. A three-year path needs more time reserve than a one-year path. We can assume this to be a linear function, so we will set aside a percentage of the total path length as time reserves. As our standard factor, we will assume 10 percent. Therefore, a 10-month path needs 1 month of reserve, whereas a 3-year path needs about 100 days of reserve (3.6 months). This percentage certainly does not come from rigorous calculations. It is more of an empirically based estimate to provide a point of reference.

The other determining factor, technical complexity, can cause vast differences in the need for time reserves. Building a bridge across a river involves relatively little technical complexity when compared to building a sophisticated new spacecraft. The spacecraft project should have correspondingly greater time reserves, though the exact amount is at best a rational guess.

The availability of certain resources affects time reserve in much the same way as technical complexity. A series of tasks that require personnel with rare skills and/or extensive training will need more reserve than a series of tasks that require more commonplace skills. A path that depends on a long, complex delivery cycle for specialized material inputs should accommodate this complexity with increased time reserves.

We can give no simple value for adequate time reserves. Based on the best information we can gather, we must make educated estimates. Even then, we cannot automatically establish the time reserves we deem adequate in our paths. Individual project conditions impose limits on the work accomplished between two points in time. We encounter these same problems in defining adequate cost reserves, however, but we still can define some value for this figure. This is simply part of the chore and the uncertainty of managing projects.

To put it simply, any positive reserve is better than none. Any cost and time reserves in our project, even though they may be inadequate, provide us tools to influence the outcome positively.

THE LEVEL OF MODEL DETAIL

Before beginning to develop the project plan, we must also decide at what level we should build the project model, that is, where to define the working level. A model that includes too much detail can create more practical problems than it solves. Each level of task decomposition increases the number of data records by 2 to 10 times. This increases the accuracy of the model, but also the labor-intensive work of tracking and maintaining the records. Like many other processes, this reaches a point of diminishing returns. We define the point at which the benefits of the increased accuracy of the detail cease to offset the additional workload of the larger database as the working level. At this practical level, we have adequate resources to maintain and analyze the database, and enough detail to recognize significant problems.

While the basic purpose of project modeling is to identify problems in future paths, we must consider the magnitude of those problems. The accuracy of our estimates for activity interfaces, constraint lags, and activity durations is just not great enough to identify minute problems. But only the larger problems really affect the project's outcome as a whole. No 3-year project suffers severe damage from a 1-day problem, but a 30-day problem could badly disrupt it. This revelation, along with the practical trade-off between detail and accuracy, suggests that we not build our project model at the lowest levels of detail.

This does not, however, give an exact indication of the correct working level at which to model. Some useful guidelines can assist us in finding the proper working level. One of these guidelines is derived from the key principle that we model to control interfaces or constraints of the project more than its activities.

This is one of the fundamental differences that separates modeling methods from other forms of scheduling. It is also the key to identifying the model working level.

The project managers must determine which interfaces they must control. This begins with interfaces across departments, functions, or locations. For example, an engineer may hand a drawing to another engineer in his or her department. The project management should not have to concern itself with this interface. This is clearly the job of the department manager. However, when this same engineer hands the drawing to another department, or the engineer needs information from another department to complete the drawing, these interfaces require wider project-level management and control.

These *significant interfaces* become the constraints of the working-level model. Since activities start and finish with constraints from predecessors or successors, the significant interfaces define the boundaries (start and finish) of the activities of our model. We derive our activities from the constraints, and not the other way around. We call this group of activities and constraints the *model of significant interfaces,* which is equivalent to the working-level model.

This leads, by the way, to a very interesting sidebar. Since many functions follow very common patterns from project to project, we can build generic model segments from the standard activities, events, and constraints for each function. Then, rather than defining new tasks and constraints for each effort, we can customize a generic model segment for the particular function from a library to match a specific series of tasks. This would not only simplify and accelerate data entry, but it would also standardize model structures of the various functions of the project.

The overall span of the project provides another guideline to define the working level. Task spans in a three- to five-year project will be much longer than the spans of the tasks in a 25-day missile launch cycle. As a rule of thumb, durations of individual detail tasks should cover about 5 percent of the total project length. Thus, activities in a three-year project should have range from 30 to 60 days long. Activities in a 25-day launch cycle should range from one-half day to a day and a half long.

This figure is derived from the possible level of accuracy. Activity durations of 5 percent will reveal significant problems in a project without miring us in minute detail.

One more consideration guides the definition of the working level, the near term versus the far term. We frequently understand near-term tasks much more clearly than far-term tasks. In fact many of the far-term tasks will be defined by effort accomplished in the near term. The model should reflect this disparity. Simply put, we model near-term tasks in more detail than far-term tasks because we understand the near term better. We always measure the near term versus far term relative to the current point in time or status date. As the project progresses, the near-term window moves, as well, enveloping activities previously considered in the far term. As this work enters the near term, we should come to better understand its detail, allowing us to decompose it into smaller, better-defined tasks and constraints. This is called the *rolling wave decomposition of detail.*

In summary, we determine the appropriate working level at which to model the project detail by considering three factors:

1. The interfaces that require management and control; these become the boundaries of activities.
2. The spans of working-level tasks in relation to the total project length; 5 percent serves as a rule of thumb.
3. The proximity of tasks to the status date; near-term tasks should be defined in more detail than far-term tasks. The near-term window advances with the project's progress.

These three guidelines should identify the working level at which to reconcile the conflicts between the top-down and bottom-up models, yielding an adequate project plan and schedule. In compromising between enough detail to identify significant problems and the manageability of the database, it is always better to sacrifice accuracy for control; that is, we should always keep our model within our resources for management and control. The increased accuracy of more detailed data does little good if adequate resources (labor, etc.) are unavailable to analyze and maintain it.

Problems can arise because the working-level model does not provide the minute detail necessary to schedule down to the smallest tasks. Maintaining the model at higher levels of detail does not remove the requirement to manage the project's intricate details. At these lower levels, however, we schedule, rather than model. Simply scheduling, tracking and controlling work effort is not nearly as difficult or time consuming as modeling, yet it is adequate to maintain the relationship between the detail activities and the working-level model activities. This is a facet of what is called *vertical integration*, as discussed in Chapter 13.

BASELINE SCHEDULE

The completion of the preproject development of the plan produces a schedule of the working-level activities of the project. This schedule shows how events meet objectives and how activities contribute to the achievement of these events. This is the project's initiation schedule.

Once the project starts, status measurements for events and activities alter the schedule, and the original plan can quickly become unrecognizable. To prevent this, we store the original plan, and continually relate the status of tasks back to it. We call this stored original plan the *baseline schedule*.

Stated another way, the baseline schedule is the schedule against which we measure the project. Based on this definition, it could contain virtually any schedule, but we can reasonably narrow this down to a few choices. Our modeling methods allow us to relate the baseline schedule data to parameters in the model.

Some practitioners believe that the original plan's late dates (LS and LF) should become the baseline dates. In a sense, the late dates are the worst-case scenario, so we might rationally measure performance relative to them. In this case, the early dates would track the project's schedule and reflect status, while the late dates would track the baseline. The PTR values would indicate both available time reserves and the span of any slips or pull-ins relative to the baseline.

Unfortunately this scheme very quickly degenerates into a zero-PTR management technique. Projects tend not to perform tasks earlier than scheduled, but rather on schedule or late. When the schedule is the Late Start or Late Finish of a task, the tendency will be to meet these dates and reduce PTR to zero. Since PTR measures the time reserve, not of the task itself, but of all paths on which the task lies, one task's consumption of its PTR robs all of its successor tasks of their time reserves.

This would certainly simplify performance tracking, because all dates would suddenly become sacred. Any missed schedule date would mean a missed schedule objective, or heavy expenditures on time reserve recovery methods like overtime, extra resources, task sequence alterations, etc. To avoid a conflict between our schedule performance measurement technique and the principles of time reserve management, we avoid specifying all baseline dates as late dates.

Remember, however, that a project plan developed through reconciliation of top-down and bottom-up plans schedules tasks by their early dates. This plan is developed in the beginning of a project before any status assessment can disrupt it. It represents a compromise that provides for achieving all schedule objectives while maintaining adequate time reserves to accommodate unanticipated, or incorrectly anticipated, occurrences. We should take this original plan as the baseline schedule and measure the performance of the project against its early dates. Unfortunately, a problem arises with this approach, as well.

Let us look at an example to see how this process works in practice. Figure 12-1 shows a simple model of four tasks. The finish of OMEGA represents a project objective, or milestone. If we take the early dates as the baseline schedule, then OMEGA finishes 12 days early. The purpose of this 12 days of time reserve is not to deliver OMEGA early, however. It is, rather, a resource to help us make the event's required schedule (July 29). Since this Late Finish date is the date that we intend to complete the path and achieve the objective, it should function as the baseline date of this single event.

Figure 12-1

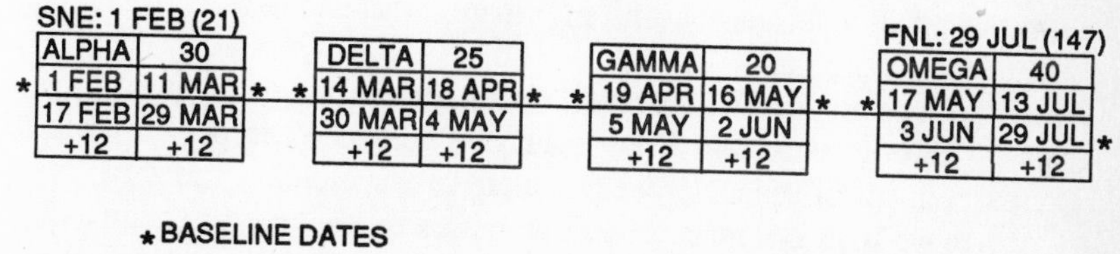

If, however, we measure all of the tasks to a baseline of their late dates, then we have a zero-reserve schedule. This does not work either. To resolve this conflict, we should measure all of the events up to and including the start of OMEGA by their early dates (ES and EF). We should baseline the finish of OMEGA by its late date. This creates a baseline schedule with a mixture of some early dates, some late dates, and maybe even some that fall in between. As a general rule:

> Baseline schedules should measure major project objectives (events), especially any customer delivery dates,

against their *late* dates. They should measure the work nodes (tasks) of the project and lower-level events against their *early* dates.

This mixture provides us a reasonable set of dates against which to measure project progress. The baseline information appears in each activity record as a pair of noncalculated date fields called, appropriately, the *baseline start* (BES) and the *baseline finish* (BEF) of a task. Once we have established the baseline dates of the project, we set these baseline fields to these values. Comparing status dates to these dates indicates project health relative to the original, baseline plan.

Suppose, for example, that a task was originally scheduled to be completed on June 30, 1995. A projected or actual finish date later than June 30, 1995 represents a slip. The magnitude of the slip, which equals the number of days between the actual/projected date and the baseline date, is called the *span of the slip* (SOS). Likewise, a projected or actual finish date that precedes the baseline date represents a pull-in. The magnitude of the pull-in, which equals the number of days between the actual/projected date and the baseline date, is called the *span of the pull-in* (SOP).

The span of any slip or pull-in on project events indicates the schedule health of the project. If a large percentage of activities' finishes show slips, this indicates poor project performance. If any key project milestones, which have late dates as baseline dates, show any slip, this indicates negative Path Time Reserve on top of poor schedule performance. Comparing status performance to the baseline schedule yields very useful information that can help us isolate both general and specific problems.

We cannot overemphasize, however, that this alone is not adequate project management. Unfortunately, too many managers use schedule performance as the sole measurement of schedule health—a tell-tale symptom of reactive management. When we limit our management processes to reacting to schedule slips or pull-ins, we focus only on the near term, reacting to today's problems, and ignoring their potential downstream impacts. We do not manage our project, we react to it—the very definition of reactive management. The alternative to this, proactive methods, will be explored more thoroughly later in this chapter.

Having determined the baseline schedule dates, we should maintain them fixed and unchanged. Only a significant modification to the work scope of the task or its interfaces should cause a change in these dates. Some managers change the project baseline schedule to hide schedule nonperformance, rationalizing "if we cannot make the schedule, change the schedule to what we can make." Slipping the baseline to coincide with actual accomplishment of work does minimize deviations between schedule and performance, but it makes these two parameters useless. Someone has to maintain their courage and resist the calls of ego in order to maintain rational planning and reasonable performance measurement.

Schedule performance measured against the original plan can provide very useful data for our analysis of project health. This should be portrayed along with the calculated data, as illustrated in Figure 12-2.

Figure 12-2

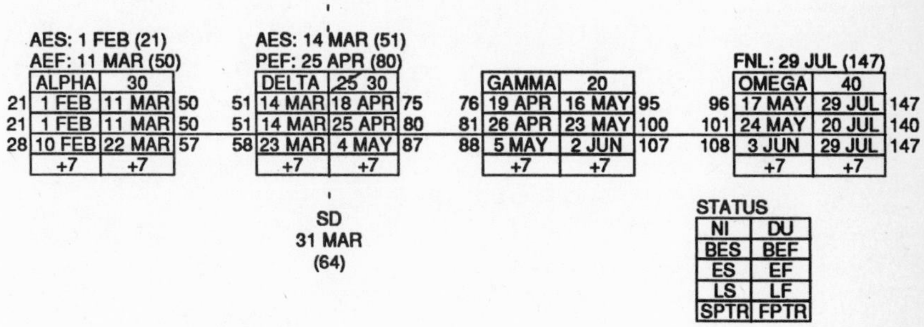

The node is laid out as before, except that the uncalculated baseline dates appear above the early dates to facilitate comparison. Status data fields also appear above the nodes where they apply. This maintains simplicity, as not all nodes have status information, and different fields contain actual and projected data. These nodes are beginning to report a great many attributes, reflecting the requirement in practical scheduling and modeling for many parameters, and therefore fields, per task.

MODELING IN THE ON-GOING PROJECT

Before initiating the project, we need to reassess the role of the CPM modeling processes. In the preproject phase, the CPM processes, along with other planning disciplines, helped us to develop a viable project schedule. The forward pass functioned as a set-forward scheduler and the backward pass functioned as a set-back scheduler. Together, they established a relationship between when we can accomplish the work of our project versus when we must accomplish it. Through an iterative process, we not only resolved the discrepancies between these two schedules, but we also added some positive time reserve to give us some flexibility with which to respond to unexpected problems that will undoubtedly arise. At this phase of the project, the CPM process proves very valid and useful, primarily in generating an initiation schedule.

Once the project begins with a viable schedule in place, however, what good is a process that is used primarily to develop a schedule? Therefore, if we are going to continue using the CPM process, its role must change.

Further examination of Figure 12-2 begins to reveal a role for this process in an on-going project. Comparing the calculated data to the baseline data gives new, richer information. We can expand our analysis beyond comparing the late dates of each task to its early dates, to also compare both the late dates and the early dates to the baseline dates. The difference between the late dates and early dates provides the current value of PTR at the point of measurement. The difference between late dates and baseline dates provides the original value of PTR at that point. The difference between the early dates and the baseline dates is the span of any slip or pull-in, since the early dates reflect current project progress. These relationships are illustrated in Figure 12-3.

Figure 12-3

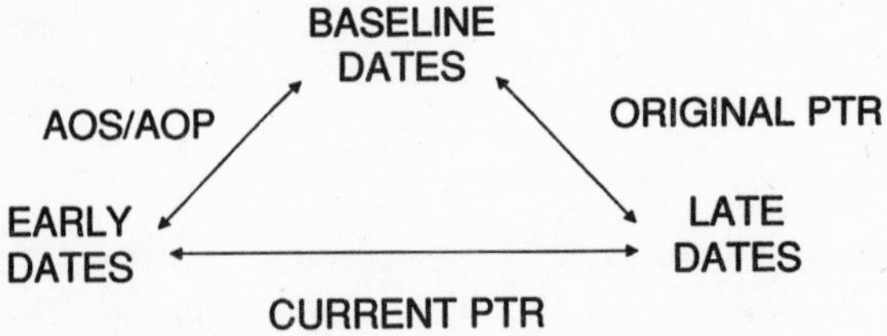

For an example of this concept, look again at Figure 12-2. Task ALPHA started and finished according to the baseline plan, so it did not affect any successor activities or any project time reserves. Task DELTA, however, slipped its finish. The span of the slip is five days (BEF – EF = 75 – 80 = –5). This slip has lengthened the path by five days, reducing PTR by the same amount. The remaining path length is 76 days (March 31 to July 13), and only 83 days remain for its completion (March 31 to July 19). The slip in time reserve from 12 days to 7 reflects our rate of erosion (ROE) of the path's time reserves. Such a loss in reserve (5 days) at the beginning of the path would represent a severe reduction, whereas the same loss toward the end of the path may be acceptable since reserve is available for use over the life of the path. What does it represent in our example?

To detect a problem, we need to compare the rate of time reserve erosion (ROE) to the the rate of progress (ROP) of the project. This relationship opens the door to proactive time analysis, as it allows us to analyze the impact of performance or nonperformance on the successors of the task.

In Figure 12-2, the project has lost 5/12 or 42 percent of its time reserves (5 of the original 12 days) in the first 37 percent of the project. We measure this current rate of project progress by dividing the amount of time elapsed between the project's start date and the status date (44 days) by the current duration of the total path work (120 days): 44/120 = 0.37. Since the rate of PTR erosion roughly matches the rate of project progress, the slip should not cause a serious concern.

We can expand upon this principle by measuring time reserve erosion on several key events in the project as a function of time (e.g., month-to-month). We choose major events because they tend to become path funnels; many paths generally culminate at or pass through them. Time reserve measured at these points (events) reflects the worst-case or least-time-reserve situation of all of the paths that end at or pass through them. By focusing on just a few major events, we can measure the net reserve of many paths, and isolate the project's significant PTR problems.

Figure 12-4

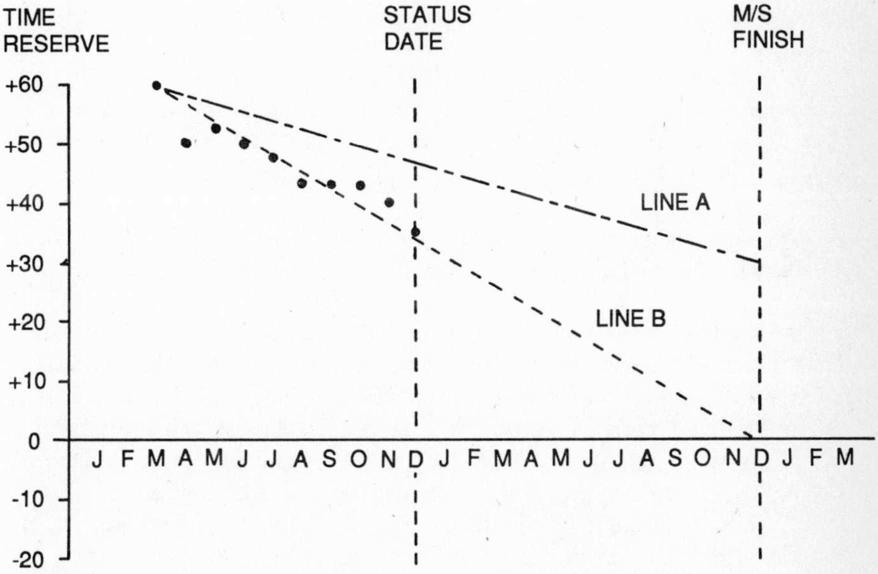

Figure 12-4 illustrates how we would reflect this measurement. This trend analysis measures the time reserve erosion at one event. We started measuring PTR at this event in March 2000, and have recorded the value for each subsequent month through the end of the year, the status date. The event is due to finish at the end of 2001. Initial measurements reflect +60 days of PTR. We certainly do not need to complete the event more than two months (60 working days) early, but since this event is internal to the project, we do need to pass some time reserve on to its successors. We have arbitrarily decided on an amount of 30 days, so we must actually complete the event some time in mid-November (30 working days prior to the end of December). We can, therefore, allow an erosion of time reserve from 60 days to 30 days over the life of the predecessor paths of this event. This acceptable rate of erosion is reflected in line A.

Plotting the values of PTR for the first 10 months gives us 10 points. Entering these coordinates into a least-squares regression function, we can determine an equation of a line that best fits them (line B). Line B lies below line A, which tells us that the actual rate of PTR erosion exceeds the acceptable rate. As of the status date, we have PTR of 35 days when we should have no less than 46 days (shown as the intersection of line A and the status date). Projecting the actual rate of erosion line beyond the status date reflects an ever-increasing gap. At our current rate of erosion, we will just make the schedule date of the event, but its successors will receive no time reserves.

This example project has a problem. Time reserve erosion due to slipped tasks threatens schedule objectives more than a year downstream. It does not threaten our immediate objective date (Dec. 31, 2001), but it does threaten the successors of this event. Our analysis enables us to project this far into the future.

It is a valid analysis because Path Time Reserve (PTR) is always the difference between the *remaining* time available (TA) and the *remaining* cumulative sequence of work (CSP). PTR shows the likely condition of the residual project, so it allows us to base our management decisions on the downstream condition—that portion of the project yet to be accomplished. The projected measurement of erosion warns us far in advance, during which time we can respond to the trend, perhaps recovering to an acceptable rate of PTR erosion.

Tracking other milestones' PTR may reveal greater or lesser problems. This allows us to rank one problem against another and decide priorities of our responses. Those with the most significant erosion relative to progression are of the highest priority. A detailed analysis must accompany each evaluation to identify the particular paths that cause or contribute to each problem. This helps us focus on the specific problems.

We now have a very powerful conceptual tool to help manage our projects. It is certainly not perfect or infallible, but it is a real-time, proactive methodology that can support rational decisions in practical implementation.

STATUS IN THE MODEL

Chapter 9 discussed the various methods for assessing the status of project models at great length. The many variations make this process seem very complex and convoluted. In the context of practical applications, we can simplify all of these concepts to some degree. First, let us review the list of the various mechanisms that show the model's status.

Actual Start (AES). The date on which work on a task began.

Actual Finish (AEF). The date on which a task finished.

Status Date (SD). The date on which status was assessed. It is used in the forward pass calculations with percentage complete and/or remaining duration.

Percentage Complete (PC). An estimated percentage of the amount of an on-going task that has been completed as of the status date. This allows computation of the length of the remaining work.

Remaining Duration (RDU). An estimate of the remaining time span between the status date and the finish of an on-going task.

Projected Start (PES). An estimate of the date on which a future task will start. It functions like a start-no-earlier-than (SNE) date in forward pass Early Start calculations.

Projected Finish (PEF). An estimate of the date on which a future task or an on-going task will finish. It functions like a finish-no-earlier-than (FNE) date in forward pass Early Finish calculations.

Start-No-Earlier-than (SNE). A date that acts on a node like a predecessor of its start to affect its Early Start (ES) date calculations.

Finish-No-Earlier-than (FNE). A date that acts on a node like a predecessor of its finish to affect its Early Finish (EF) date calculations.

If we tried to maintain status by all of these methods, we would have to manage as many as eight fields of data for each activity. Yet, only two of these fields give actual status information; the other six indicate projected or on-going task status. The measurement of projected status introduces the excessive complication. There is no reason to work through a variety of methods to accomplish the same thing. For simplicity's sake and to standardize our process, we will eliminate some of these fields.

We will use actual start and finish dates as they are, since they are already standard and simple. We need only update two date fields (AES and AEF) with actual calendar dates. The actual start (AES) date then becomes the Early Start

(ES) of the node regardless of any other predecessors (either SNE dates or constraints). The actual finish (AEF) date becomes the Early Finish (EF) of the node regardless of any other predecessors. We will adjust task durations to equal the difference between the task's actual start and its actual finish on the appropriate calendar. This keeps the late date calculations in agreement with the actual condition.

In three situations, data on completed tasks creates a contradiction or update-out-of-sequence (UOS) condition:

1. A task with an AEF value, but no AES value has an out-of-sequence error of the third type (see Chapter 11).
2. A task with an actual start date later than its actual finish date has an out-of-sequence error of the fourth type (see Chapter 11).
3. Any task with either an actual start (AES) or actual finish (AEF) date later than the status date (SD) has an out-of-sequence error.

Each of these should be identified and flagged along with other UOS problems in a data validation step prior to CPM processing.

An on-going task has both an actual date and a projected date. Tasks that have not started may have either a projected start and/or finish. We assess the status of the completed portion from its actual start date, just like any other actual start date. In assessing the projected status, however, we can simplify our previously discussed process, in effect making it identical to the actual date process in that it will require only the projected start (PES) and finish (PEF) dates. Rather than deriving a task's projected finish date from a new span (DU), a percentage complete (PC) value, or a remaining duration (RDU) value, we will provide this information directly to the process. PC or RDU values give us nothing that projected dates do not, and the projected dates require no global point of reference (the status date).

We will use projected start and finish dates in lieu of other methods for another primary reason, though. Projected dates represent specific performance commitments by project personnel, and not some value derived from a computerized modeling process. Therefore, they inspire a sense of ownership in personnel who are accountable for their accomplishment.

Projected start and finish dates bring another advantage since they eliminate the need for separate fields for start-no-earlier-than (SNE) and finish-no-earlier-than (FNE) dates. SNE and FNE fields serve exactly the same purpose as PES and PEF fields, making them redundant. We will still assign date successors or FNL dates as late date targets, as these dates should not be confused with FNE or PEF dates.

In total we have cut our requirements for status fields in half, from eight to four fields. We have also eliminated the need for separate SNE and FNE fields. In addition, we record actual and projected status by a consistent method, using dates in both cases.

This method also removes the influence of the status date (SD) on EF date calculations. It still serves as a point of reference, however, and for detecting and flagging certain update-out-of-sequence conditions.

Users should retain percentage complete and remaining duration processes in a utility of the software. They can still yield benefits in off-line analysis.

Now that we have settled on our method for assessing projected status, we can establish its rules and disciplines. These disciplines govern our estimates of the finishes of on-going tasks or the start and finish of any other task that begins after the status date.

Projected Start Date Rules

1. Every start node must have a projected start (PES), actual start (AES) date or SSTR (start specified time reserve). Data validation will report any start node that does not have one of these three values as a hanging start (SNF = HS). If both dates are present on a node, the actual start date takes precedence.
2. For internal nodes (i.e., nonstart nodes) we model date predecessors of their starts with PES dates.
3. Anytime a node has a projected start that differs from its baseline start (BES), this date should be entered into the node's PES field.
4. Any task with a baseline start (BES) date prior to the status date (SD) that lacks an actual start (AES) date must show a PES date later than the status date. Data validation will flag an update-out-of-sequence condition on any node that has not started (shows no AES) and either lacks a PES or shows a PES date earlier than the status date.
5. The modeling process treats PES dates as start-no-earlier-than (SNE) dates. They directly affect the calculation of the node's ES date, in accordance with the latest early date rule.
6. If the process finds a PES date on a node, but no update for the node's finish, it does not recalculate the node's duration. Instead, it adds the original duration to the resultant Early Start date to find the Early Finish date. If, therefore, the projected start date of a task slips and we wish to retain the original finish date, we must enter it as a projected finish (PEF) date on the task to force a new duration calculation, as explained further below.

Projected Finish Date Rules

1. Anytime a node has a projected finish date that differs from its baseline finish date (BEF), this date should be entered into the node's PEF field.
2. If the process finds both a PEF date and an update for the start of the node (AES or PES), it recalculates the task's duration on the appropriate calendar by one of the following equations:

   a. If the node has an AES: DU = PEF – AES + 1.

   b. If the node lacks an AES, but has a PES: DU = PEF – PES + 1

   This changes the task's duration for both early and late date calculations.
3. Any task with a baseline finish (BEF) date prior to the status date (SD) that does not have an actual finish (AEF) date must show a PEF date later than the status date. Data validation will report a UOS condition for any task that has not actually finished and either lacks a PEF date or shows a PEF date earlier than the status date.
4. The modeling process treats PEF dates as finish-no-earlier-than (FNE) dates. Recalculation of the duration by adding the Early Start date to the duration (and subtracting one day) could very well corroborate the EF. This depends on what determines the task's ES date.

An example will help illustrate how the status assessment process (either actual or projected) relates to the modeling process. Figure 12-5 shows four nodes, three of which contain status information: one task is completed, one is on-going, and one has not yet begun. We are taking status as of April 22 (the status date).

Task ALPHA, the completed task, shows status information including an actual start date (AES = Feb. 1) and an actual finish date (AEF = March 18). These dates become its Early Start and Early Finish, respectively. The process first recalculates the duration of ALPHA:

DU = AEF – AES + 1
= day 55 (March 18) – day 21 (Feb. 1) + 1 = 35 days

This way the duration of the task is compatible with the actual status and this allows for the correct backward pass calculations.

Task DELTA, the on-going task, shows an actual start date (AES = March 28) and a projected finish date (PEF = May 9). The AES date becomes the task's Early Start, while the PEF date, as the latest predecessor of the task's finish, becomes the Early Finish. The process first recalculates the duration:

Figure 12-5

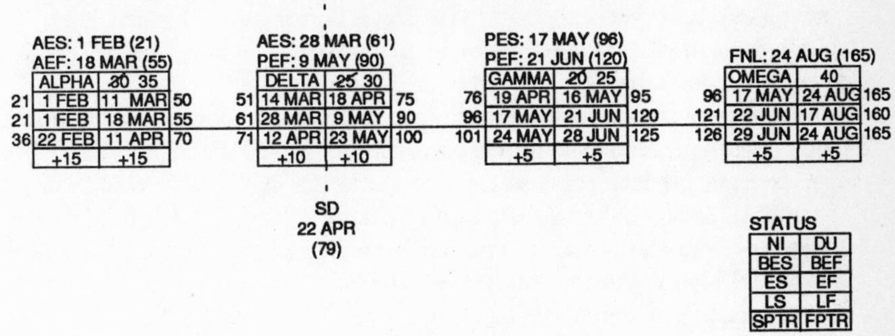

$$DU = PEF - AES + 1$$
$$= \text{day } 90 \text{ (May 9)} - \text{day } 61 \text{ (March 28)} + 1 = 30 \text{ days}$$

The task's span added to its ES and the finish date predecessor (PEF) yield the same Early Finish date. Since the Early Start date of an on-going task is always an actual date, this equation always yields the EF value, except when another predecessor of the node's finish (an FF constraint) provides a later date. The new duration keeps the backward pass calculations consistent with the status information.

Task GAMMA shows only projected status information: a projected start date (PES = May 17) and a projected finish date (PEF = June 21). These dates become the task's Early Start and Early Finish dates in the forward pass because they are the latest predecessors of its start and finish, respectively. The process recalculates the duration from the projected dates.

$$DU = PEF - PES + 1$$
$$= \text{day } 120 \text{ (June 21)} - \text{day } 96 \text{ (May 17)} + 1 = 25 \text{ days}$$

This duration keeps backward pass calculations consistent with the current status information. In this example, the projected dates all become the node's early dates, because they are the latest predecessors. Later predecessors might have overridden the ES or EF of GAMMA.

THE SCHEDULE-MODEL RELATIONSHIP

Figure 12-5 not only illustrates the various ways to assess status, but also one of the primary conflicts between modeling and project control. The Early Start and Early Finish dates of tasks ALPHA, DELTA, and GAMMA reflect the status of these tasks. The Early Start and Early Finish of task OMEGA (and any of its successors), however, do not provide status information. Instead, they reflect the *potential* impact of the slips of its predecessors, as any changes in the early dates of a node's predecessors may affect its early dates.

To evaluate the status assessment process as a project control tool, we must ask a fundamental question: Should we consider the rippled dates derived by the forward pass from predecessors' status information as new status information, or as potential impacts on the downstream schedule? Clearly, status information and estimates of potential impact are not the same thing. Status information provides the model alterations to simulate accomplishment or expected progress. Rippled early dates provide an estimate of potential impacts resulting from the status. This is the proper understanding of this disparity.

> We should not consider schedule slips or pull-ins derived from CPM calculations as status until the project managers have done all they can to recover any adverse schedule impacts.

The process of identifying problems and then devising potential resolutions takes time. During this period, *the project schedule should not change.* To remain useful, a project schedule must maintain a consistent and stable character. It should not bounce back and forth in response to our efforts to resolve problems. Only after the project has exhausted all opportunities for problem resolution should its schedule reflect any residual impacts.

If we try to schedule and track task performance by the early dates, the schedule changes with every impact. Yet for the modeling processes to work properly, they must reflect status information and its effects on successors. This incompatibility between stable schedule tracking and modeling dynamic processes creates the primary obstacle to successful modeling processes for ongoing projects.

To overcome this obstacle, we must understand that the forward pass is not the proper mechanism for schedule tracking and control. It helps us to analyze potential impact, but we must find some other means to maintain schedules. Therefore, we must separate the schedule tracking and control data from the early date fields. This separation cannot be complete, as these two functions remain closely related and each affects the other.

We must constantly update the forward pass with current schedule data in response to status and change. For this reason, we need to maintain all schedule and analytical data in a single modelfile. Since the two primary functions of schedule data are to track and report progress and update the model, we can simply append this information to each node.

We need to add no new fields because the method we have chosen to assess status also satisfies tracking and control requirements. The six status and baseline fields (BES, BEF, AES, AEF, PES, and PEF) provide all of the current schedule information for tasks and events. The baseline dates for every activity also specify the current schedule start and finish of each activity unless the other status fields (AES, AEF, PES, and PEF) reflect other values, in which case these become the most current schedule. A task's current schedule follows a progression. It is the actual dates (AES or AEF) if they are present, then it is the projected dates (PES or PEF) if they are present, and finally the baseline dates (BES or BEF) when neither actual nor projected dates are present. Altogether, this cumulative date data (baseline, projected, and actual dates) represents the current schedule. It provides all the information we need for schedule tracking, reporting, and subsequent program control. Chapter 15 will examine how to display this information effectively.

The status-to-model relationships have us simulate the project's current schedule in the activity (node) parameters, from which the forward pass process derives early dates. These early dates reflect:

1. Actual dates for any task that has either started or finished.
2. Any projected dates that are not overridden by another predecessor (by the latest early date rule).
3. The potential impacts on a task as the result of status assessment or other changes to its predecessors.

In an on-going project, the early dates reflect the current schedule and any rippling resulting from this. We can measure the potential impact in two ways. First, we can determine the impact on the schedule by comparing the early dates of a task to its current schedule dates. If a task's early dates precede the current schedule dates, we could pull the task in. If the task's early dates are later than the current schedule dates, the task could slip. For example, the start and finish of Task OMEGA from Figure 12-5 have slipped 25 days due to the cumulative slip of its predecessors: BES – ES = 96 – 121 = –25, yet the current schedule data of this task reflect no such impact.

We can measure the second, and more important, form of impact on time reserves by examining the PTR values of each node before and after the assessment of status. For example, the slips of the predecessors of task OMEGA eroded the PTR from +30 (baseline condition) to +5, an 83-percent reduction (25/30 = .83). Comparing this PTR loss to the project's progress gives a rate of erosion. The 83-percent reduction has occurred after completion of only the first 45 percent of the project (59/130 = .45). Losing 83 percent of the path's time reserves in accomplishing only 45 percent of the work indicates an excessive rate of erosion. If this trend continues, the project will miss at least one schedule objective. It is in trouble, even though it currently has positive time reserves.

Comparisons like these provide us valuable information about potential impacts on task completion in the form of individual activity schedule changes, and on the overall project in the form of PTR erosion. By these techniques, we can measure the effects of status and change on the balance of the project. The modeling processes serve a different purpose in the on-going project environment than they served before the project started, even though the fundamental processes did not change.

We have added a fourth function, current schedule, to the comparison and reconciliation process. This complicates the methodology somewhat, but it allows us to satisfy the requirements of both schedule control and proactive modeling analysis in a single modelfile. It provides the methodology we need to put the CPM processes to work on an on-going project.

Figure 12-6 illustrates the scheme of comparisons that we must make for each task in the project to conduct a thorough analysis. At the top of the figure, the relationship between the baseline dates and the current schedule reflects the amount of slip or pull-in of the task, revealing schedule performance. On the left side of the diagram, the relation between the baseline schedule and the late dates provides the baseline time reserve, while the comparison along the bottom of the diagram, late dates to early dates, provides the current time reserve.

Figure 12-6

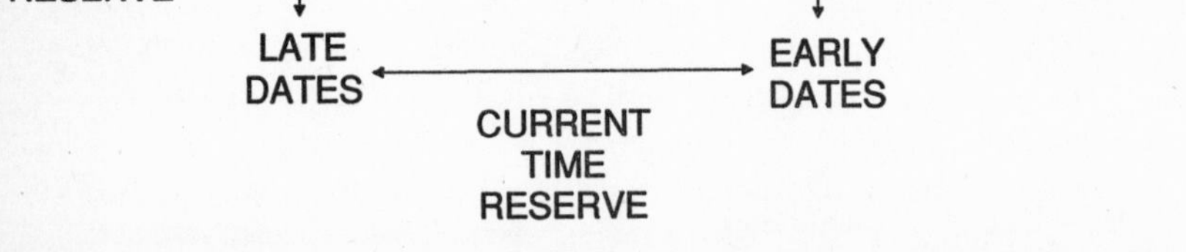

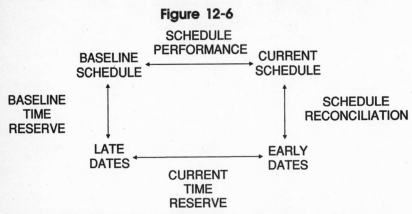

The right side of the diagram reflects the comparison of the current schedule of the task to its model schedule. If the node shows projected dates for the task, the early dates will either confirm them or reflect a conflict. For activities that indicate no current status assessment, any disparity in this comparison reflects the potential impact of the status of its predecessors. It is also possible to have projected status on a task and still have it impacted by the status of other predecessors.

Understanding these relationships and controlling them will allow us, in turn, to exercise considerable control over the destiny of our project.

EXAMPLES OF RECONCILIATION

To more thoroughly understand the information we can obtain from comparing one set of parameters to another, we need to examine each relationship in detail.

Schedule Performance Comparisons

Comparing the baseline schedule to the current schedule provides the schedule performance measurement.

Current Schedule Date Earlier than the Baseline Date. This indicates that the projected or actual schedule has been pulled-in by the difference between these two dates (the SOP).

Current Schedule Date Later than the Baseline Date. This indicates that the projected or actual schedule has slipped by the difference between these two dates (the SOS).

Analytical Comparisons

Comparing early dates to late dates provides the current PTR value at each event. A negative value indicates that the project is in jeopardy of missing at least one schedule objective. A positive value indicates that the early date schedule will meet all schedule requirements. Even positive PTR values can indicate problems if they are diminishing faster than the project is progressing.

Schedule-to Model Reconciliation

Comparing late dates to the baseline schedule gives the baseline PTR value at each event. This is not necessarily the original PTR value. It may reflect changes since the original date in either the successors' cumulative path lengths or the time available to accomplish these paths or both. The backward pass and any subsequent PTR calculations reveal the effects of these changes.

Comparing the current time reserve to the baseline time reserve yields the effective amount of time reserve erosion. This is the amount of erosion of the current path lengths and date parameters relative to the baseline condition. Do not confuse this with the PTR tracking process, referred to earlier.

Comparing early dates to status information establishes the schedule-to-model relationship. The dates on the current schedule (the status) should match the early dates. As change and status perturb the model, however, this comparison will reflect conflicts between the schedule and the model. If the model date precedes the current schedule (baseline, projected, actual), then the task is being delayed by something other than its predecessors. The model indicates that no predecessor prevents performance of the task earlier than currently scheduled. This results in a delay with consequent time reserve erosion. The difference between the early dates and the current schedule dates tells us how much time reserve the project will lose if the current schedule holds. Hopefully this outside factor can be corrected and the task's schedule pulled-in to match the model, preserving time reserves.

A model date later than the current schedule date indicates one of two things. Either the model conflicts with the ability to perform the task as early as currently projected because a predecessor(s) will prevent it, or the activity will be potentially slipped as a result of slips or changes to its predecessor(s). In either case, we must adjust the work to change the terms of the time reserve equation (PTR = TA – CSP) to alleviate any problems in the downstream schedules. As we resolve our problems, the model's early date will shift back to the left and realign with the current schedule. We must make constant adjustments to the project to keep the early side of the model aligned with the schedule.

These are the general guidelines that structure the analysis processes. We can now look at some examples to see how the schedule, analytical, and reconciliation processes work in practice. Figure 12-7 shows two tasks: OUT and BAK. Task OUT is on-going and Task BAK is awaiting the completion of OUT and another predecessor so it can begin. OUT started according to the original plan (AES – BES = 108 – 108 = 0), but it took 20 days longer to complete than expected (PEF – BEF = 147 – 127 = 20). Thus, the task's duration increased by 20 days, which becomes the SOS. The cumulative span of the path or paths on which OUT lies has also increased by the span of the slip. Path Time Reserve has eroded from 50 days (LF – BEF = 177 – 127 = 50) to 30 days (LF – EF = 177 – 147 = 30) at this point. The current schedule does reconcile with the early dates (AES – ES = 108 – 108 = 0 and PEF – EF = 147 – 147 = 0).

Task BAK, a successor of OUT, could suffer due to this slip. When we reconcile Task BAK's current schedule to its early dates, we find that its start (ES) and finish (EF) dates do show an impact, but only 10 days (ES – PES = 158 – 148 = 10 and EF – BEF = 182 – 172 = 10). This 10 days matches the amount of time reserve erosion at this point [(178 – 148) – (178 – 158) = 10]. Upon examination, we can see that BAK suffers only the 10-day delay because of the effects of its other, unseen predecessor.

This illustrates that slips do not always delay the successors of the slipped task by the amount of the slip. Successor tasks can have many other predecessors, including date predecessors (PES dates), that could override the effects of any one slip. Only after we determine the latest early date at each node end

Figure 12-7

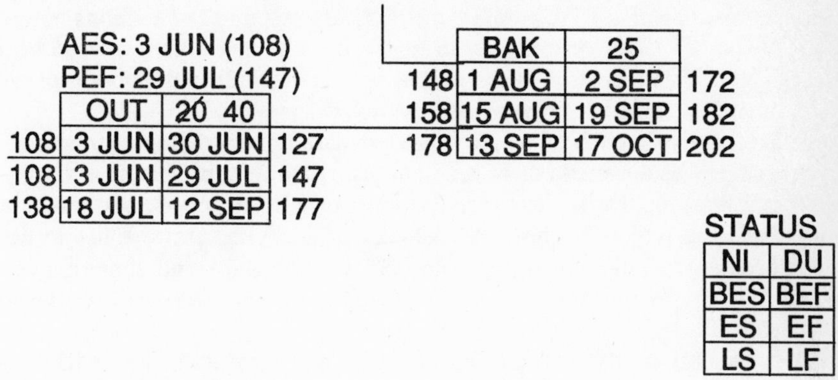

(event) in the forward pass can we see the actual effects of a pull-in, slip, or other change. In this case, our analysis determines that the 20-day slip of OUT will delay its successor by 10 days. Since this is a severe reduction in PTR (33%), we should look for ways to avoid or mitigate this impact.

Figure 12-8 further explores this process. The finish of Task 3B has slipped 30 days (March 25 to May 9). The effects of this slip diminish as we proceed down the path to successors, though. We see this by comparing each task's Early Start and Early Finish date to its current schedule start and finish dates. The start and finish of Task SS slip by only 20 days; the start and finish of Task 2B slip by only 10 days; the start and finish of Task 1B do not slip at all. Cumulative effects from the other predecessors of these tasks make up for the slip of 3B by the start of 1B.

Figure 12-8

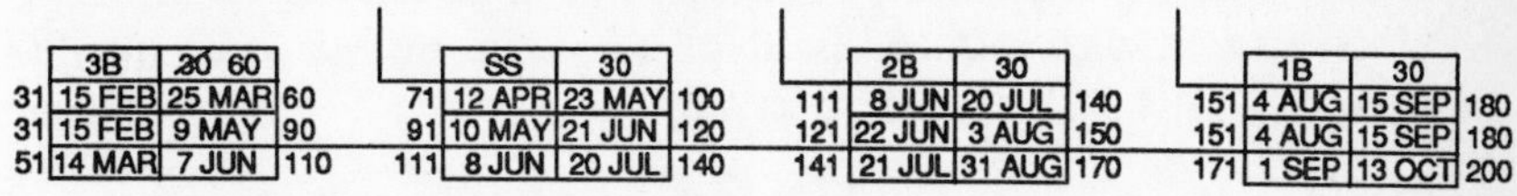

Once we understand the techniques of isolating problems by comparing and reconciling parameters, we must learn how to avoid or mitigate any adverse impacts. This process is not new, however. As we measure impacts by their effects on Path Time Reserves, we can look to the basic time reserve equation for the solutions:

$$PTR = TA - CSP$$

We can either increase the time available (TA) to a path or decrease the cumulative length of a path (CSP) to recover the downstream schedule, just as we can recover negative PTR by these methods. By reacting to situations before the PTR becomes small or even negative, we have a longer path in which to make changes. This increases our available options, including the option of allowing some PTR erosion, provided we control its rate.

This proactive problem resolution process based on modeling analysis significantly improves our methods of problem resolution and mitigation. By monitoring time reserves, we see trouble, not when or after it occurs, but before it occurs. Once the project begins, we maintain the modeling process not to develop schedules, but instead to assess the effects of change and status on our project.

> The model of the on-going project helps us identify and measure *potential* downstream problems (schedule and PTR) caused by slips, pull-ins and other changes, and to test the effectiveness of potential solutions.

We will find that the savings from these proactive decision processes more than offset the time and expense of the modeling effort.

ANOMALY STATUS CONDITIONS

Now that we have derived our working methodology, we can further examine the effects of simulating status in our model. Figure 12-9 shows two tasks, the first (LEFT) with an actual start date (AES), and the second (RITE) with a projected start date (PES). The absence of projected finish dates in both cases forces us to answer a question before we can simulate status in the model. From a schedule perspective, task LEFT is currently scheduled to start on February 15 (its AES date) and it is scheduled to finish on March 11 (its BEF date). Likewise, the current scheduled start and finish of task RITE are March 21 (its PES date) and April 25 (its BEF date), respectively.

Figure 12-9

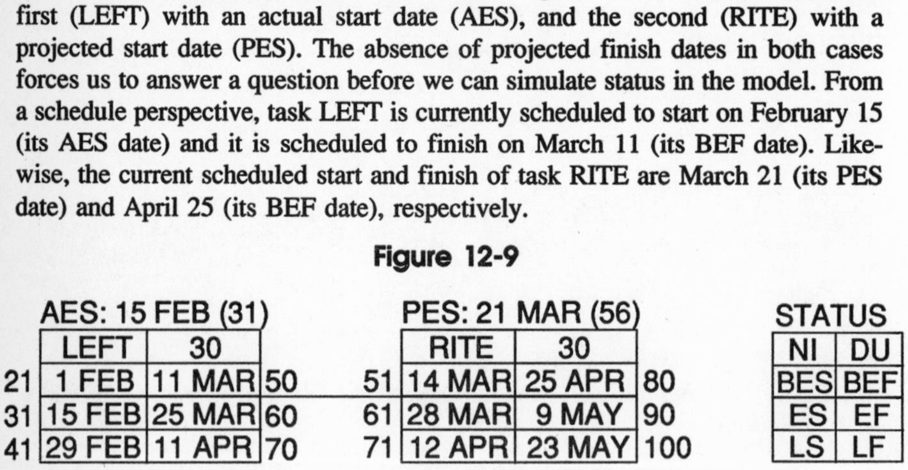

But does the model process yield the same dates? Should we maintain a task's baseline finish date as its Early Finish date, or should we maintain its duration constant and recalculate a new EF date? Neither option works in all cases. Sometimes we will want to hold the original finish date; sometimes we will want to maintain the duration to determine a new finish date. Therefore, we need the ability to do either. The process that preserves these options requires that we specify a new projected finish date (PEF) or an actual finish date (AEF) for the task before recalculating a new duration. Then we control when a date is used and when the duration is used.

When we set no projected finish date on a task, its original duration will enter into its Early Finish date calculation. To hold the task's baseline finish date (or any other date for that matter), we simply enter that date in the projected finish (PEF) date field. This may seem to contradict the principle that a task's baseline dates function is its schedule unless changed during status assessment, but this is only true for the schedule. This rule is in conflict with the principle of modeling philosophy that tasks take time to accomplish (they have durations). Therefore, as the task's schedule changes, we must overtly control and alter the duration of each task.

Figure 12-9 did not specify either task's projected finish date, so the process did not recalculate their durations. The 10-day slip of the start of Task LEFT delayed its finish by the same amount (EF – BEF = 60 – 50 = 10). In the calculation of the Early Start of RITE, the slip of LEFT's finish overrode RITE's projected start date (pushing it forward from 56 to 61). The slip of the start of LEFT potentially delays both its finish and its successor's start and finish. This also cuts time reserve in half, from 20 (LS – BES = 71 – 51 = 20) to 10 (LS – ES = 71 – 61 = 10).

Figure 12-10

AES: 15 FEB (31)
PEF: 11 MAR (50)

	LEFT	~~30~~ 20	
21	1 FEB	11 MAR	50
31	15 FEB	11 MAR	50
56	21 MAR	18 APR	75

PES: 21 MAR (56)
PEF: 25 APR (80)

	RITE	~~30~~ 25	
51	14 MAR	25 APR	80
56	21 MAR	25 APR	80
76	19 APR	23 MAY	100

To hold the original or baseline finish dates of these tasks in the model, we need to specify these dates as projected finish dates (PEF), as in Figure 12-10. Inserting these baseline dates as projected finish dates causes the process to recalculate the duration of each task, affecting both the forward pass (early dates) and the backward pass (late dates). This causes both tasks to absorb the slip of the start. Most of us hold this kind of optimistic myth, that even though we start a task late, we can still complete it on time. Many projects have marched behind such a fallacious banner into deep trouble.

Hopefully, our status procedure will keep us aware of this problem. At least it will show us the extent of our handicap by divulging how much duration we have conceded (i.e., 10 days in LEFT, 5 days in RITE).

To summarize, if a task shows no projected finish (PEF) date or actual finish (AEF) date, the process will not recalculate its duration. Baseline finish (BEF) dates do not function as finish-no-earlier-than (FNE) dates in the modeling process.

If we reverse the situation, assuming tasks have projected finish dates, but no projected or actual start dates, we reach a similar conclusion. As Figure 12-11 shows, the projected finish of Task KNACK is specified, but not its start. The scheduled start date of KNACK is March 14, the baseline start date. Is this also the task's Early Start date?

Figure 12-11

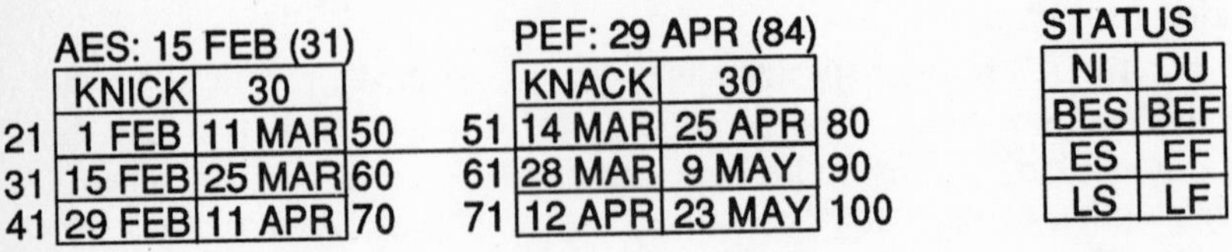

The process does not recalculate duration for a node with a PEF date, but no status information about its start. The projected finish date functions as a finish-no-earlier-than (FNE) date, but the duration of the task will not corroborate this.

In the model, the baseline start date does not function as a start-no-earlier-than (SNE) date. Therefore, the Early Start of the task will depend solely on its true predecessors and the latest early date rule. The forward pass will also determine the Early Finish date of the task. Predecessors of the finish of KNACK include its start and its projected finish date. Its Early Finish date will therefore be the later of:

$$EF = ES + DU - 1 = 61 + 30 - 1 = 90$$

or

$$EF = PEF = \text{April } 29 = 84$$

The 90th workday is the later of the two and becomes the Early Finish date of KNACK. In our reconciliation, we should note that the task's projected finish conflicts with the interfaces of the model.

In order to recalculate the task's duration and for the baseline start (BES) date to function as a start-no-earlier-than (SNE) date, we must enter it as projected start (PES) date. This would enter it directly into the Early Start (ES) date calculation, and through this into the duration recalculation. Even this will not guarantee that the ES will match either the BES or the PES. The latest early date rule may intervene. To illustrate this, let us enter the baseline start (BES) date of Task KNACK (March 14 or day 51) as its projected start (PES) date and examine the results. First, the process would recalculate the task's duration:

$$DU = PEF - PES + 1$$
$$= \text{day } 84 \text{ (April 29)} - \text{day } 51 \text{ (March 14)} + 1 = 34 \text{ days}$$

The forward pass would give an Early Start date for the node (its latest predecessor) of:

$$ES = EF \text{ of KNICK} + Lag + 1 = 60 + 0 + 1 = 61$$

or

$$ES = PES = \text{March } 14 = 51$$

The 61st workday, the latest predecessor, overrides the projected start date. Status remains inconsistent with the model.

The forward pass calculates KNACK's Early Finish date as the result of the latest predecessor, as well. In this case, the duration of the task has been changed to agree with status information:

$$EF = ES + DU - 1 = 61 + 34 - 1 = 94$$

or

$$EF = PEF = \text{April } 29 = 84$$

The latest date, day 94, is the Early Finish date of KNACK. Again, status conflicts with the model due to the impact of KNICK on KNACK's Early Start date, shifting it from March 14 (day 51) to March 28 (day 61).

To summarize, as for finish dates, if a task lists no projected start (PES) or actual start (AES) date, the process will not recalculate its duration. Baseline start (BES) dates do not function as start-no-earlier-than (SNE) dates.

Figure 12-12 shows a projected start date equal to the task's BES date entered on the successor (NITE). Of all predecessors of NITE, its PES is the latest early date and becomes its Early Start date.

Figure 12-12

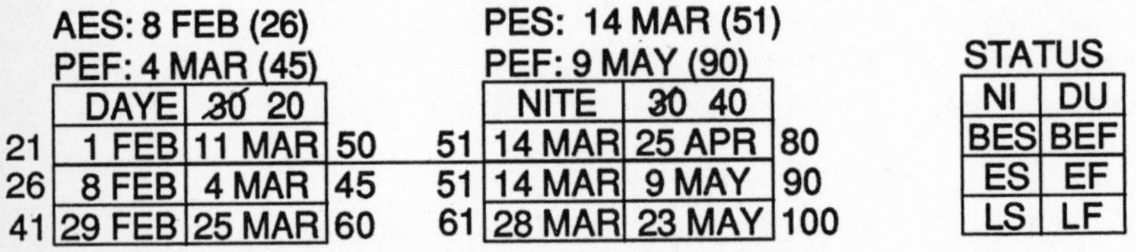

The process will recalculate the task's duration in response to status information on both the start and the finish:

$$DU = PEF - PES + 1 = 90 - 51 + 1 = 40$$

Since the task's duration is the difference between the two projected dates, and the Early Start date of the task is the projected start (PES) date, then the calculated Early Finish date of the task will equal the projected finish (PEF) date.

$$EF = ES + DU - 1 = 51 + 40 - 1 = 90$$

or

$$EF = PEF = \text{May } 9 = 90$$

Because the task's Early Start date matches its projected start date (PES), the calculation of the task's Early Finish simply rearranges the duration equation (where ES = PES).

This series of examples shows that, when we assess status, we should always measure and report the current condition of both the start and the finish. This keeps status information in the form of a pair of dates (AES and AEF, AES and PEF, or PES and PEF). In doing this we are forced to reconsider the task's span every time we update it, because the process will alter it to correlate to the status information. It also injects status dates directly into the early date calculations. The only exception is when we elect to hold the task's duration constant and recalculate the finish date based on a start date and the duration.

These may seem like mere details, but we cannot overlook them. A great many parameters are required to clearly demonstrate the current condition (status and change) of the project model. It is all of these details that actually allow us to harness these processes and make them project management tools.

WHAT-IF ANALYSIS

What-if analysis is a contingency exercise by which to test the results of a potential scenario. What if a critical supplier were to go on strike? What if a product were to fail a crucial test? What if the engineering release were 90 days late? What if our funding were cut 30 percent or delayed six months?

The proactive project manager constantly asks these what-if questions to try to anticipate the many contingencies that may arise during the course of a project. Many of these projections never come to pass, but we can never be sure just what will happen in the future. In this uncertainty, we must concern ourselves with any potential as well as the actual threats to the outcome of our project. Our management philosophy dictates that we plan for these possibilities in order to avoid the worst of them, or at least mitigate their consequences. This is one more area where the modeling processes can be extremely useful.

If we examine a what-if question in the context of a model, we can view the worst tragedy as no more than a change that has not yet happened. We evaluate a what-if question in virtually the same way as we accommodate change. Like everything else, however, we must maintain discipline in order to ensure our results are valid.

Since a what-if question supposes a potential situation, we do not wish to alter our working project. It must continue to facilitate the chores of real status and change analysis. If we alter the model to simulate a contingency condition, we cannot be sure that we have returned it to its original state at the end of the exercise. Therefore, we always perform the what-if exercise on a separate copy of our project model. Most software applications provide a utility to easily clone data for such a purpose. We can change this copy as much as necessary, and simply dispose of it when we are finished. We need not try to return it to its previous condition.

The discipline of what-if analysis also requires that we document our process thoroughly. A what-if exercise run in January could easily yield very different results from the same exercise run six months later. Thus, we must maintain a record of the configuration (status and baseline) of the model for a specific exercise (e.g., "What-if analysis run against project model status as of June 3 and baselined from the March 16 replan"). Also, we must document the contingency's conditions and assumptions, which are necessary to define the specific alterations of the evaluation. Finally we need to define and document the specific alterations to the modelfile. This includes addition of new activities, changes to existing activities, removal of activities, and any alterations, additions, or deletions to the constraints.

With these disciplines in place, we make the alterations, additions, and deletions required to model the contingency in the appropriate copy of the project modelfile and run the CPM processes. We can then assess the impacts of these changes by comparing the newly calculated information (early dates, late dates, and time reserves) to the current model. This resembles Figure 12-6 with an additional set of calculated dates. From the differences, we can find and understand the impacts of the changes by analysis like that discussed throughout this chapter. Then we can make a decision as to our course of action or contingent course of action.

In one final step, we should package all of the information from this process, log it, and store it. In this way, we maintain a history of our contingency analysis, much of which we may find useful at a later point in the project.

SUMMARY

The Preproject Schedule-Model Relationship

In the preproject phase, the modeling processes serve to help us generate a viable project schedule. This process involves several methods and techniques. For instance, we must decide whether to use top-down or bottom-up planning. The principal advantage of the top-down approach is its development of the detail tasks from the overall project perspective, ensuring consideration of all objectives and the work required to meet them. Its principal disadvantage is that the durations of the detail tasks is determined from the overall project span, without regard for the nature of the work.

The principal advantage of the bottom-up approach is its derivation of the durations of tasks from the scope of the work, the available resources, and other pertinent factors by the work performers. Its principal disadvantage is its failure to consider the project's schedule objectives and its lack of a mechanism to ensure that it covers the entire scope of the project.

In practice, we use both the top-down and bottom-up approaches. We decompose structures, objectives, and requirements from the top down. We assess spans, lags, and constraints from the bottom up.

In the model we simulate the bottom-up project data (task durations and interfaces determined by the nature of the work) in the forward pass, and the top-down project data (the objectives imposed by outside requirements) in the backward pass. We then reconcile the resulting early and late dates until we reach a viable, acceptable plan.

We should establish positive time reserve in all paths of our project to provide a margin to accommodate errors and problems. To decide how much reserve is enough, we consider:

1. Total path length—We should try to establish about 10 percent of the path's length as time reserve.
2. Risks, technical complexity, resource concerns, etc.—We should increase the amount of reserve to allow for these factors.

The working level of the model should be determined by significant project interfaces. The following guidelines lead us to this level:

1. Define the interfaces that the project needs to manage and control, then develop the activities from these boundaries.
2. Define each working-level, or detail, task with a span of about 5 percent of the total project length.
3. Decompose near-term tasks to a greater level of detail than those in the far term.

Once we develop the initial plan, we define the project's baseline schedule from late dates for major project milestones and early dates for all tasks and lower-level milestones.

Modeling in the On-Going Project

The role of the model changes after the project begins. In the on-going project, the modeling processes serve to analyze the potential impact of status and change. Potential impact can be measured as:

1. Delays in successor tasks reflected in their early dates (ES and EF) compared to their current schedules.
2. The erosion of Path Time Reserves. This can be measured at every event (node end) or just at selected major events (milestones).

Status information appears in the model as actual (AES or AEF) or projected (PES or PEF) dates. Whenever we update the status of a task, we should update both its start and finish through changes in these fields. We impose actual dates (AES and AEF) on the forward pass calculations, whereas projected dates (PES and PEF) function as date predecessors.

We measure current schedule status based on a progression of dates. As the current schedule of an activity, we look first to the actual date. If we find none, we look to the projected date. If we find neither actual nor projected dates, we look finally to the baseline date.

The process recalculates the duration of a task if both its start and finish reflect status dates:

With both AES and AEF: DU = AEF – AES + 1
With no AEF, but both PEF and AES: DU = PEF – AES + 1
With no AEF or AES, but both PES and PEF: DU = PEF – PES + 1

If either end of the node reflects no status information, the process does not recalculate duration. We do not use the baseline dates as projected dates in the model. This is a compromise to provide the user the greatest flexibility.

The rippling of a task's Early Start and/or Early Finish date as a result of a status update or a change of its predecessors should not be considered as status information, but rather as a potential impact.

We separate the modelfile into a schedule side (listing baseline, projected, and actual dates) and an analytical side (listing early and late dates). To analyze impact, we compare these dates.

What-if analysis is a contingency exercise that tests the impact of a potential scenario. The principal steps of a what-if exercise include:

1. Making a duplicate copy of the working model on which to perform the analysis.
2. Thoroughly documenting the conditions of the exercise, including:
   A. Modelfile status/baseline configuration.
   B. Ground rules and assumptions.
   C. Specific changes to the modelfile to simulate the contingency.
3. Simulating the contingency on the copy of the model and running it through the CPM processes.
4. Performing the analysis by comparing the calculated data back to the current model.
5. Establishing position and initiating contingency plans.
6. Collecting, logging, and storing all information for each what-if exercise for future reference.

CHAPTER 13

Summarization and Vertical Integration

We will find that because of the ease of computer modeling the size of our database increases very quickly. This can create problems for both schedule control and analysis due to the awkwardness of dealing with a large project all at once; there is just too much data to handle. The sort and selection processes reduce this problem to some degree, but they still handle and display the data at its lowest level.

We need a means of condensing our information, especially to satisfy management reporting requirements. Typically, management is a hierarchy that parallels the functional or organizational structure of their projects. Managers at each level in this pyramid need to keep abreast of the progress of their specific piece of the project. At the lowest levels of the management structure, a segment of the detail data is the right information. Upper-levels managers, on the other hand, need to oversee their wider responsibilities without having to absorb large volumes of detailed data. This need for the appropriate amount of information at each level in the hierarchy is the management information problem.

As an important practical requirement, our modeling processes must include a method for solving this problem. We will call this method *summarization.* It is a process that condenses detail for reporting model information up the management hierarchies.

HAMMOCK STRUCTURES

The summarization process is one where we represent a great amount of detail concisely. We must combine many model elements (activities, events, and constraints) into groups and report net data for each. The first attempt at this process, called *hammock structures,* came out of the Arrow methodologies. Recently this method has been modified to work in the Precedence scheme.

To understand this method, let us look at an example. Figure 13-1 depicts a short series of tasks: ALPHA through OMEGA. We can represent these four tasks as one task by defining a summary node that begins with the start of task ALPHA and ends with the finish of task OMEGA. To do this, we create summary node ALEGA, which has special constraint relationships with task ALPHA and task OMEGA. We call this combination of two relationships and a special node a *hammock* because, in effect, it spans a series of tasks from distinct points at each end.

Figure 13-1

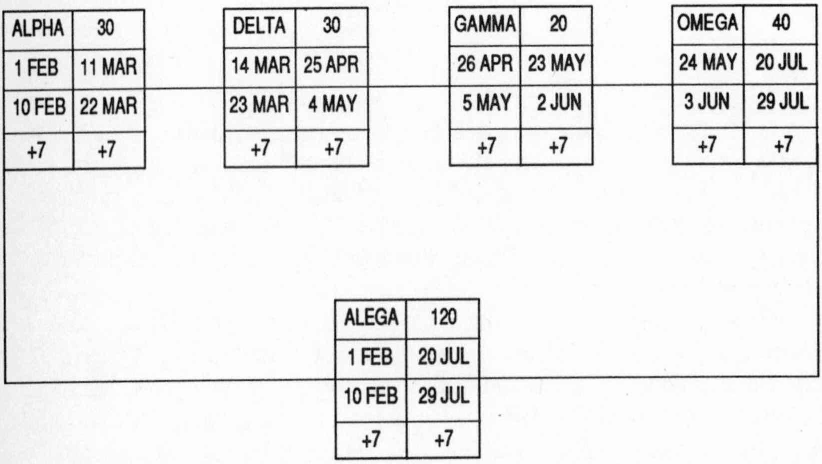

The hammock node resembles other activity records in some ways: it has a unique identifier, a calendar, and a description. The sort and selection structures discussed in Chapter 11 can be appended to it like any other node.

However, the hammock node must also be flagged as such to identify its special qualities. For instance, it must identify the nodes that define its start and finish. These two end nodes provide all the hammock node's date data through a process. All of its start dates (Early Start, Late Start, baseline start, actual start, and projected start) come directly from the same fields of the designated start node. All of its finish dates (Early Finish, Late Finish, baseline finish, actual finish, and projected finish) come directly from the same fields of the designated finish node.

Hammock node data could also capture the Path Time Reserve of its start and finish nodes, but we must do this carefully. These values represent the time reserves measured at these two points, giving no information about the nodes that lie in between. The worst-case time reserve of all the nodes between these two ends may not appear at the ends (see Appendix E). This limits the utility of the time reserve values in hammock nodes.

From the ES and EF dates derived from the start and finish nodes, we can calculate the duration of the hammock using the date math equation. This value would include the effects of any external constraint stretching or delays revealed in the forward pass, but none from the backward pass. This establishes a single activity record that summarizes data from several activities between its start and finish parameters.

If we summarize all of the activities of a project in a series of hammocks, we create a smaller number of special nodes that reflect the entire project, but in less detail. The Arrow technique provides a model-like structure in that the activities are linked together. This is somewhat deceptive though. The hammock data represents the project model, but the hammock structure in itself is not a CPM equivalent to the detail model (see Appendix E). We cannot extract the hammock model data, perform the CPM processes, and generate identical data to the detail model. This is a common misconception regarding hammocks. Precedence does not even produce the visual effect of a model composed of hammock mechanisms without additional constraints linking hammocks. The hammock process has more serious problems, however.

First, the structure itself can be deceptive. Figure 13-2 illustrates two levels of a hammock structure; one summarizing the model into three hammocks (intermediate), the other a single hammock. Because we were not careful in establishing the structure of the hammocks, the sum of the durations of the intermediate hammocks (30 + 30 + 30 = 90) does not equal the duration of the whole hammock (70) because it counts the durations of two tasks (C and E) twice. Thus, we must take care that our hammock summary does not yield a false picture, in this case accounting for tasks more than once. There is no mechanism to prevent us from doing this.

Likewise, the hammock structure could just as easily omit several activities and therefore portray only a portion of the project. Keep in mind also that we have grossly oversimplified the practical process of establishing hammock structures by limiting our example to a single path consisting of only seven activities. Creating a hammock structure that accurately summarizes a modelfile of 1,000 activities or more is a much more complicated effort and both types of errors are easy to make.

Hammock mechanisms also require additional model structure (nodes, relationships, etc.), adding another level of separate records of data to the already complex modelfile.

Figure 13-2

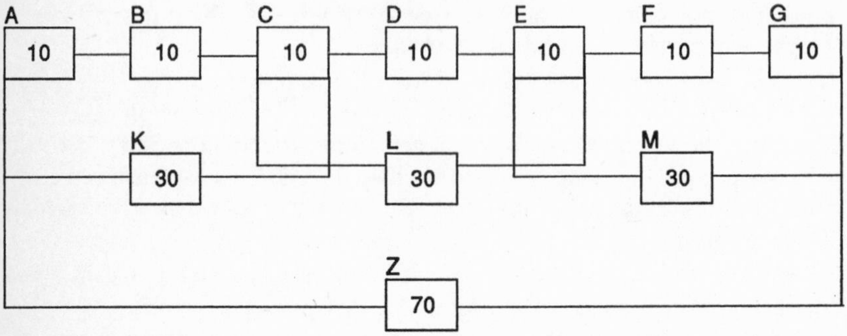

Yet another problem with hammocks arises when we need to summarize our project by more than a single criterion (i.e., function, or organization, or product, or location, etc.). We can build a hierarchy of hammocks like that in Figure 13-2, but it is difficult to overlay a multicriteria hammock structure onto a project.

Because of these reasons plus those discussed in Appendix E, hammocks have too many discrepancies to adequately meet our needs. To avoid all of these problems, we will leave the hammock between trees in the backyard where it better serves its purpose. We will find a better tool to meet our summarization needs.

CODE FIELD SUMMARIZATION

Activity Summarization

Fortunately, there is a much better and easier way to summarize a project model. This process has many similarities to hammocks but avoids its pitfalls. Figure 13-3 illustrates a small project model in which the nodes carry both product and functional structure criteria. This structure data appears in two fields: the PROD field (for *product*) and the FNCT field (for *function*). We could summarize our data with either criteria but will first summarize model data for individual products (PROD), creating an activity to represent each product: the Widget, the Gizmo, the Whozits, and the Assembly. The example shows only calculated data, but the process accommodates the schedule fields of the records (baseline, projected, and actual dates), as well.

Since the early dates reveal the schedules of tasks, the summary schedule of the Widget runs from February 1, the first finish, to July 20, the last finish. To define these summary activities, we do not define their start and finish nodes

Figure 13-3

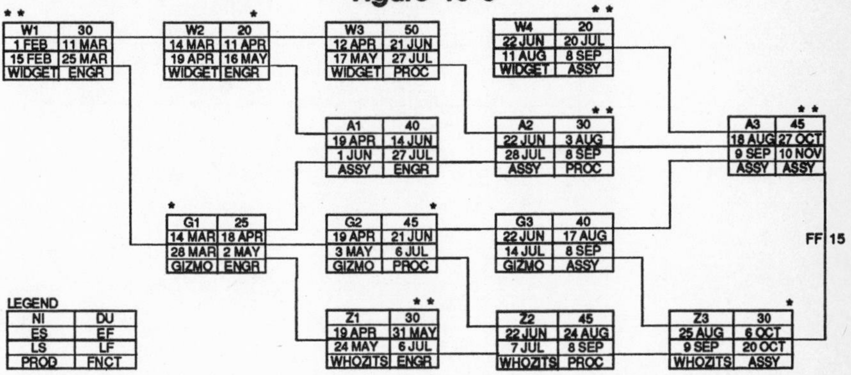

but rather determine the earliest start date and the latest finish date of all the tasks in each group. These two dates become the start and finish dates, respectively, of the corresponding summary tasks. We do not need to identify the first and last node, or add records to our modelfile.

By this method, the summary schedule of the Gizmo would run from March 14 to August 17, that of the Whozits would run from April 19 to October 6, and that of the Assembly would run from April 19 to October 27.

This data shows up best in a Gantt barchart, as explained in Chapter 15. We will use this format now to illustrate our summary data better. Figure 13-4 depicts the summary schedule of the four-product project in the Gantt barchart format. Simply, the bars span the expected period of performance of the work of each summary task and is depicted relative to the calendar strip.

The example combines the information for four groups into four summary activities. To maintain this information properly, we create a summary datafile separate from, but related to, the modelfile. This file assigns the criteria values (e.g., Widget, Gizmo, Whozits, etc.) as the node or record identifiers. It can also assign other structural and descriptive information to each record in appropriate fields. Each record also contains the six standard schedule date fields (BES, PES, AES, BEF, PEF, and AEF) and four analytical date fields (ES, LS, EF, and LF). However, all date data comes from the detailed modelfile through the summarization processes (i.e., the earliest start dates and the latest finish dates of each group). Like the hammock mechanism, this does impose some additional data input and upkeep requirements, but it does not increase model complexity. We store all of this summary data in a separate data structure, the summary datafile.

In this summarization method, we could very well utilize the structures that were established for sort and selection criteria, saving us from adding anything

Figure 13-4

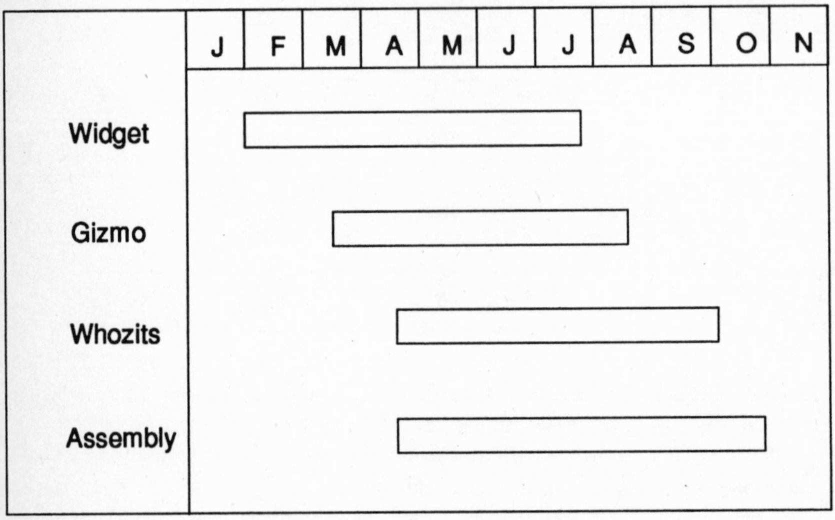

to the database to accomplish summarization. In some cases, however, the summarization criteria is different from the sort and select criteria, necessitating additional field coding. We can also summarize the project by different criteria (i.e., one time by function, the next time by product) simply by changing the field by which we group activities for the summarization process. A summary datafile would be established for each criteria.

The summary activity definition process can be defined as follows:

1. The user chooses a criterion of the summary by selecting a field or subfield (e.g., PROD). This identifies the values for the software's summarization process.
2. The process then divides the project activities into groups of nodes with matching values in the specified summarization field (e.g., Widget, Gizmo, Whozits, and Assembly).
3. From each group, the process selects the earliest activity's start date and the latest finish date of all date types (baseline, projected, actual, early and late) to arrive at the summary schedule for that group.
4. The process stores this information in a summary datafile, defining each summary activity as a record.

Earlier we pointed out two of the hammock scheme's structural pitfalls: structural overlaps (Figure 13-2) and structural gaps. Summarizing data based on code fields easily overcomes both of these problems. A code field or subfield can hold only a single value in each record. Both *Widget* and *Gizmo* will not fit into the PROD field of one record. This prevents the process from including one activity in more than one group during summarization.

A record could have no value in a field or subfield, causing the summarization process to overlook it. We can easily add a verification step to the process to ensure that every node in the modelfile has a value (other than null) in the summary field. This would preclude missing any nodes in the summarization process and the total of summary activities represents the entire detail project. In this way, the code field summarization method enables us to summarize data while avoiding the hammock's pitfalls.

Event Summarization

In the second phase of code field summarization we summarize events. The project example in Figure 13-3 includes 13 nodes. There are two events per node, one for each node's start and finish, which total to 26 events. A summary schedule need not reflect details of all 26. Rather, it should show a subset of these events. The process of event summarization is a selection of the more prominent events.

We choose to include the start of node W1, the finish of W2, and the finish of W4 from the events of Widget tasks, only three events out of the possible eight. We append the appropriate information from these three events to the Widget activity record in our summary datafile to designate them as milestones.

In the same manner, we go through the entire database flagging the events that we want to show in the summary. Figure 13-3 indicates the selected events with asterisks (*). In the modelfile, this as accomplished through coding in the SEF and/or FEF fields. Graphically, we can portray the information we added to the summary datafile as event symbols, geometric shapes that appear above the appropriate summary activity bars. (This is another graphic device explained thoroughly in Chapter 15.) Textual descriptions from the detail record (in the SED field for a node's start event, and the FED field for its finish event) appear beside each milestone. Figure 13-5 depicts our example subset of information.

Our combined activity and events processes establish a complete method of summarizing these two model elements. Such a scheme provides extensive built-in flexibility, as we will now examine.

Varying the Summary Criteria

Data in Figure 13-3 also included codes for the functional criteria of the project (in the FNCT field). We could summarize project data by this criterion, as well.

Figure 13-5

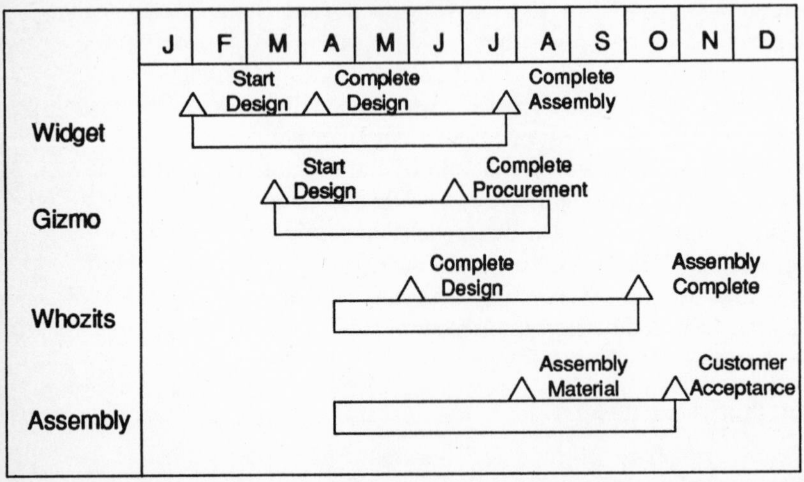

In this case, the process would divide the database into three groups, one for each of the values in the FNCT field: Engineering, Procurement, and Assembly. The engineering group contains five activities (W1, W2, A1, G1, and Z1), among which, the earliest Early Start is February 1 and the latest Early Finish is June 14. These become the summary schedule dates of the Engineering activity. We could arrive at summary schedules for Procurement and Assembly by the same method. These schedules all appear in Figure 13-6.

Figure 13-6

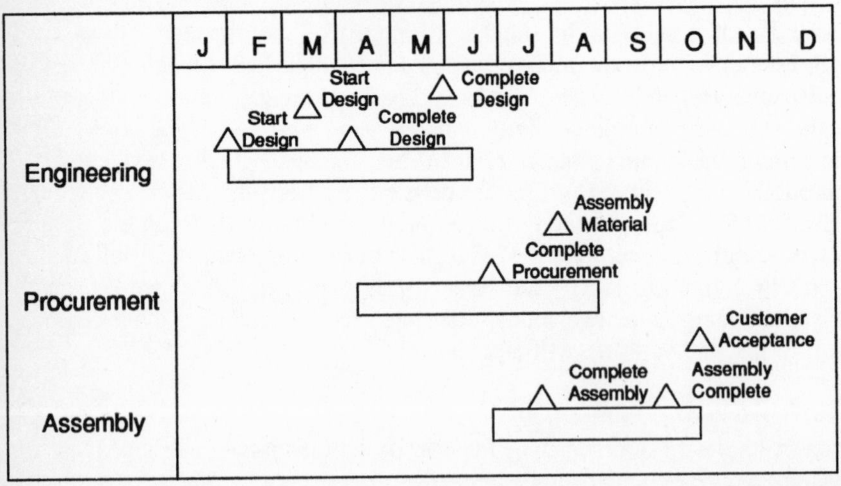

The same events that we had previously selected also appear in this graphic output product. Each appears on the bar appropriate for its function (FNCT) code value (e.g., the start of W1 appears with Engineering, the finish of A2 with Procurement, the finish of Z3 with Assembly, etc.).

One problem becomes apparent in the textual titles of the milestones. The SED and FED values adequately describe the events in Figure 13-5 because other information supplements them. The bar or summary activity title provides complementary definition. When we change the summary criterion from PROD to FNCT, however, we change the orientation of our summary activities. The same SED and FED values become inadequate because the summary activity description does not complement the events in this orientation. For example, the first milestone above the Engineering bar in Figure 13-6 does not indicate the specific START DESIGN to which it refers (Widget).

To correct this problem, we might manage to code the event descriptions (SED/FED fields) with more universal standalone values, but it is difficult to find succinct values that fit all summarization criteria. A better technique combines the values from two or three fields for textual description. In the schedule summarized by function, we could append the PROD field to the SED/FED field, giving milestone tiles above the Engineering summary bar of: WIDGET START DESIGN, GIZMO START DESIGN, WIDGET COMPLETE DESIGN, and WHOZITS COMPLETE DESIGN. These headings more adequately describe these milestones.

We can take this a step further and reduce the SED and FED values to the words *START* or *COMPLETE.* In the product-based summary (Figure 13-5), the milestone titles combine the SED/FED fields with the FNCT fields. This would yield milestone descriptions like: START ENGR, COMPLETE ENGR, COMPLETE ASSY, COMPLETE PROD, etc. These values, coupled with the titles of the appropriate bars (Widget, Gizmo, Whozits, and Assembly), provide adequate and succinct milestone descriptions.

In the function-based summary, we would combine the PROD field with the SED/FED field, yielding milestone descriptions like: WIDGET START, WIDGET COMPLETE, GIZMO START, GIZMO COMPLETE. Combining these with the summary bar titles (Engineering, Procurement, and Assembly) gives adequate milestone descriptions. This solution best solves the milestone identification problem because it provides simple, flexible, and concise descriptions. It does require a sophisticated level of discipline.

Summary Level Milestones

In addition to marking individual activity initiation and completion with events, we often need events that mark initiation or completion of groups of activities. A report might require that we note the completion of all Engineering, or Procurement, or Assembly work. The finish of node A3 marks the completion of the project, as well as the completion of all Assembly work, so it meets one of these needs. No other single event will suffice, however, to mark the completion of either Engineering or Procurement. The completion of A1 is the last event associated with Engineering, but as the project progresses (status), it might not always be the last finish of Engineering. If the finish of activity Z1 were to slip 20 days, it would become the last finish of Engineering. Thus, we cannot say that the finish of the last Engineering activity occurs with the completion of either A1 or Z1; it is the completion of both events. All other Engineering activities are predecessors of these two events, so no other activity should finish later.

This determination of the event that occurs last should bring to mind the process for dealing with multiple predecessors in the forward pass, the latest early date rule. If the finishes of nodes A1 and Z1 were both predecessors of another node, then the forward pass would sort out which event would occur last through the latest early date rule. We could create an events node (DU = 0) constrained by all of the potential Engineering activity finishes (A1 and Z1) and another that is constrained by all of the potential Procurement activity finishes (A2 and Z2). The finish dates (EFs) of these two events nodes would always reflect the current latest finish of Engineering and Procurement activity, as illustrated in Figure 13-7. Every node contains two events (the start end and the finish end), and we generally need to consider only one of them for events nodes. In this case, we need the event at the finish end of the node as we are measuring finishes. We simply ignore the data at the start end of the events nodes.

Currently, the Early Finish date of node E1, June 14, also marks the conclusion of all Engineering activity. If Task Z1 were to slip 20 days to June 21, then the Early Finish of E1 would pick up this date in the forward pass as it is the last finish of its predecessors. Similarly the finish of node P1 currently shows the finish of all Procurement work as August 24. If node A2 were to slip 20 days, then the Early Finish of P1 would reflect this date. Through the forward pass process, the Early Finish of E1 will always mark the completion of all Engineering activity and the Early Finish of P1 will always mark the completion of all Procurement activity.

The creation of special summarization events nodes coincidentally creates a problem. Since no relationship drives the Late Finish dates of nodes E1 and P1, these nodes become hanging finishes unless we tie them back into the model. If we do nothing, these dates will default to their Early Finish dates, which will taint the late date calculations of all their predecessors. Through the earliest late date rule, the backward pass would override any predecessor's positive time reserves to zero.

Figure 13-7

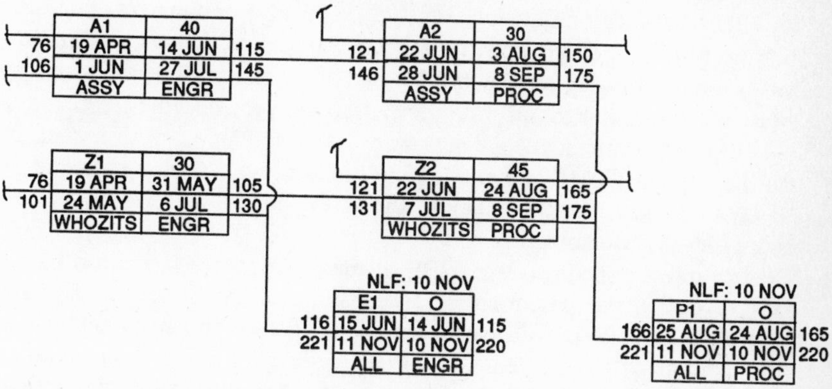

We cannot leave these nodes hanging. We must either define downstream constraints for them or assign proper date targets. We created these events merely for reporting purposes, however, so they have no appropriate successors. We cannot usually, therefore, constrain them to any downstream node so we must assign them date successors that will prevent them from influencing their predecessors' late dates and subsequent time reserve calculations. In our example, we simply assigned November 10, the required project finish date, as an FNL date on each node. Now the meaningless late dates of these nodes will not adversely affect the late date calculation of their predecessors. We can simply ignore them. Remember that we established these nodes only for their Early Finish dates; they are not part of the working model.

There is an alternative way to determine summary events. These usually measure the same data as our activity summaries. In our summary datafile, we could create events to represent the start and finish of each summary activity, deriving all dates and structural information for these subrecords from their parent summary activities. In this way, we could define, for example, the completion of the Engineering, Procurement, and Assembly activities. To these records we can also append any additional textual or structural data that is necessary. This method would only accommodate events at either the start or finish of the summary activities. Interim events would have to be derived from the model.

Hierarchical Summary Structures

So far we have described a summarization process for the detail data and one higher level. However, typical management structures include several levels and we need to develop schedules that correspond to each. The code structure summary process can accommodate multiple levels of a hierarchy through manipulation of the code values of the summarization fields. Figure 13-3 allowed summary by either product or function, but we could also summarize the entire project in the form of a single activity. It would appear as a single bar starting on February 1 and finishing on October 27. For this small project, the summary schedule seems somewhat trivial. In a more practical project with 1,000 activities, we could first summarize our data into 100 activities at an intermediate level and then again into 10 activities at the highest level. A step factor of 10 is not always practical, but the example shows the correct difference between one schedule level and another.

Certain project structures, like WBS or management trees, impose hierarchy codes to define the parent-child relationships of the structure. In other words, the code associated with each element of the significant structure also defines the parents of the element (e.g., WBS code 1.2.3.1 is a child of WBS code 1.2.3, which is a child of WBS code 1.2). If we summarize on the various subfields of this structure, we create four levels of summary data. This type of structure can range from four to nine levels deep, providing very powerful support for a multiple summary structure.

We can develop a simple hierarchy, however, from the criteria we have already used. We might summarize Engineering department data in categories like Electrical, Mechanical, and Structural Engineering, then further summarize this data for Engineering in total. At the first level of summarization, three activities would report the cumulative schedules for Electrical, Mechanical, and Structural Engineering. At the highest level of summarization, a single activity would report all of the Engineering work.

We should structure our code values to accommodate this scheme, perhaps defining the following codes: ELECENG, MECHENG, and STRUENG. Summarizing data based on these field values would give schedules for the three intermediate summary activities. To arrive at a single schedule for all of Engineering, we could summarize data based on the subfield of the last three characters of the field value (ENG). This would combine all of the tasks of the previous summary process into a single group. We control the process by specifying part or all of the field in our summarization process. This also ensures that we include all of the work that makes up the intermediate activities in the overall activity.

Along with this tiered activity summarization, we need a way to summarize hierarchies of events. Referring again to Figure 13-3, we originally selected nine events on which to include data at the intermediate level. We now select a subset of five of these nine events, the start of W1, and the finish of W4, Z1, A2, and A3, for the next higher summary. The double asterisks in the figure differentiate these events from the four intermediate events. In the model, we identify events for various tiers of the summary by adding codes that specify a level of schedule (SEF and FEF fields). We might code the events we want to summarize at the highest level with the number *1*, the events at the next lower level with the number *2*, the events at the next level with *3*, and so on until the events at the lowest level of detail have no such code (e.g., the finish of W1, the start of W2, the start and finish of W3, etc.). When we summarize, we simply specify which level of events we wish to portray. This numerical designation allows us to include level 1 and level 2 milestones in a level 2 schedule, levels 1, 2, and 3 in a level 3, etc.

Vertical Traceability

Whenever we create a hierarchy of schedules, whether with 2 levels or 10, we must always maintain *schedule traceability*, that is, we must keep all schedules in correlation with one another. The date of any event at the highest schedule level must match the date of the same event at all lower schedule levels. If we must choose between accurate schedules or total consistency (traceability), we must choose consistency. It is more important to keep all levels of the project on the same page, than it is to have some on the correct page. Historically, schedule traceability has severely challenged project managers.

One of the reasons we prefer field code summarization over other methods is its benefits of traceability. Field code summarization reports the same information about any activity or event at the highest levels of detail that it reports at all other levels because all the information comes from a single database. Information reported at any level comes directly from the detail level. Any changes are made at the detail level so all schedules derived from this data match exactly. Only by drawing all data from a single database can we count on achieving these results.

This guarantee of perfect traceability requires that we maintain reasonable discipline. It is possible to produce a schedule, change the database, and then produce another schedule. Such shortcomings in discipline destroy traceability between these two schedules.

Our summary process also passes status data up the hierarchy, including the earliest actual or projected start date and the latest actual or projected finish date. (Status reporting is discussed later in the chapter.) This information keeps our summary status for project tasks and events in agreement with the detail schedule. This allows project managers to isolate problems by examining only selected areas at the detail levels. The project managers can see substantial problems clearly without having to engage in a complete detail-level analysis.

CONSTRAINT SUMMARIZATION

Thus far, we have explained the summarization process only for schedule elements (activities and events). We have not as yet considered constraints. We can also summarize this model element, though we must maintain a proper perspective. We do not summarize constraints to create a summary level model, but rather to enhance our other summary information. We will add constraint data from the detail level to the summaries of activities and events to make those summaries more comprehensive.

Constraint summarization resembles the selection process for event summarization more than the net results calculations of activity summarization. To examine the process in detail, we will return to Figure 13-3. This simple model has 19 detail constraints. We need to reduce this number and still provide useful summary information. First, we must specify a criterion for summarization. We might like to summarize the schedule by product (PROD), for example. Examine the detail activities and constraints in the Gizmo summary activity. The constraints fall into two types. The first, like the constraint between tasks G1 and G2, are internal to the summary task. In the second type of constraint, like that between tasks W1 and G1, one of the nodes, in this case the predecessor, is external to the summary task. The constraint between G1 and A1 also has one node, the successor, that is not part of the summary task.

Constraints for which both the predecessor and successor lie within the summary task would be entirely under the control of that task's manager. These constraints would not add much value to a summary schedule. The constraints for which either a predecessor or successor falls outside the summary task indicate a shared responsibility as they interface one product to another. These constraints would attract the interest of the managers of both summary tasks. Therefore we will select for summary constraints those for which the predecessor and the successor fall within different summary activities. In our first example, the summary product-oriented schedule includes 10 constraints: W1-G1, W2-A1, W3-A2, W4-A3, G1-A1, G3-A3, G1-Z1, G2-Z2, G3-Z3, and Z3-A3.

We reduced the number of constraints only from 19 to 10 for the summary, but in a more practical example, summary tasks will likely contain a larger proportion of the constraints within their boundaries, leaving fewer constraints to summarize. We will reduce the number of constraints through summarization to the same degree that we reduce the number of activities through summarization.

The constraint summarization process in effect accentuates interfaces with ends that occur in different summary tasks. Just as summary tasks and events affect the project more significantly than those at the detail level, the summarized constraints have stronger effects than detail constraints. These interfaces cross between major project structures and this makes them more important. If we can assume that a level of management corresponds to each summary level, then movement up the schedule hierarchy isolates the constraints that concern successive levels of management.

Having selected the summary constraints, we must next relate them to the appropriate summary activities. To do this, we must choose a point of reference. We must decide whether to summarize a constraint relative to the successor or the predecessor. Otherwise, we will add each constraint to the summary product twice: once for the predecessor and once for the successor. Summarizing the constraint on the successor side is called *input relativity,* as the interface resembles an input to the summary task. Summarizing the constraint on the predecessor side is called *output relativity,* as the interface resembles outputs from the summary task. For our initial example, we will summarize constraints on the predecessor side, or input relativity.

The date information we display will tell us when the predecessor of the constraint is needed versus when it becomes available. A finish-to-start constraint like W1-G1 is needed on the Early Start date of the successor (March 14). The predecessor becomes available on its own Early Finish date (EF + lag + 1 = March 11 + 0 + 1 = March 14). We convert this date to the next workday morning by adding one day for easier comparison. In this case, the predecessor becomes available on the same day that the successor needs it (March 14). This means that no free reserve separates the predecessor and successor of the constraint. If the predecessor task (W1) were to slip, it would immediately affect the G1, Gizmo task. This is important information for the summary schedule to track.

We can display this data graphically in the Gantt barchart format, as in Figure 13-8. First we draw an arrow symbol beneath the appropriate summary bar (Gizmo) at the point at which the predecessor is needed (March 14). We attach the arrow symbol to a vertical line that marks the point at which the predecessor becomes available. When these two dates are the same, as in our first example, only the arrow symbol is drawn. This indicates a lack of free reserve between the predecessor and the successor of the constraint. The textual description of the predecessor activity appears next to the arrow to describe the input.

The finish-to-start constraint into the Assembly, W4-A3, has two separate dates. In this case, the predecessor is needed on August 18, the Early Start date of the constraint's successor (A3). The predecessor becomes available on July 21 (EF + lag + 1 = July 20 + 0 + 1 = July 21), the beginning of the workday after the Early Finish date of the predecessor (W4). In this case, the predecessor is available 20 days before the successor needs it, leaving 20 days of free reserve between them.

On the barchart, this is illustrated as an arrow below the Assembly activity (the successor summary task) on the predecessor's need date, August 18. A vertical line attached to this symbol reflects the predecessor's availability, July 21. The title of the predecessor activity next to the need arrow defines the input (Widget Assembly). Figure 13-8 shows immediately that the Widget Assembly predecessor will become available to Assembly almost a month before it is needed. The Gizmo Assembly, on the other hand, will not become available until the day Assembly needs it. The manager of Assembly should monitor the progress of the Gizmo much more closely than the Widget.

Other constraint types (FF and SS) with lag values require a different process. In the finish-to-finish constraint Z3-A3, A3 is the successor. We will summarize the constraint information relative to to the summary Assembly task. The availability of the predecessor (Z3) remains its Early Finish date, October 7 (EF + lag + 1 = Oct. 6 + 0 + 1 = Oct. 7). Its scheduled need date is no longer the Early Finish date of the successor, however, because the constraint includes a lag of 15 days. The scheduled need date is therefore 15 days prior to the Early Finish date of A3, or October 7 (EF – lag + 1 = day 210 – 15 + 1 = day 196 or Oct. 7). We need to add a day in both equations to convert the dates to the start of the remaining portion of the task. In this example, the need and availability dates are the same, so again no free reserve separates the finish of the predecessor (Z3) from the start of the remaining portion of the successor (A3).

In the barchart, the arrow appears (WHOZITS ASSEMBLY) at the predecessor need date, or the start date of the remaining portion of the successor represented by the lag on the constraint (October 7). No vertical line connects it to the predecessor's availability date because the two dates are the same, indicating an absence of free reserve. Figure 13-8 shows no indication of the constraint type of this interface, but this does not matter because it accurately represents the interface.

Though the example includes no start-to-start constraint, the summarization process resembles that of the finish-to-finish constraint. In the SS relationship, the need date of the predecessor is the schedule or Early Start date of the successor. An arrow will appear below the successor's summary task on this date. The availability date of the predecessor is its own Early Start date plus the lag on the constraint (ES + lag). Since this type of constraint ties completion of an initial portion of the predecessor (represented by the lag on the constraint) to the start of the successor, we need not to adjust these dates for the start-finish convention. Both are start dates. The graph would show a vertical line on this date with a horizontal line attaching it to the arrow, provided the two dates are different. Any difference between these two date values will be the free reserve between the predecessor and the successor.

Figure 13-8

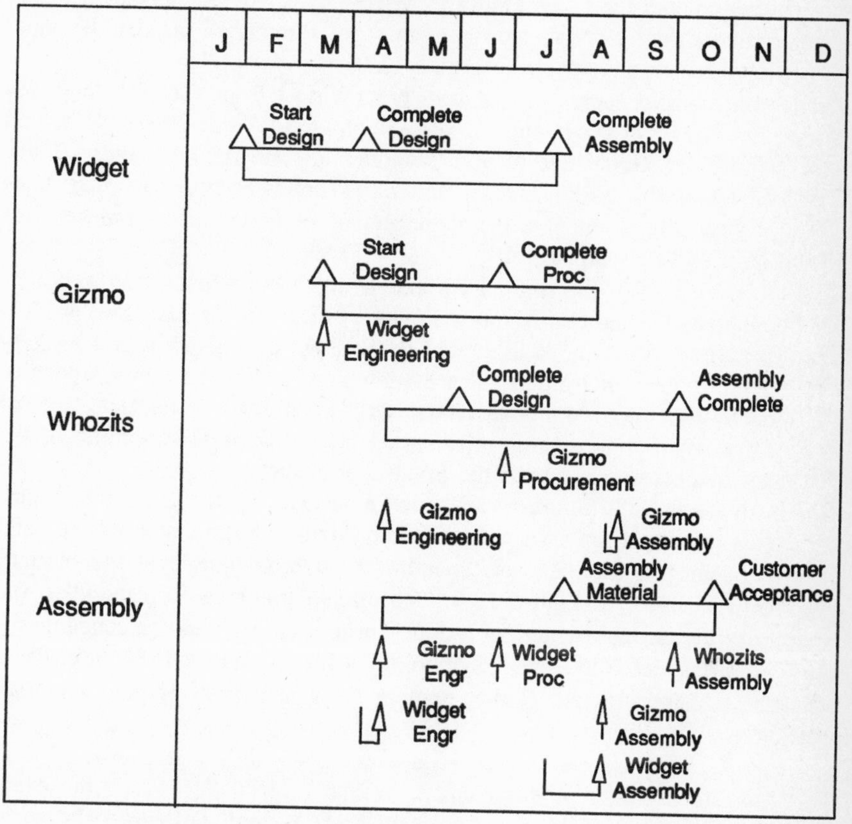

Changing the Summarization Criteria

The process summarizes constraint information based upon the field (subfield) by which we choose to group detail activities into summary tasks. If we change the criterion for the summary, then the process would select different constraints. We based our initial summary on the values in the product (PROD) field, but now we will change this to the function (FNCT) field.

In this situation, the summary tasks are: ENGINEERING, PROCUREMENT, and ASSEMBLY. The process will select and summarize constraints that cross functions, like those between Engineering and Procurement, or Procurement and Assembly. Out of the 19 constraints, 8 meet this criterion: W2-W3, W3-W4, A1-A2, A2-A3, G1-G2, G2-G3, Z1-Z2, and Z2-Z3. All of these are finish-to-start constraints, so their need dates are the scheduled starts (ESs) of the successors and their availability dates are the scheduled finishes (EFs) of the predecessors (plus any lag plus 1). Again, we summarize the data by input relativity, and the results appear in Figure 13-9.

No free reserve separates W2 and W3 (Widget Engr #2), W3 and W4 (Widget Proc), G1 and G2 (Gizmo Engr), G2 and G3 (Gizmo Proc), or Z2 and Z3 (Whozits Proc). Some free reserve separates A1 and A2 (Assembly Engr), A2 and A3 (Assembly Proc), and Z1 and Z2 (Whozits Engr). If the schedules of any of these inputs change, the managers of other functions can see this information at the summary level.

As the structure of the summary schedule changes its focus from products to functions, the selected constraints also change from inputs into each product to inputs into each function. In Figure 13-9, the input data below each summary bar reports on the external inputs to each summary functional activity. Specifically, the need date (arrow symbol) versus the availability date (vertical line) of each external input indicate this information. The activity descriptions of the constraints' predecessors describe the imput that is shown.

Above the bar, the figure shows some modifications to the reported summary events to correct previous concerns. First, two milestone symbols (Δ and Ⓐ) now appear. The triangles correspond to the intermediate-level events highlighted with asterisks in Figure 13-3. The circled triangles correspond to the highest-level events highlighted with double asterisks. This adds a schedule hierarchy significance to the event symbology which provides for quick visual recognition. Chapter 15 will further explore the significance of this and other symbology.

Second, labels associated with each event now follow the combined field scheme discussed earlier in this chapter. We have elected to attach the event description (SED/FED) field to the product (PROD) field. This properly correlates the events' labels to the current summarization structure (FNCT field).

Output Relativity

If we summarize the constraint information in terms of output rather than input relativity, we track the external outputs from each summary task instead of tracking external inputs into them. Basically, we summarize the same information that we did with input relativity, but now we show the cross-functional constraints as output attached to the predecessor summary activity rather than input attached to the successor summary activity. We still report the need for the constraint's predecessor versus its availability.

In finish-to-start constraints, the need dates are the scheduled starts (ESs) of the constraints' successors and the availability dates are the scheduled finishes (EFs) of the constraints' predecessors plus any lags on the constraints. The

Figure 13-9

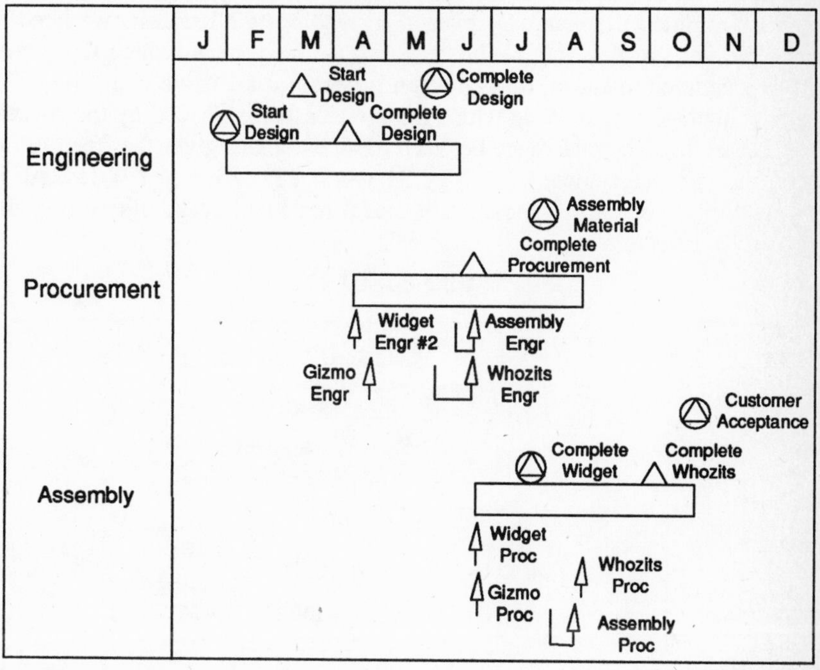

need dates of finish-to-finish constraints are the scheduled finishes (EFs) of the constraints' successors less the lags, and the availability dates are the scheduled finishes (EFs) of the constraints' predecessors (plus 1). Finally, the need dates of start-to-start constraints are the scheduled starts (ESs) of the constraints' successors and the availability dates are the scheduled starts (ESs) of the constraints' predecessors plus any lags. These equations match identically those for input relativity.

Graphically, output relativity does create some problems. Input seems to fit well under the bar, as though the data comes up from below. Output, conversely, seems to fit best above the bar, as though the data emerges out of the bar. If we did this, the outputs would compete with the summary events for the same space. This leaves us with two reasonable choices: either we do not depict the summary events with outputs or we reflect the summary outputs beneath the bar, just as we did with inputs, but point the arrows down (⇓). We must also label each output with the description of the constraint's successor activity, as opposed to the predecessor description from input relativity. We need a definition of the destination of the interface rather than its source.

Figure 13-10 illustrates a summary of constraint information in reference to functions with output relativity. We have placed the output information above the bar and omitted the summary events. In reading this barchart, we need to interpret the vertical line (e.g., Whozits Proc) as the point at which the predecessor is scheduled to finish, that is, when it becomes available (e.g., May 31). The arrow indicates the point at which the predecessor is needed by the successor (e.g., June 22). The difference between these two dates gives the free reserve between the two tasks (June 22 – May 31 – 1 = 121 – 105 – 1 = 15). As in input relativity, a lone arrow indicates identical need and availability dates, and a lack of free reserve.

Figure 13-10

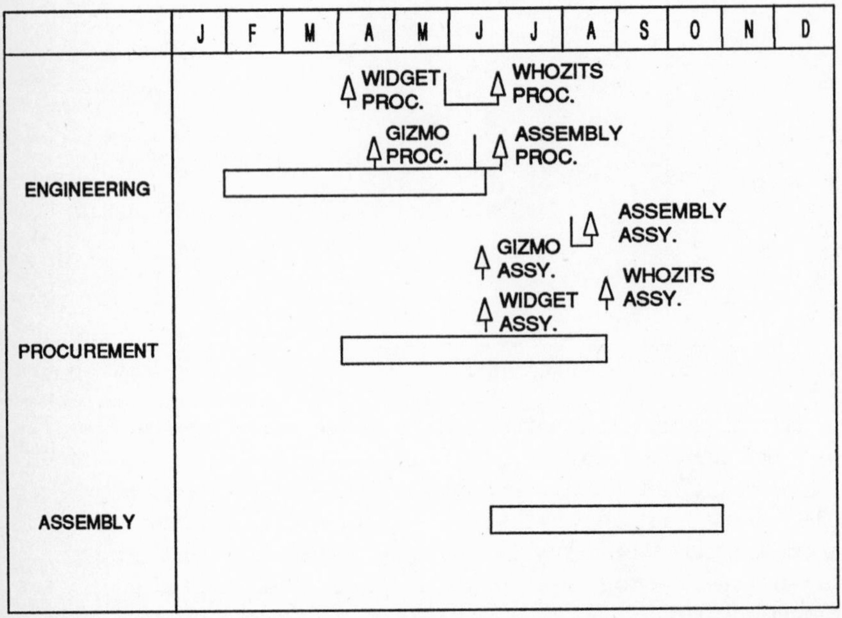

In summarizing constraints, we select out of the modelfile those interfaces between different summary activities as either external predecessors (for input relativity) or successors (for output relativity). We are showing the relationships of the external inputs to and/or outputs from each summary task, including the dates they are needed and dates they will become available. Information about status appears immediately on the summary schedules. This adds a unique and useful component to our status and control mechanisms, as we can see clearly and quickly how external factors affect our ability to accomplish the work of our project.

Together, the summaries of activity, event, and constraint information provide a very comprehensive schedule summary. Each summary element contributes vital information to help project managers detect schedule problems. The summary activities reveal the overall schedule of the tasks, and provide a subjective understanding of schedule performance. Summary events show specifically which milestones the project has accomplished on schedule, which have slipped or been pulled-in and by how much, and the prospects for those that remain in the future. This provides an objective measurement of schedule performance. Finally, summaries of the principal project interfaces, those that cross between the major project structures, reflect the status of any external interface to each summary task.

At the beginning of this chapter, we set out to develop a process to inform upper management about the schedule performance of their particular pieces of the project in order to help them identify problems that threaten their schedules without having to wade through all of the detail data. Field code summarization helps us accomplish this goal very effectively. Once managers have isolated a problem, they can selectively review specific portions of the project at the detail level to get a more complete understanding of causes and effects, avoiding unnecessary detail elsewhere. This is the purpose and power of the *summarization processes—problem identification at higher schedule levels, then selective isolation of the specific problem at the detail level.*

Replacing Calculated Data with Schedule Data

Summarization provides information for reporting, not for modeling or analysis. Activity summarization divides the project into structural groups (e.g., product, function, etc.) and reports certain data from each. Along with the summarized schedule and status data, the analytical data derived from this process includes the earliest Early Start, the latest Early Finish, the earliest Late Start, and the latest Late Finish dates for each group of detail tasks. But we must analyze this calculated information with great care. The earliest Early Start and the earliest Late Start of a group of tasks might come from different activities, as might the latest Early Finish and the latest Late Finish dates. Any time reserve computed from this data could be very misleading, and certainly would not represent an overall condition of the detail data.

Further, time reserve problems in a group of tasks may not necessarily show up in the first or last nodes. Serious problems in paths that cross the project structures in the middle would not show up at all at the summarized activity ends. Although summarization does provide a very powerful scheduling and control mechanism, neither hammocks nor field code summarization support accurate time reserve analysis. This is best done at the detail level through the sort and selection processes.

For these and other reasons, we find the summarization of schedule dates is the most beneficial, including baseline dates (BES and BEF), projected dates (PES and PEF), and actual dates (AES and AEF). This produces a clear picture of the baseline data measured against actual and projected performance, for schedule tracking and control purposes at all levels of the project management hierarchy.

SCHEDULING BELOW THE MODEL LEVEL

Chapter 12 defined the proper level at which to model as the level of significant interfaces where the work crosses between project structures, plus any other principal project interfaces. We must, however, schedule work in more detail than this, at least in the near term. To meet this need, we do not drive the model to lower levels, but rather we develop a complementary scheduling and tracking methodology for use below the model level. This method will schedule this detail work and report on its status, but not attempt to model it. This is because we can maintain schedule data with considerably less effort than modeling requires, and modeling at the lowest levels of project detail returns only very marginal value for the investment of effort. Although we will not define constraints between such detailed tasks, we will still develop very useful detail schedule mechanisms.

In doing this, the most important consideration is maintaining schedule traceability from the model to these lower-level schedules. This requires that we integrate the detail data with the model data, even though each will reside in a separate data structure. Therefore, every task at the detail level must be a part of a single task in the model. This establishes a parent-child relationship between each detail task (child) and a model activity (parent). Many children can be derived from each parent, but each child can derive from only one parent, creating a *many-to-one relationship*. To define this interface between model and detail schedule, each detail task must specify its parent activity through a field. Each model task has a unique node identification (NI) code, making this an ideal mechanism to designate exact relationships.

Figure 13-11 illustrates this principle. In this example, detail tasks 1 through 8 are components of the model task ABAA. We indicate this relationship by entering the parent activity's NI code (ABAA) in a designated field of each of the eight children tasks. Once we have related all of the tasks in the detail schedule datafile to the activities in the model, then we accomplish the traceability process.

Schedule traceability requires compatible dates between related tasks. We compare these dates from the top down. This dictates that the schedule of the detail tasks must start on or after the model task's start, and finish on or before the model task's finish.

Figure 13-11

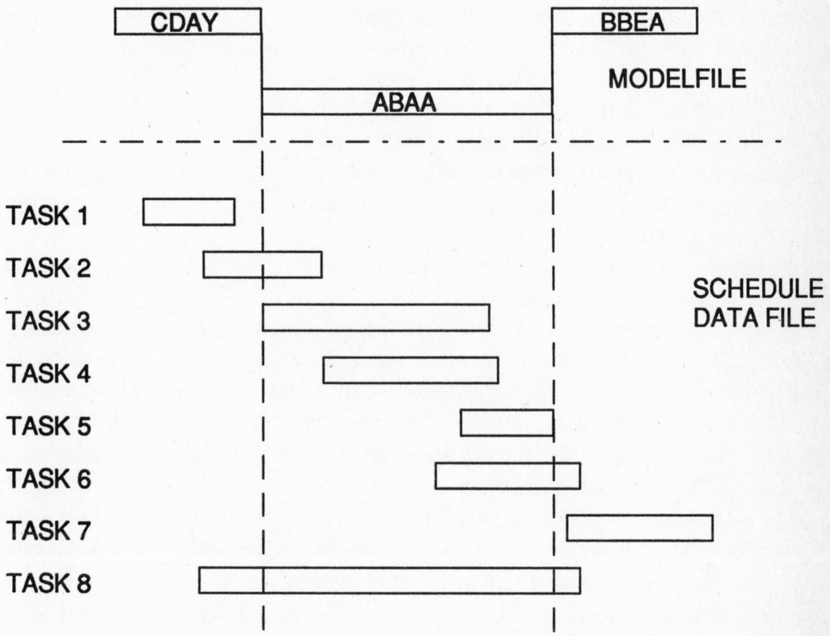

Tasks 3, 4, and 5 in Figure 13-11 comply with this, but the other tasks do not. Specifically, Task 1 starts and finishes before ABAA even starts, Task 2 starts before ABAA starts, Task 6 finishes after ABAA finishes, Task 7 starts and finishes after ABAA finishes, and Task 8 starts before ABAA starts and finishes after ABAA finishes. We cannot trace the dates for these tasks to the schedules of their parent model task.

In integrating these databases, we cannot assume that one always drives the other. Resolving conflicts between dates requires a knowledge of which values, those in the detail schedule or those in the model, are correct. This requires investigation and decisions, both of which exceed the simple capabilities of our detail scheduling process. This is the role of human intelligence.

To restore traceability to nontraceable records, the process should begin by identifying the condition in a report rather than automatically trying to impose one database on the other. Having isolated the disparities, we can investigate and resolve the differences.

Since both sets of data can also contain project structure fields, we can conduct traceability checks on this information. For example, if a single function performs both a model activity and its associated detail tasks, we can compare the value in the appropriate code field of the model activity to the same data in the associated detail tasks for consistency. Again, we will identify the disparities rather than automatically replacing any value in one database with that from the other. We may find valid exceptions.

To satisfy our detail schedule system requirements, we need the data records in the detail schedule task file to contain fields that (1) describe the record and relate it to the parent model activity, (2) relate the record to project structures, and (3) define the schedule data of the task. These fields are described below and summarized in Figure 13-12.

To satisfy the first requirement, we need three fields. First, an alphanumeric field stores a unique identifier for each task to facilitate record identification for data maintenance. Second, a textual field describes the work. The third field contains the parent activity's node identifier code. We can combine the third and first field to form a single unique task identifier. The first portion of this field would repeat the model activity node identifier, followed by an additional two, three, or four characters to uniquely identify the detail task (e.g., ABAA01, ABAA02, ABAA03, etc.).

The fields that relate detail tasks to project structures are the same fields we use on the model activity record to describe the project structures. These are peculiar to the project and its management environment, but would include things like: function, location, organization, WBS, etc. The values used in these fields must match the corresponding values used in the model to maintain traceability.

The fields that contain schedule data repeat those from the schedule side of the model record: baseline, projected, and actual dates. As we are dealing with tasks, fields must store data for both the start and finish of each. We must also expand our data traceability checks to include validation of baseline, projected, and actual dates to the model activities.

Baseline data requires only a straightforward traceability check for consistency between the dates in the two databases, with a report of any discrepancies. We should do this when we originally develop the data, and then routinely to maintain traceability. Checking status data for traceability is another matter. We could simply check for consistency of projected and actual dates, but we must consider that usually we assess status at the lowest level of detail. Rather than independently updating the model and the detail schedule, we can pass the information in the detail schedule datafile directly into the model. We maintain traceability in such cases by automatically imposing dates from one database on the other. However, as many detail tasks correspond to one model activity, we

Figure 13-12
Detail Schedule Records

Field Name	*Field Type*	*Field Description*
Identification Parameters		
TI	Alphanumeric	Unique task identifier
NI	Alphanumeric	Parent activity identifier
AD	Text	Task description
Status and Control Parameters		
BES	Date	Baseline start date of task
PES	Date	Projected start date of task
AES	Date	Actual start date of task
BEF	Date	Baseline finish date of task
PEF	Date	Projected finish date of task
AEF	Date	Actual finish date of task
Structure Parameters (Example)		
S1	Alphanumeric	Structure field 1
S2	Alphanumeric	Structure field 2
S3	Alphanumeric	Structure field 3
S4	Alphanumeric	Structure field 4
S5	Text	Structure field 5
S6	Text	Structure field 6

pass only a portion of the detail-level status data to the model, as defined by a consistent process.

1. A task's current schedule follows a consistent progression. If an actual date is present, this is the current schedule. If an actual date is not present but a projected date is, then this is the current schedule. If neither an actual or projected date is present, the current schedule is the baseline date.
2. To assess status for detail tasks, we first divide the detail tasks into groups that share a common parent. We will call these groups *detail groups*.
3. If any task in a detail group has an actual start date, we know that the parent model task has also actually started. To keep the actual start date

of the parent activity in agreement with those of the detail tasks, we pass the earliest actual start date from each detail group to the model as the actual start date of the parent activity.

4. The parent model activity actually finishes when the last of its detail task children actually finishes. We impose the latest actual finish date of the detail group when all detail activities are finished as the actual finish date of the parent task.
5. If any of the children tasks of a group have actually started, but not all of the children have actually finished, then the parent activity is ongoing. We impose the earliest actual start date of the group as the actual start date of the model activity as before. The model activity's finish date will come from either projected or baseline data of the detail tasks. We compare the latest projected finish date of the detail group to its latest baseline finish date as well as the parent task's baseline finish. If they agree, then the parent's baseline finish date remains its current schedule date. If they do not agree, then the latest projected or baseline finish date of the detail group (which may be earlier or later than the model activity's baseline finish) becomes the projected finish date of the parent.
6. If none of the detail tasks have actual starts or finishes, then all status dates are either projected or will be assumed to retain the baseline. In each detail group we compare the earliest current schedule start date (projected) to the earliest baseline start date and the latest current schedule finish date (projected) to the latest baseline finish date. If the earliest current schedule start date of the detail group differs from the earliest baseline start date of the parent task, this current schedule date becomes the projected start date of the parent task. Likewise, if the latest current schedule finish date of the detail group differs from the latest baseline finish date of the parent task, then this current schedule date becomes the projected finish date of the parent task.

By this process, we pass the most appropriate status dates on to the parent model activity. Figure 13-13 illustrates the general idea of these principles.

Integrating these databases in this way requires very rigorous discipline, based on the principle that the sum of the parts equals the whole. In real-world schedule control, this is not always true. For instance, managers commonly schedule a group of detail tasks to finish earlier than the schedule at the next higher level requires. The difference is a form of time reserve to ensure that the overall activity makes its schedule. However, should dates set earlier than the schedule define status at the next level? Sometimes not.

Even deliberate inconsistencies like this do not invalidate the schedule-to-schedule integration mechanism. The process reconciles the baseline schedules of the two databases, then imposes status information from the detail tasks on the model, but it can allow some latitude in the "automatic" processes. We could impose the baseline of the model on the detail, then reconcile the status between the two data files. It does not matter so much that we impose one database on the other or reconcile the data, as long as, by either mechanism, we keep schedule data consistent in both databases.

Figure 13-13

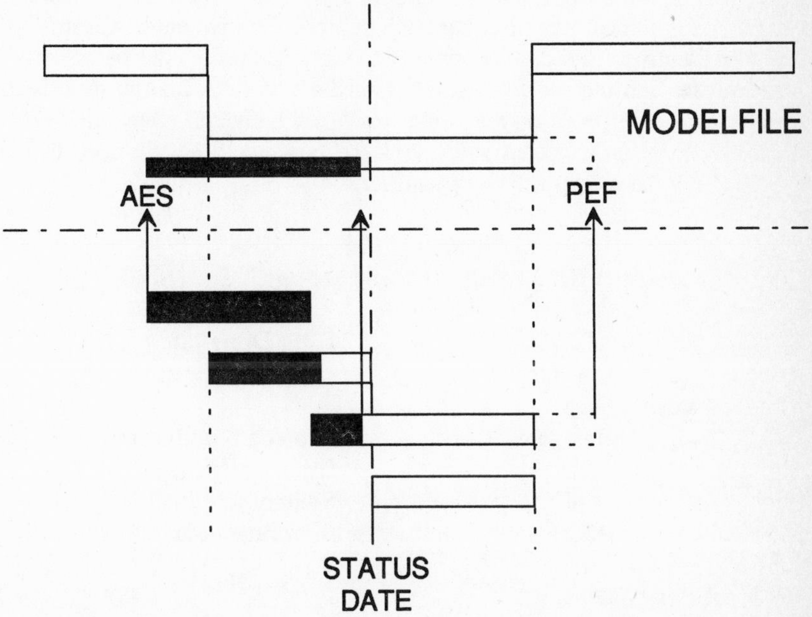

Just as we have detail-level activities, we also have detail-level events. The first of the two types of detail-level events, the independent event, is a separate record in the datafile. The second type, associated with a detail task, is a subrecord of the task record.

We track all events as points in time without any duration. This is the basic difference between task records and independent event records. We can store the data in the same record format, with structure and identification fields exactly like the task record. In the date fields, however, differences arise, as we maintain only one set of dates for an event. If the event measures a start, its record includes the start date fields: baseline (BES), projected (PES), and actual (AES). If the event measures a finish, its record includes the finish date fields: baseline (BEF), projected (PEF), and actual (AEF).

Events that are associated with a detail task share its identity and structure, but must also reflect additional unique information pertinent to each event. Figure 13-14 lists these parameters. In an event subrecord, we do not need to repeat any fields that are common to the task, although we could add a generic structure field or two for sort and selection of the events. Events do require some additional identification, however.

Since we generally find few events (less than 10) associated with an individual detail task, we do not need an additional identifier for each. Instead, we add a number to the task identifier for each subrecord as we enter it, forming a unique event identifier (the combination of the task identifier, the parent model activity identifier and the sequence number of the event). A textual description is also necessary to provide an adequate event label. Events associated with a task are part of the same model activity as the task itself, so we need add no separate field for the model activity identifier.

Figure 13-14
Subrecord for Events Associated with Activities

Field Name	*Field Type*	*Field Description*
Identification Parameters		
SRN	Integer	Subrecord number
ED	Text	Event description
SYM	Alpha 2	Symbol designation
BIA	Alpha 1	Start/finish bias
Date Control Parameters		
EBD	Date	Event baseline date
EPD	Date	Event projected date
EAD	Date	Event actual date
PER	Integer (positive)	Percentage of task
SFS	Integer	Span from start

There are three different ways in which we can specify the schedule of an event associated with a detail task. These appear in Figure 13-15. First, we can directly define the event's schedule by assigning a date in the baseline date field (EBD), as illustrated by the events associated with Task C. Event C2's date coincides with the start of the activity (Task C) and event C4's date coincides with its finish. C3, an interim event, occurs during the task's progress (on April 2). Events C1 and C5 occur before and after the task performance, respectively. These events are related to the task, but not directly produced as a result of its work. The date in the event subrecord's EBD field controls their position.

In the second method of date specification, we define a span of time as an offset from the task's start date (SFS), as illustrated by the events associated with Task B. Event B2 coincides with the start of the task (offset 0), while event B3 occurs 10 days after the start of the task. Positive values (+10, +20, +30) reflect events to the right of (subsequent to) the task's start and negative values (–10) reflect events to the left of (prior to) the task's start. Again, we can specify events that range beyond (both before and after) the duration of the work, illustrated by events B1 and B5. If we defer or pull-in the start date of the task, the schedules associated with each of these events will change accordingly.

Figure 13-15

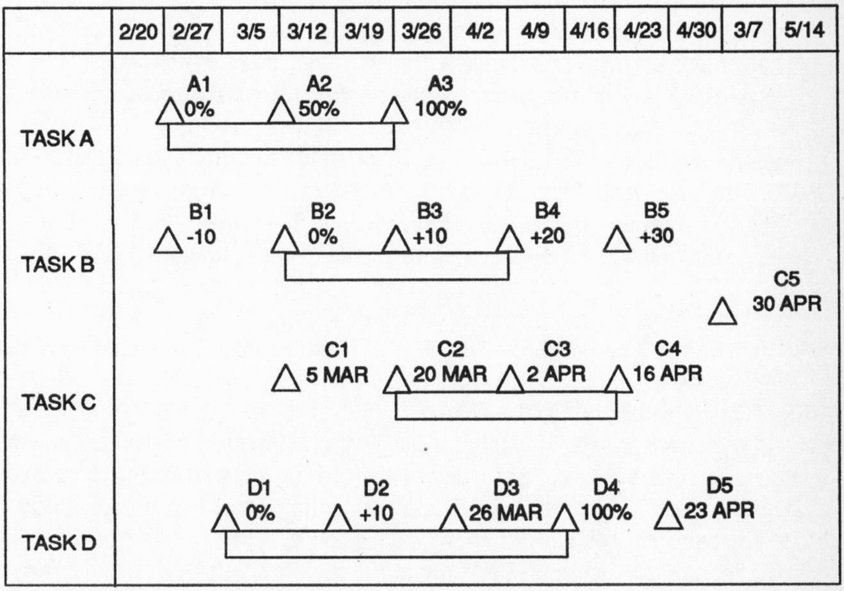

The final date specification technique defines points in time as percentages of the work accomplished (PER), as illustrated by the events associated with Task A. The total span of this task, defined as the difference between the detail activity's baseline dates (BEF – BES + 1), equals 100 percent of the work. The start date of the task (A1) occurs at 0 percent, and the task's finish (A3) occurs at 100 percent, in this case four weeks. Values between 0 and 100 (A2) define points during the work of the activity. Event A2, a 50-percent milestone, occurs after two weeks of work ($.5 \times 4 = 2$). Adding this duration to the task's start date yields a date (March 5) where the event is expected to occur. This method cannot specify milestones before or after the span of the work since they are directly related to task performance. Again, a slip or pull-in to the start date of the task changes the schedule of these events.

The user specifies one of these date specification mechanisms for each event. In determining the appropriate schedule, the process will also calculate the values for the other methods' parameters. For instance, the process would determine that event C3, which was assigned a date, begins after an offset of +10 work days from the start after completion of 50 percent of the activity. We could specify different date setting methods for different events that relate to the same task, as illustrated by the events associated with Task D. D1 and D4 are defined in terms of a percentage of the work, D2 is defined as an offset, and D3 and D5 are assigned specific schedule dates.

When we assess status for events, we assign dates to either the EPD field for a projected date or the EAD field for an actual date. These dates then aid schedule tracking and performance measurement, and we can transfer them upward to show the status of model-level tasks.

Besides these, we must define two other fields to complete an event definition. A graphic symbology designator, SYM, carries a user-input value that correlates to a specific shape. All of the events in Figure 13-15 appeared as triangles, indicating identical values in all of the event subrecords' SYM fields. A different value in this field would generate a different geometric shape.

The final field biases the event as either a start (S) or finish (F). For instance, the event B2 could define March 5 as a finish date or March 6 as a start date. The *S* bias seems most appropriate, as the event correlates to the start of the task. On the other hand, the *F* bias would best fit event B4 as this corresponds to the task's finish. The appropriate bias of B3 would depend on whether it measures the end of the second week of the task or the start of the third week.

The data produced by the detail task scheduling process provides a critical tool for total project schedule tracking and control, while freeing us from the overly cumbersome process of modeling at the lowest levels of detail. Removing this burden is a key factor in making the modeling processes more practical. The diverse benefits of this type of schedule control mechanism are limited only by the user's imagination. It can provide many kinds of information, not all of which have to be traceable to a model. The process will work as a standalone schedule control mechanism, as well. This tool can either complement project modeling or operate independently, and the project planner will want to keep it handy at all times. The data from this method would need access to a report writer and a Gantt barchart writer (Chapter 15) for effective reporting.

SUMMARY

Hammocks

A hammock or hammocking is a summarization mechanism that uses additional model structure and is part of the CPM process. A special node called a hammock is established with a user defined start node and finish node. The hammock takes its start data from the start node and its finish data from the finish node. Although hammock structures can be useful for reporting summary data—they are not in themselves a model that duplicates the detail.

They also have serious limitations. Their structuring adds clutter to the modelfile and requires great care to avoid missing or duplicating portions of the project. Also, it organizes information by only a single criterion when most projects need reports from multiple perspectives.

Code Field Summarization

Code field summarization is a process based on the coded values in the modelfile structure fields and is usually performed separately from the CPM processes. A field or subfield is specified prior to initiation of the summarization process. All three model elements (activities, events, and constraints) can be summarized relative to the specified criteria. Through these methods, vertical schedule traceability can be guaranteed.

All nodes of the modelfile are divided up into groups each designated by a unique value in the summary field or subfield. Each of these node groups is a summary activity. The process establishes the start and finish dates of each summary activity as the group's earliest start (BES, PES, AES, ES, and LS) and the latest finish (BEF, PEF, AEF, EF, and LF) dates for each date type. This information is all stored in a summary datafile and can be depicted on reports or graphic products. Since structure fields are designed to hold one value, it is very unlikely for a node to be included more than once or missed altogether.

Events are summarized by selecting and designating key events (milestones) for reporting from the total event pool (two events per node). A method can be employed to signify events for various levels of summary reporting (SEF and FEF fields). Events are related to summary activities by virtue of the value in their summary field (subfield).

Constraint summarization is a process of selecting the constraints that cross the summary structure. If the value in the summary field of the predecessor is different from the value in the successor, this is designated a summary constraint. Each constraint can be depicted in relation to either the predecessor or successor summary activity. Their depiction enhances the summary schedule (activity and events) by depicting inputs into or outputs from each summary task.

The summary process can take advantage of multiple coding fields by generating a summarized structure for every field and subfield in the modelfile. The actual summarizations performed are controlled by the user. This is a complete, structured process that satisfies most all reasonable requirements of data summarization and vertical traceability.

Scheduling below the Model Level

In the near term, it is almost always necessary to schedule and control work at a lower level of detail (greater granularity) than the project model. To accommodate this, we need a method of scheduling and tracking that does not require the extensive effort of modeling.

This process is a list (datafile) of detail tasks and events that retains identification, structural, and schedule information. All data is supplied and maintained by the user. The database accommodates communication, traceability to the model (if desired), and multiple (traceable) product outputs (Gantt barcharts, tabular reports, and management graphs).

All detail tasks can be linked to a parent model task for vertical traceability (down the hierarchy) and status reporting (up the hierarchy). Events are established as either independent (separate records) or dependent (task subrecords) as necessary.

The entire process can be used as a standalone schedule tracking and control tool, or as a complementary application to modeling that provides a project's top-to-bottom schedule, control, and proactive analysis.

CHAPTER 14

Horizontal Integration

We need to complement vertical summarization with an even more challenging process to simplify management of large projects. We need to break a large modelfile into more practically manageable segments. Typically a project is managed as a group of distinct functional areas and it is along these same lines that we need to fragment the model. This accommodates the information needs of the distinct functional areas that typically complete various portions of the work.

The typical structure of modelfiles can further complicate this problem. Many software packages limit access to the data of a single modelfile to one user at a time for record entry, alteration, and deletion. This alone forces us to fragment the database structure so multiple users can simultaneously work on the total project, simulating multiple access.

Even if the software did allow for simultaneous access into a single modelfile, problems would remain. Joint access to a database creates a discipline problem in controlling who does what to which data at what moment. Too many users can spoil the database. We generally find that multiple access quickly plunges the database into anarchy, with groups of simultaneous users feuding about who is at fault for errors, omissions, and other problems.

Before long most users conclude that one individual should have sole responsibility for the data in each modelfile, while allowing personnel within functional areas access to information about their parts of the project. For all of these reasons, we must learn to manage a fragmented or distributed database structure.

We will call the partial model based on the database of a particular segment of the project a *submodel,* which is a model with interfaces into and out of other models. The various submodels are separated by the boundaries of authority in the organizational structure. Dividing the project model into submodels lets each organizational group deal with its own responsibilities within the project, while the sum total of all the submodels is the total project.

We cannot, however, completely separate the activities in any one group from the activities in the other group. Even though modelfile structures distinguish the work into submodels, the effort in one submodel could still depend upon the work done in other submodels. This creates a requirement for interfaces or constraints that cross from one model structure to another, while still maintaining the integrity of the submodels. We call the process of accommodating the submodel interfaces or constraints *horizontal integration.*

Since work in one modelfile constrains work in others, we need the ability to perform the CPM processes treating these constraints as any normal constraint while still maintaining the integrity of the individual submodels. Several schemes have been devised to do this; unfortunately, some of them do not work properly.

FILE PASSING

The first method for sharing data attacks the problem in a simple fashion; it attempts to pass the necessary information from one modelfile to the other. We call this process *file passing,* as computer files are the medium we use to transfer the data.

Figure 14-1

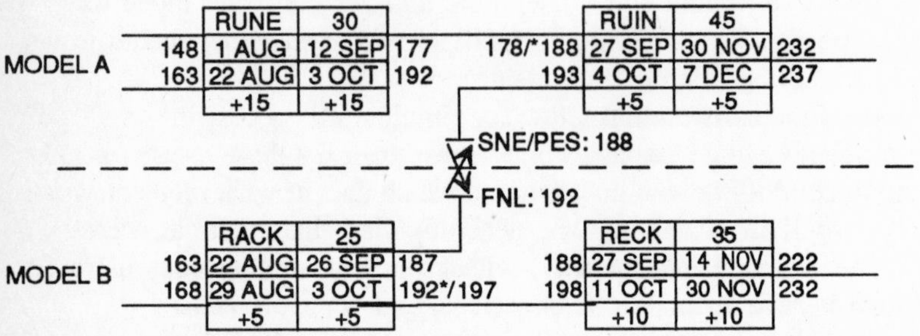

Figure 14-1 illustrates model segments in two separate submodels (A and B). The start of Task RUIN in Model A depends upon the finish of Task RACK in Model B. Practically, this means something like the initiation of the procurement of a certain component (a task in the Procurement submodel) requires the completion of the engineering design of the component (a task in the Engineering submodel). This constraint crosses the boundary between functional submodels. The Early Finish of Task RACK crosses the interface to constrain the start of Task RUIN. More specifically, we need to process RACK's Early Finish date through the constraint equation:

$$\text{ES of SN} = \text{EF of PN} + \text{lag} + 1$$

The value that we actually need to start RUIN includes the lag of the constraint plus one time unit for the start-finish date convention. Our example yields a date of September 27. To simulate the interface, we can process the Early Finish date of Task RACK through the forward pass constraint equation and then enter that date as an SNE date for Task RUIN. (Note that we enter both SNE dates and projected dates on any node in the PES date field.) The normal forward pass process can then correctly assess all of Task RUIN's predecessors and determine the latest one which would become its Early Start date. This date may or may not end up as the SNE/PES date, but the calculation does consider the cross submodel constraint.

The CPM process goes both ways through a constraint, and we must also look at the effects of this constraint relationship through the backward pass, as well. This requires a Late Start date for Task RUIN to determine the Late Finish date of Task RACK, since the backward pass determines dates for an activity from those of its successors. Again we must process this data through the constraint, this time using the finish-to-start backward pass equation:

$$\text{LF of PN} = \text{LS of SN} - \text{lag} - 1$$

To find the Late Finish date of RACK we deduct the lag of the constraint and a time unit for the date convention from the Late Start of RUIN. Entering this date as an FNL date for Task RACK allows the normal backward pass process to determine the earliest successor requirement, which becomes its Late Finish date. This date may or may not end up as the FNL date, but the calculation considers the cross submodel constraint.

To summarize the file passing method, then, we enter the processed Early Finish date of Task RACK into the SNE date predecessor (PES) field of Task RUIN, and the processed Late Start date of task RUIN into the date successor field (FNL) of Task RACK. This passes the necessary forward pass data from Model B to Model A and backward pass data from Model A to Model B.

We first identify the external interface in both submodels and then process the necessary data through the appropriate constraint equation. As a simple way to accomplish all of this, we can insert an events node (DU = 0) in both modelfiles at the point of interface, as illustrated in Figure 14-2. These nodes have identical node identifier codes (INF1), which combine, along with a few other parameters to be introduced later in the chapter, to identify a specific interface. These function as a single event or point in time that resides in both submodels.

Figure 14-2

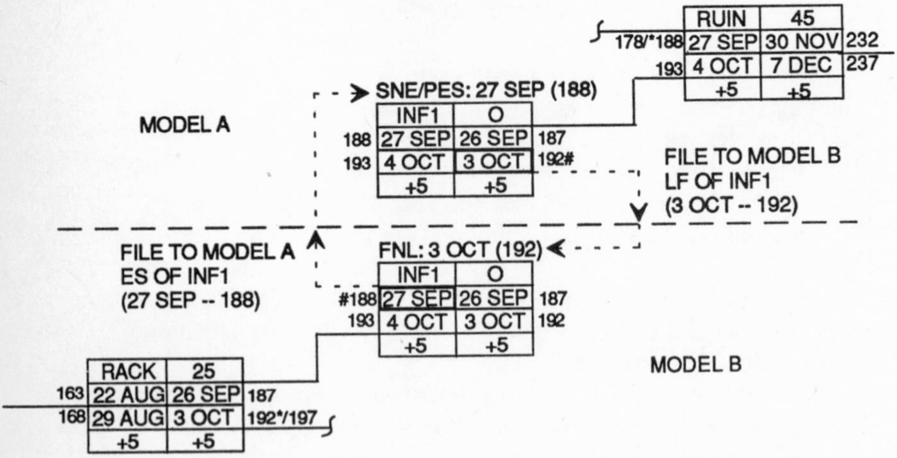

Through the normal forward pass process, the Early Start date of node INF1 in Model B (Sept. 27) has passed through the constraint from Task RACK. We take the ES date of this node, pass it across the interface and enter this value into the SNE date (in the PES field) of its counterpart events node in Model A (INF1). The zero duration of this node assures that the correct date will pass on through the forward pass from the events node to the successor activity node, Task RUIN.

Likewise, the Late Finish date of the events node INF1 in Model A (October 3) also passes back through the constraint from RUIN. We take the LF date of this node, pass it back across the interface and enter this value into the FNL date of its counterpart events node in Model B (INF1). The zero duration of this events node assures that the Late Finish date passes unchanged back to the predecessor, Task RUIN, through the backward pass.

We should note that any lag on the constraint should appear in only one of the submodels or it will be counted twice in each pass. This can be done to add some time reserve between two submodels, but as a general rule any lag value on these constraints should appear only once. Since the node identifiers provide no direct means of identifying which interface node is the predecessor and which is the successor, we add codes to do this. This is important to define the flow of the relationship.

To review, first we create events nodes in both models with an identical node identifier. The file passing process inserts the Early Start date of the predecessor interface node as an SNE date in the PES field of the corresponding successor node in the other submodel. The file passing process then inserts the Late Finish date of the successor interface node as an FNL parameter of the corresponding predecessor node in the other submodel. Performing new forward and backward passes in each model provides a correct picture of the cross-structure effect.

Unfortunately, this process can fall apart. It works well for a single interface from one model to another. In typical models, however, many constraints cross between organizational or functional interfaces. Work flows virtually continually back and forth between functions. Therefore, we need to reexamine the file passing process under these circumstances to ensure that it still works properly. Figure 14-3 illustrates a very simple multiple interface. As a convenience, we have elected to enter only numeric values into date calculations.

The cross-submodel constraint from TBS to CBS is identical to those illustrated in Figures 14-1 and 14-2. First, we process Model B through the forward pass, then we pull the Early Finish of Task TBS in Model B across the structure for the Early Start calculation of Task CBS. The appropriate date (day 110) would typically become the SNE date parameter (in the PES field) of node (CBS). A forward pass on Model A would drive CBS's Early Start and Finish dates.

Only then can we address the second cross-structural interface, which depends upon the first. The constraint in this case binds Task CBS in Model A and task HBO in Model B. After we have determined the Early Start and Early Finish dates of Task CBS, we pass the EF value back across the structure to find the Early Start date of Task HBO. Again the file passing process inserts this date value (day 135) as an SNE date parameter (in the PES field) of node HBO. Another forward pass in model B generates a correct Early Start date of Task HBO which reflects the effects of both cross-structural interfaces.

Finally, we need the Early Finish date of Task HBO to determine the Early Start date of Task PBS. We pass this date back across and rerun the forward pass on Model A.

Figure 14-3

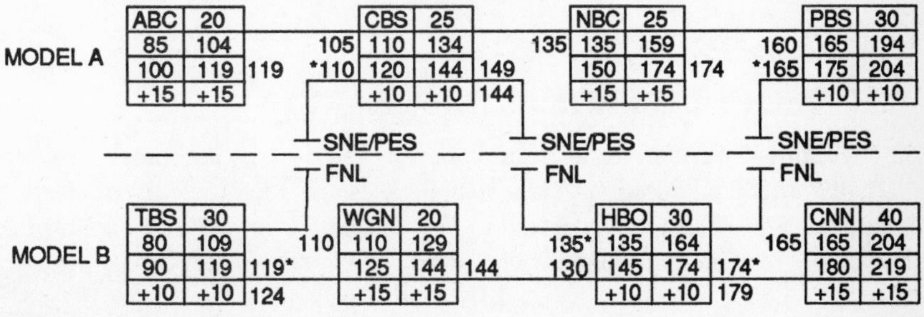

Three interfaces required three separate file passes and two subsequent forward passes through each model. We would have to calculate, pass, and recalculate late dates an equal number of times. This is all very convoluted and time consuming.

We cannot make all of the file passes at one time because each interface affects calculations to its right in the forward pass and to its left in the backward pass. For instance, the Early Finish of HBO at the time of the first forward pass on Model B (workday 159) does not include the effects of the finish of Task TBS on Task CBS. Likewise, the interface value from Task TBS changes the Early Start date of CBS. The file passing process requires that we pass a date, process the forward pass, pass the next interface date, reprocess the forward pass, and so on until we cover all of the interfaces. The number of iterations depends upon the number of cross-model interfaces and their sequence relative to one another. This could be as few as one and as many as the number of cross model interfaces. To be sure that we account for all of the interfaces in the proper sequence, we will have to perform this process once for each cross-model interface.

Figure 14-3 represents a very simple example. A more typical project might have 10 to 15 individual submodels, each with 30 to 50 interfaces to other submodels. We could be passing dates and rerunning CPM processes for several hundred iterations to ensure correct results. This severe problem makes this method generally impractical.

It may still prove useful in proper situations, however. If several somewhat independent components come together in a final assembly, they should create minimal cross-model interfaces. Figure 14-4 illustrates an abbreviated model for final assembly of a satellite. Each of the start events nodes in this submodel corresponds to a finish events node in another submodel. That is, the finished products in the component submodels become inputs to the final assembly submodel. The single cross-model interface would probably allow us to overcome the deficiencies of the file passing process in this special situation, although the assumption of minimal parallel interfaces between the various components is tenuous. Certainly, at a minimum, many design considerations would cross back and forth. A case could be made, however, that the process could work in this scenario.

NESTED DECOMPOSITIONAL INTEGRATION

Another common method of horizontal integration adapts the hammock structure method of vertical integration. This method uses the hierarchical structure to accommodate the horizontal integration. We can best examine and explain this method through an example like Figure 14-5. The overall project model includes

Figure 14-4

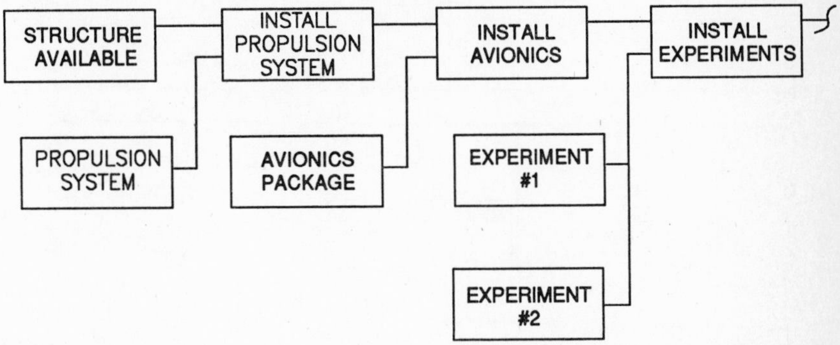

the three activities (A, B, and C) at Level 1. We decompose each of these high-level activities into detail tasks, events, and interfaces. This creates Level 2, which is represented by three submodels: A (A1-A7), B (B1-B10), and C (C1-C10). Further breaking down each of the tasks in these models into more detailed tasks and interfaces establishes Level 3, which is only partially displayed (one submodel each) due to space limitations. Decomposition of activities A6, B4, and C5 has created submodels: A6 (A61-A66), B4 (B41-B45), and C5 (C51-C57).

In this method all cross-model interfaces are normal constraints at the next higher level. A constraint between submodel A6 and A7 in Level 3 is the constraint A6-A7 in Level 2. On the surface, this appears to be a very neat, clean method as it integrates tasks horizontally in a very structured fashion. Each decomposed model appears to be quite autonomous, and the number of records in each submodel remains small as they must by definition combine to form a single activity.

However, even the three levels in the very simple example generate a large number of submodels. This is not a great problem if we maintain discipline. For instance, the node identifier code structure indentures submodels as derived from next higher level activities. For example, nodes B41-B45 all sum to node B4 at the next level, and nodes B1 through B10 are all components of Task B at the next level. This hierarchical node coding structure clearly reflects vertical relationships, which is an example of the kind of discipline necessary to make this method workable. As an integration mechanism this scheme seems almost too good to be true, and unfortunately it is.

This method imposes the same problems as the hammock scheme for vertical integration. As we decompose activities down through a hierarchical structure, the interfaces between the submodels pass into and out of many interme-

Figure 14-5

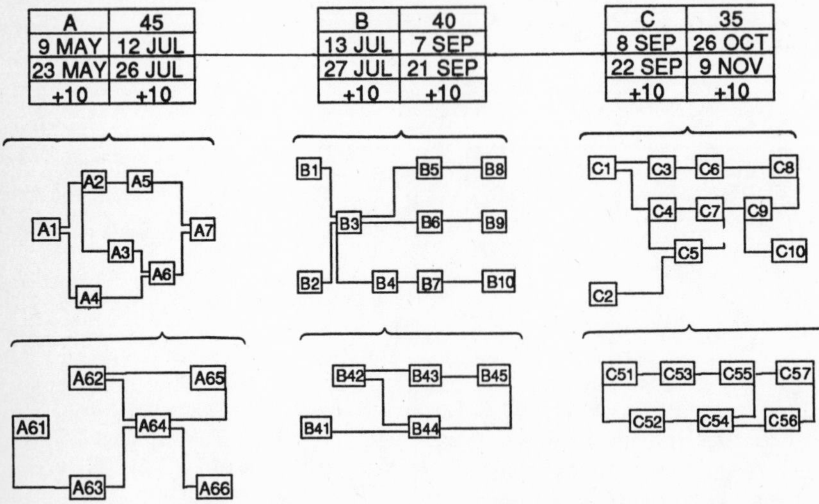

diate nodes, as well as the first and last nodes. That is, once we decompose our project from Level 1 to Level 2, we find that the interface between Models A and B does not necessarily join nodes A7 and B1 or B2 (the last and first nodes in these two submodels). Interfaces could join A6 and B3, A5 and B1, A6 and B4 or any number of other combinations. It is not practically possible to represent the cross-submodel interfaces of one level by a single finish-to-start constraint at the previous level. This is the same reason we discarded the hammock structure and replaced it with the coding structure as a summarization mechanism.

MULTIMODEL INTEGRATION

Instead of dealing with each cross-model interface individually, we need to perform the CPM process on the entire group of submodels, at once taking into account their cross-model interfaces and retaining the analytically derived values (early/late dates and time reserve calculations) on each activity in a single forward and backward pass flow. There is more than one way to achieve this rather difficult goal, but each requires these steps:

1. Combine all of the submodels together.
2. Simulate the cross-model interfaces between them.

3. Perform a total CPM process.
4. Simulate the effects of the entire model at each cross-model interface point.
5. Break the model back into autonomous submodels.

In this way, we take into account all of the detail interfaces of the entire project model while retaining submodel autonomy. We can still simulate the effects of the whole model on each autonomous piece, horizontally integrating data without the previously discussed pitfalls. The general procedure has been adapted into several specific techniques.

Submodel Integration Using Cross-Model Constraints

One technique simply loads all of the submodels together into one database structure and then adds a file that contains the cross-model constraints. This would in fact constitute a total project model yielding completely integrated date and time reserve calculations. Including all of the interfaces at once in a single model eliminates the need for multiple file passing.

A problem arises, however, in step 4, in simulating the effects of the whole model on each cross-model interface. This step is necessary because it enables us to break the total project back into the various submodels, each reflecting the effects of the total project. We simulate the effects of the entire model on each submodel through date targets.

This requires an additional process to extract certain values from the cross-model constraint table, and insert them as date predecessors or successors on appropriate nodes in the submodels. These imposed dates carry the overall project effects to each submodel.

For each constraint in the table, we calculate a finish date in the forward pass and a start date in the backward pass. The forward pass finish date calculation adds the lag of each cross-model constraint to the Early Finish date of the predecessor, then adds one time unit. This date matches the Early Start date processed through the constraint. Entering it as an SNE date (in the PES field) on the constraint's successor models the effects of a predecessor node in one submodel on its successor node in the other submodel.

The backward pass start date calculation subtracts the lag of each cross-model constraint and one more time unit from the Late Start date of the successor, providing the same Late Finish for the predecessor as that processed through the constraint. Entering this date as an FNL date on the constraint's predecessor models the effects of a successor node in one submodel on its predecessor node in the other submodel.

Completing these calculations for all cross-model constraints models the effects of the total project on each submodel. Removing the cross-model constraints to break the submodels apart leaves the residual effect of the total project on each submodel. These dates will remain valid until changes are made to the data in any of the actual modelfiles.

Figure 14-6 graphically illustrates the principles of this process. It shows segments of two submodels (A and B), with a single cross-model interface between Tasks NORTH and SOUTH. In our first step, we combine the two independent models (A and B), and then add the cross-model constraints, in this case the finish-to-start interface between NORTH and SOUTH. We run the CPM processes for the total model and then execute the horizontal integration step.

In Figure 14-6, the forward pass establishes an Early Finish date of May 16 for NORTH (the constraint's predecessor), which, through the equation EF + lag + 1 = May 17, drives the Early Start of SOUTH, it is the latest predecessor. We insert this date value into Task SOUTH's SNE (PES) date field. The backward pass establishes a Late Start date of May 3 for SOUTH (the constraint's successor), which, through the equation LS – lag – 1 = May 2, drives the Late Finish of NORTH, it is the earliest successor. We insert this value into the FNL date field of the predecessor of the constraint, Task NORTH.

Figure 14-6

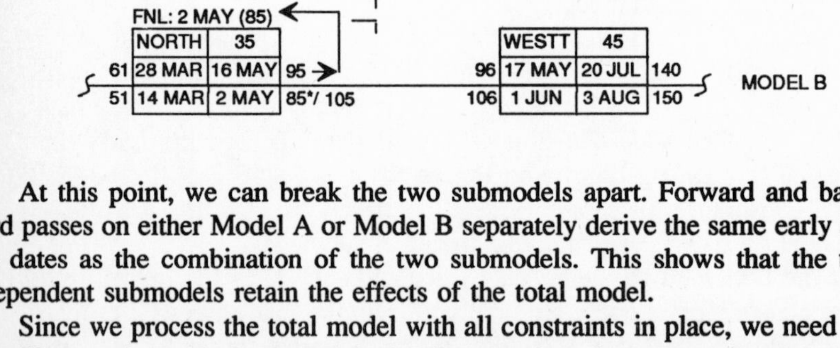

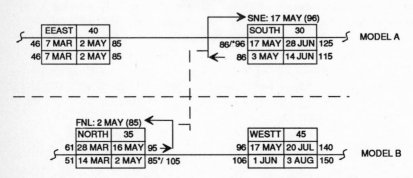

At this point, we can break the two submodels apart. Forward and backward passes on either Model A or Model B separately derive the same early and late dates as the combination of the two submodels. This shows that the two independent submodels retain the effects of the total model.

Since we process the total model with all constraints in place, we need not pass files continuously back and forth. We complete the process in one set of passes. This fairly simple, clean process has only one drawback. In their independent states, the two submodels provide no indication of the cross-model interface other than the imposed dates, which might exist on the nodes for any number of other reasons. Model analysis requires understanding of what external constraints affect our project at what points. Therefore, we need to modify the process slightly to add visibility of the external interfaces in all of the submodels.

Submodel Integration Using Common Nodes

The final process takes bits and pieces from all of the previously discussed mechanisms to accomplish all essential steps of horizontal integration while also identifying the sources of the interfaces in each submodel. Clearly identifying the source of every interface facilitates analysis by allowing each submodel to stand alone as a project element. With both the effects and the sources of the external constraints clearly visible, we can analyze each individual submodel with attention to its relationship to the rest of the project. We can narrow our analytical focus to each individual submodel, and yet clearly see the interfaces that pass into or out of it. In total, our analysis of all of the individual submodels constitutes an end-to-end analysis.

The method by which we do this resembles the cross-model constraint process discussed earlier, except that we simulate the external constraint in both submodels by terminating each interface at an events node. We include the cross-model constraint in both of the interfacing projects. The cross-model interface actually passes through these special events nodes which will be called *interface nodes.*

Our horizontal integration process includes nine steps:

1. Define the cross-model interfaces (constraints) by identifying all predecessor and successor activities and the modelfile names of the submodels in which they reside.
2. For each interface, establish corresponding events nodes in both models that share an identical node identifier. Code each of these nodes with a value in a special interface parameter (INI) that labels it as an interface node (I).
3. On the predecessor side of the interface, the interface events node is a finish node (i.e., one with no successors). The latest immediate predecessor task within the first submodel defines its Early Finish date. The process will label these nodes as interface predecessors (P) in another special parameter field (INT).
4. On the successor side of the interface, the interface events node is a start node (i.e., one with no predecessors). The latest immediate successor task within the second submodel defines its Late Start date. The process will label these models as interface successors (S) in another special parameter field (INT).

Figure 14-7

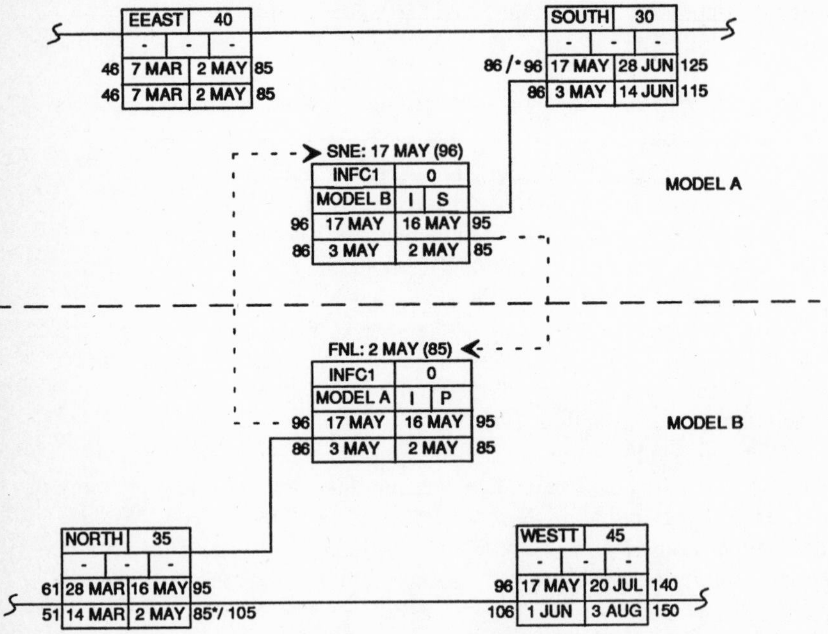

We will cover Steps 5 through 9 after reviewing Figure 14-7 to see how this process works.

The two events nodes that indicate a single cross-model interface carry identical node identifier codes (e.g., INFC1) establishing a common event in each of the submodels. This pairing creates the link of each interface.

Together, any and all predecessor activities in Submodel NORTH define the Early Finish of the first interface events node INFC1 in Submodel B through the finish-to-start constraint NORTH-INFC1. The second interface events node has an identical identifier INFC1 in the successor Submodel A. It affects the starts of successor tasks in Submodel A through the finish-to-start constraint INFC1-SOUTH.

The events nodes INFC1 in the cross-model constraint between Tasks NORTH and SOUTH function just like a single constraint between the two tasks. The events nodes will retain the effects of the interface, however, once the submodels break apart. The special parameter INT distinguishes interface nodes from all other nodes, as illustrated in the second row, middle column where the value I appears on the two interface nodes, while blanks (null values) appear on all other nodes. The predecessor interface node may have more than one predecessor and the successor interface node may have more than one successor. The normal rules of multiple predecessor or successor relationships (latest early date and earliest late date) apply in these cases.

As part of our process, we must identify which date will be retained on which interface node. The disciplined method for setting up interface nodes in steps 1 through 4 allows the process to label any events node that meets the conditions described in the third step as a predecessor interface node (INT = P). Likewise, it can label any interface events node that meets the conditions described in the fourth step as a successor (INT = S). This eliminates a step the user would have to perform, which is to identify the directional flow of the interface.

In our example, this parameter (INT) appears in the second row, far-right column. The interface node in Model B that meets the conditions of step 3 shows a *P*, while the interface node in Model A that meets the conditions of step 4 shows an *S*. All other nodes reflect no values as they meet neither the conditions of step 3 or 4. These labels define the direction of flow of the cross-model constraint from Model B to Model A. This determines where the imposed dates will go.

In step 5, we merge all of the submodels together into one modelfile and perform a CPM process on the entire project. The special identification label (I) causes the integration process to allow nodes with identical node identifiers to exist in the same modelfile. When running the entire model (the collective submodels) through the CPM time analysis, it treats interface nodes (INI = I) with identical node identifiers as a single node. In effect, it merges them together and leaves each pair of interface nodes with identical calculated data.

Once the total CPM process is complete, the process imposes certain date parameters on each of the interface nodes to retain the effects of the total project. In step 6, it imposes the calculated Late Finish date (e.g., May 2) of the combined interface node as an FNL date on the predecessor interface node (INT = P). In step 7, it imposes the calculated Early Start date (e.g., May 17) of the combined interface node as an SNE date in the PES field of the successor interface node (INT = S).

When the process accomplishes these steps for all interface node pairs in the project, it can rebreak the model into its distinct submodels. In any subsequent CPM time analysis of Model A, the constraint INFC1-SOUTH (through the SNE date) has an identical effect to the cross-model constraint NORTH-SOUTH. Likewise, the constraint NORTH-INFC1 in Model B (through the FNL date) also has an identical effect to the cross-model constraint NORTH-SOUTH. We have retained the effects of the total project on each submodel, until, of course, the data in either is changed.

The interface nodes reflect the cross-model constraints in both modelfiles, but this only partially completes interface identification. We still lack any indication as to the source of each interface. To correct this, we simply add step 8 to our process. After imposing the dates on each interface node, we also insert certain values into special fields on each of these nodes:

a. We insert the modelfile name of each predecessor interface node into a special field (IFS) of its corresponding successor interface node.
b. We insert the modelfile name of each successor interface node into the same special field (IFS) of its corresponding predecessor interface node.

In effect, the identical nodes exchange modelfile names. After splitting apart, the interface nodes still indicate the other side of each interface through this value. This value appears in the two interface nodes of our example in the second row, the far-left column (MODELA in Model B, and MODELB in Model A).

In step 9, the process breaks the total project back down into individual models. Each independent model clearly identifies and simulates the cross-model constraints and the process has successfully been completed.

Our horizontal integration process requires three special parameters on each interface node; the user sets one, and the process sets the other two:

Interface Node Identifier (INI): User-defined with value *I* or no value

Interface Node Type (INT): Process-defined with value P for predecessor or S for successor

Interface Source (IFS): Process-defined with source modelfile name(s)

These fields will reflect values for all interface nodes and null values for all other nodes. The user identifies nodes involved in cross-model constraints through the INI parameter; the process recognizes this and sets the appropriate values in the other two fields during the horizontal integration process.

Through this horizontal integration process, we can simulate all cross-model interfaces, conduct a total project CPM time analysis, and then simulate the effects of the total project in each submodel. In addition, we have transcended these essential objectives of horizontal integration by identifying the source of the interface in each interface events node. This yields a very thorough, yet simple method of accommodating the complex requirements of horizontal integration.

We achieve these results less by applying the process than by maintaining user discipline. Even more than most other processes, this scheme requires considerable organization and coordination within the project structure for successful implementation. Interfaces that cross submodels must be carefully identified and monitored. The process works properly only in the presence of proper schedule control and planning discipline. It also creates a responsibility because once we identify these interfaces—then we must manage them.

Multiple Predecessor/Successor Interface Nodes

Events quite commonly have successors or predecessors in more than one submodel. This is called a *one-to-multi* or *multi-to-one cross-model relationship.* It creates a complication that we must accommodate in our horizontal integration process without disrupting or invalidating it, and without adding excessive new complexity.

As one means of accommodating these situations, we could create a new series of node pairs to model every interface. For a one-to-multi relationship, we would create several interface events nodes in the predecessor's submodel that represent the same event, but each with a different node identifier. Each of these nodes would correspond to a duplicate successor interface node in each successor submodel.

Similarly, to model a multi-to-one relationship, we would create several interface events nodes in the successor's submodel to represent the same event, each of them having a unique node identifier. Each would correspond to a duplicate predecessor interface node in the various predecessor submodels. Creating a pair of nodes to simulate each interface, we generate a series of nodes to represent a single event.

This technique gives us the advantage of a unique identity for every interface. It does so, however, at a cost of creating many extra nodes to model a single event. A simple addition to our original process can prevent this while still taking multiple relationships into account.

Figure 14-8 shows two examples of multiple cross-model relationships. The left-hand example (IF2) shows a multi-to-one relationship in which the interface events node in Model A has two predecessors, one in Model B and the other in Model C. These two events will start on different dates, as reflected in their Early Start dates (September 19 in Model B and October 3 in Model C).

First, we relate these three events nodes together by giving them all the same node identifier, along with the appropriate interface node identifier (INI = I). The augmented process determines that this interface consists of more than one pair of nodes, indicating a multiple interface relationship. It also determines that two of these nodes are predecessors (INT = P) and one is a successor (INT = S). This invokes the additional multiple interface process.

To begin, we must determine which predecessor drives the successor. Since the successor task cannot start until completion of all predecessors, the driver would be the last one to finish. The process determines the latest Early Start date from the predecessor information (e.g., October 3) and imposes this

Figure 14-8

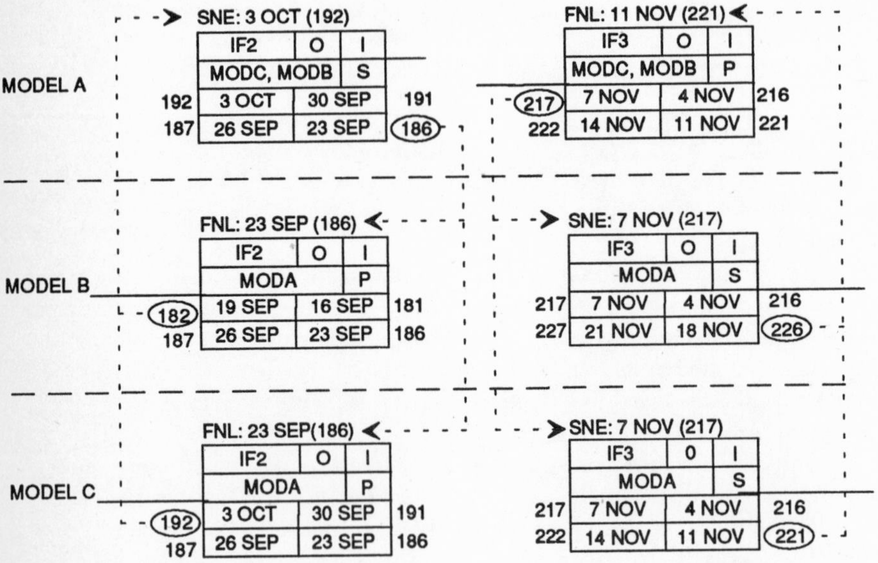

date as an SNE date in the PES field of the successor interface node in Model A. This models the net effect of the forward pass on these two constraints.

The process imposes the backward pass date (the LF of IF2 in Model A) from the lone successor node (September 23) as an FNL date on each predecessor interface node. This models the requirement imposed by an activity's successor in each of the submodels.

This completes the horizontal integration process for this group of nodes. Note that the early dates of these nodes are no longer identical (ES/EF of IF2 in Model B) as we retain the calculations made in each submodel.

The right-hand example (IF3) shows a one-to-multi relationship in which the interface node in Model A has two successors, one in Model B and the other in Model C. Again, the process first determines that the constraint involves more than a single pair of nodes. This multiple interface invokes the multiple relation process. Two of these nodes are successors (INT = S) and one is a predecessor (INT = P). The process imposes the Early Start date of the single predecessor (November 7) as an SNE date in the PES field of all successors. In the backward pass, each of the two successors has its own requirement date (November 18 in Model B and November 11 in Model C). The process must determine which successor drives the predecessor. To satisfy all successor requirements, all predecessors must finish prior to all successors' late dates. The process determines the earliest Late Finish date of the successors (November 11) and imposes it as an FNL date on the predecessor interface node (Model A). This parameter completes the requirements for determining the net effect of this group of interfaces. Note that the late dates of the three nodes are no longer identical (LS/LF of IF3 in Model B).

After setting the dates in each node, the process next exchanges the modelfile names. In our first example, it sets the modelfile name of the successor in Model A (MODA) in the interface source (IFS) parameter of both of the predecessor nodes, one in Model B and one in Model C. This is basically no different than before.

In the multi-to-one relationship, however, the process must set both of the modelfile names of B and C (MODB and MODC) into the IFS parameter of the interface node in modelfile A. To accommodate these two values, and maybe many more depending upon the interface, this parameter (IFS) must be a large text field. The process groups the modelfile names in the field, separated by either spaces or commas.

The sequence of the modelfile names depends on the date calculations. It establishes this order from the latest Early Start to the earliest Early Start [e.g., MODC first (October 3), then MODB (September 19)]. In this way, the interface node in Model A, IF2, shows that this node's Early Start date is derived from an interface in Model C, but is also constrained by an interface in Model B.

In the right-hand example, the process would insert the modelfile name of the single predecessor interface node in Model A (MODA) into the IFS field of both successors, one in Model B and one in Model C. It would need, however, to set both successor modelfile names (MODB and MODC) into the IFS field of the interface node in Model A, again ordered according to the appropriate date calculation. In this case, the order would run from the earliest Late Finish to the latest Late Finish [e.g., MODC first (November 11), then MODB (November 18)]. The interface node IF3 in Model A would then show that its Late Finish date is derived from an interface in Model C, but it is also constrained by an interface in Model B.

Since the multi-to-one relationship process affects the calculation only of the early dates and the one-to-multi relationship process affects the calculation only of late dates, it is not difficult to extend this process to cover a multi-to-multi relationship. We could, that is, have multiple predecessors and multiple successors all tied together simply by combining the two multiple relationship methods. This process would impose the latest Early Start date from the group of predecessor interfaces as an SNE date in the PES field of all successor interface nodes. It would then impose the earliest Late Finish date from the group of interface successors as an FNL date on all predecessor interface nodes. Properly exchanging modelfile names would place multiple modelfile names in all nodes, completing the process. This is complex, but acceptable provided the user can manage the discipline.

It should be apparent that the multiple interface process works identically to any other multiple predecessor or successor relationship, but for interface events. Therefore, our horizontal integration process must vary its steps slightly to accommodate the interfaces external to each model, but it remains totally consistent with the way any internal constraints are processed, taking the earliest late date, the latest early date, etc. It adds a minimum of complexity to our process, since it adds fewer nodes than other alternatives. Further, the common node identifier on all nodes involved in the multiple interfaces enhances the visibility of any given interface.

SUMMARY

Horizontal integration must follow this process in order to achieve its objectives:

1. Combine all submodels together
2. Simulate the cross-model interfaces
3. Perform a total CPM process
4. Simulate the effects of the entire model at each point of cross-model interface, such that each independent submodel will derive identical calculated data to the whole
5. Indicate the source of each interface in both submodels
6. Break the model back down into autonomous submodels

Of the several horizontal integration schemes to accomplish these goals, many do not work properly.

The file passing method maintains the individual submodels separately, passing integration data back and forth between them electronically. This will work until multiple parallel interfaces arise between the submodels. These could require multiple file passes and CPM processes in order to account for all interfaces, making this method too convoluted for practical use. This method will work when the submodels are structured in a linear fashion.

The nested decompositional integration method handles horizontal integration through a hierarchical structure. It always addresses cross-model constraints at the next higher schedule level in a manner very similar to hammock mechanisms. It fails for the same reasons that hammocks failed as a vertical integration mechanism—once a series of tasks is decomposed, the relationships at the previous level of detail lose validity.

The cross-model constraint method groups all submodels into a single modelfile, adding a file of constraints to tie the submodels together. It then runs the entire model through the CPM process. An additional process imposes appropriate dates on all nodes involved in the cross-model constraints to retain the residual effects of the entire model. The process works properly, but provides no visibility to the horizontal integration.

The most complete method of horizontal integration is accomplished through the following process:

1. Define the cross-model interfaces, identifying all predecessor and successor activities and the modelfile names of the submodels in which they reside.
2. For each interface, establish corresponding events nodes in both models that share a common node identifier. Code each of these nodes with a value in a special interface parameter (INI) that labels it as an interface node (I).
3. On the predecessor side of the interface, the interface events node is a finish node (i.e., one with no successors). The latest immediate predecessor task within the first submodel defines its Early Finish date. The process will label these nodes as interface predecessors (P) in another special parameter field (INT).
4. On the successor side of the interface, the interface events node is a start node (i.e., one with no predecessors). The latest immediate successor task within the second submodel defines its Late Start date. The process will label these nodes as interface successors (S) in another special parameter field (INT).
5. Merge all of the submodels and perform CPM calculations on the entire project.
6. Impose the LF date from each combined interface node as an FNL date on the corresponding predecessor interface node(s).
7. Impose the ES date from each combined interface node as an SNE date in the PES field of the corresponding successor interface node(s).
8. Exchange the modelfile names. Enter the project modelfile name of each successor interface node into a parameter field (IFS) in its corresponding predecessor interface node(s). Also enter the predecessor's modelfile name into a field in the successor interface node(s).

9. Break the total project back down into individual submodels. Each submodel clearly identifies and simulates the cross-model constraints.

When an event has predecessors or successors in more than one submodel, it creates a multiple relationship. A node established in each submodel with the same node identifier represents the event. The process recognizes a multiple relationship, then:

a. Imposes the latest Early Start date of the predecessors as an SNE date in the PES field of each successor
b. Imposes the earliest Late Finish date of the successors as an FNL date on each predecessor

CHAPTER 15

Output Products

Output products provide the principal means of communicating the needs and results of the modelfile and the modeling processes. This capability depends almost entirely upon the application software, so product needs strongly influence the choice among software alternatives.

Output products provide a variety of information in varying formats to address many different requirements. These styles do, however, fall into product families, each with its own formats, characteristics, and options, including tabular reports, model plots, Gantt barcharts, and a group of plots called management graphics (histograms, piecharts, and X-Y plots). This collection of reports can help managers identify problems associated with the structure or proper modeling techniques, analyze and reconcile data, communicate information about project status or other conditions, isolate problems with the work itself, and depict results of potential solutions. Output products can serve as broad a spectrum of purposes as the scope of information contained in the database. The most basic of the output product families, and the one that should have the most extensive use, is the tabular report.

TABULAR REPORTS

A tabular report displays information taken from the modelfile record by record, listing some of their parameters in regular columns across the printed page. Typically, we print such tables on 14 by 11 inch or 8½ by 11 inch paper, which limits the number of fields that can be shown. The purpose of each report also restricts the information in it. We must, therefore, screen data records and arrange them appropriately to achieve our specified purpose.

The wide variety of uses for tabular reports creates a need for very flexible software that allows us to generate reports in a broad spectrum of formats. Unfortunately, we can frequently gain flexible reporting only at the cost of complexity. This can diminish practical flexibility if the user cannot master the software's complexity. To avoid this vicious circle, the software application must provide the necessary flexibility, yet in a simple enough fashion to be useful to someone other than a software programmer.

The resulting tabular report generation software is called a *tabular report writer* (TRW). Let us examine some of its functional requirements.

Field or Parameter Control

Each activity or constraint record contains a great many parameters (see Appendix G for a list of possibilities). Even if paper-width limitations allowed a single report to list all of these parameters, limits on comprehension would make it undesirable. It is too much data to be of practical use at one time. The width of the page (up to 132 characters) and the requirement of the report establish what data needs to be portrayed. Usually this width is sufficient, provided the tabular report writer gives us adequate controls over the format of the output.

As its first requirements, the tabular report writer must easily allow the user to:

1. Specify the fields in each report
2. Specify the sequence (left to right) of these fields

In addition to these essential requirements, it should also allow the user to:

3. Specify headings for each field
4. Control the formats in which the report displays the specified fields

The first two requirements are the most important and the easiest to satisfy. The tabular report writer should ask the user to identify the fields (parameters) on the report. The user then lists the field names in the order in which they will appear across the page. One simple response provides the tabular report writer the necessary information to satisfy the first two critical requirements.

For example, we might need a report to help us reconcile our schedule to the model. This requires that we compare schedule data to forward pass data for each activity, so we need listings of the fields that contain this data. The sequence of the fields should facilitate the appropriate comparisons, grouping start dates together, then grouping finish dates, for example:

NI DU BES PES AES ES BEF PEF AEF EF

This report would list the task's node identifier and its duration (although the duration is not absolutely necessary), followed by two groups of dates, one of which gives information about the task's start (BES, PES, AES, and ES), while the other gives similar information about its finish (BEF, PEF, AEF, and EF). Each date group shows side-by-side the dates that we must reconcile. We can quickly catch the disparities, especially if we separate the start dates from the finish dates with some empty space. The sum total of all the field lengths, plus any breaks, defines the total length of the record line on the report.

The specified fields of data will appear on the report in columns. Each column must provide sufficient width to accommodate the necessary information. For example, date fields must be either eight or nine characters wide, depending on the date convention used (dd/mm/yy or dd-mmm-yy with the delimiters counted as characters). Any titles above the columns to describe the data must conform to the field format. To label the field BES as "BASELINE START DATE," we would stack the words in the column, as:

BASELINE
START
DATE

The longest word in the title (BASELINE) is eight characters long, so it fits within the column width in either date convention. We can specify where to break words in the column headings and link column titles to specific fields through a wide variety of techniques. The method is not important, as long as some method is available and simple to accomplish. All methods basically provide a table or file that matches each field name to the corresponding heading on the report, as well as format information for each title.

For some reports, we need to alter the formats of certain fields. Text fields, such as the textual activity description, appear in a variety of formats. To set the number of characters allowed for each description, we generally define this field last, which places it at the far right of the report, setting its width to match the remaining available space. This requires the capability of truncating text fields to specific lengths.

We also need the ability to control field formats to:

1. Limit the length of integer, alphanumeric or text fields
2. Specify a date format
3. Run text fields to subsequent lines (called *word wrap)*
4. Add prefixes or suffixes to field values

Since the record fields display the parameters of the model, we need to exercise maximum control over them.

Record Selection, Order, and Zoning

Chapter 11 described how a software package must allow the user to screen or select certain data, and to control the sequence of the data from top to bottom. Reports impose similar requirements.

Many reports address only portions of the model's activities, events, and constraints. In fact, a report on the total database would lose its focus in the avalanche of information, masking or obscuring the purpose of the report. Selecting only pertinent data allows us to focus our analysis, better tailoring the report to specific needs. This essential feature of our tabular report writer is called *data selection* or *data filtering*.

In this process, we impose criteria by which the software then segregates the data for the report. Specification of the segregation criteria is accomplished through values in the fields of each record. We could select certain groups from our structure fields, or schedule dates earlier than or later than a particular date, or tasks reflecting marginal or unacceptable levels of time reserve, or combinations of any of these. Satisfying complex information needs requires more than a list of filtering criteria; instead, we need a separate process specifically devoted to criteria specification.

This process consists of three steps. First, we designate the field or subfield that contains the criteria values. We then specify conditions such as equal to (=), not equal to (≠), greater than (>), less than (<), present, absent, contains, etc. Finally, we specify acceptable values, for example, the structure field value, a date value, or an integer value. At the end of these three steps, we have generated a simple selection criterion statement such as:

Tasks with a given value in a structure field: S1 = DESIGN

Tasks with schedule dates prior to a certain date: BEF < 30-JUN-96

Tasks that fall short of certain levels of PTR: FPTR < +30

Tasks with projected finish dates: PEF PRESENT

We can generate such a selection statement in several alternative ways. Regardless of how we communicate this information to the tabular report writer, it is only important that we can instruct the process to filter or select the data, and limit the report to those records that meet the specified criteria.

Complications in the criteria impose further requirements on the data selection process. For example, complex data selection may require multiple criteria values, including a list of values after the operative expression, separated by a delimiter such as a comma or a space. This delimiter character must never appear in any of the field values. To select records with the values DESIGN, PROCUREMENT, or FABRICATION in the S1 field, we would enter: S1 = DESIGN/PROCUREMENT/FABRICATION.

In another complicated situation, complex selection criteria may require more than one selection statement to fully state the needs. We might need to select all design activities (S1 = DESIGN) scheduled to finish before June 30, 2001. To address this type of requirement, we need the ability to link select statements with "AND" or "OR" operators. Adding the AND operator would solve the example problem, as: S1 = DESIGN AND BEF BEFORE 30-JUN-01. The tabular report writer would select only records that satisfy both of these criteria.

Suppose, instead, that we needed data for activities in either the procurement department (S1 = PROCUREMENT) or another structural requirement (designated PR3365 in the S2 field). The OR operative would allow us to select any data that meet either requirement, as: S1 = PROCUREMENT OR S2 = PR3365. The tabular report writer would select data that satisfied either of these criteria. The data need not satisfy both.

As the selection criteria become more and more intricate, the need arises for virtual mathematical expressions within parenthesis ( ) and brackets [ ] to control the sequence of AND and OR operators. Many common tabular reports require such flexibility, and the tabular report writer's select function must include features that provide it. Such a complex selection statement might select data for a certain type of Soon-to-Come Due report: (BES < 7/1 AND BES > 3/31) OR (BEF < 7/1 AND BEF > 3/31). All of this requirement creates a serious dilemma in that it is difficult to accommodate without complexity. This can be done, but it requires very innovative software.

Three different mechanisms of the tabular report writer control the sequence of data: page breaking, regional zoning, and data order. To distribute certain information to different people, the tabular report writer must begin each person's section of the report on a new page. This *page breaking* function, like data selection, relies on values in the fields of the records, although the process is entirely different. To break pages for reports to different departments or different work groups, these would be distinguished in a specific structure field or subfield of each activity record in the modelfile.

The process would group data records together that had matching values in the designated page break field. It would then sort these groups by the alphanumeric sequence of these values. As it printed the report, the tabular report writer would cause the printer to skip to the top of the next page each time the value in the page break field (or subfield) changed. This would separate the report into distinct sections according to the values in the user-specified page break field.

Within sections of a report, we often need to group certain records together and separate them visually from the other records. This separation could take the form of a skipped line, several skipped lines, and/or a horizontal line across the page. To enhance this separation function, the report writer might also provide a title for each of these horizontal bands, or *regional zones,* of data. This terminology gives a name to the mechanism: *regional zoning.*

It resembles page breaking, in that it groups data records that share a user-specified zone field or subfield. It then orders these groups by this value, and finally separates each group in some way (i.e., with spaces, lines, titles, etc.). Regional zoning follows page breaking, so the tabular report writer sorts the selected data first by the paging criteria, and then by the zone criteria. This organizes a report into separately paged sections, each with groups or bands of data.

Data within any regional zone may need further organization, however. The function that does this is called *data sort* or *data order.* It also requires specification of a field or subfield, but this field must contain different values in order to establish an ordering medium between the various records. The tabular report writer orders the records by values in this field. If the user specifies an alphanumeric or text field, it orders the data in alphanumeric sequence (either right or left justified). It sorts date field data chronologically, and integer or decimal fields numerically.

The possibility of identical sorting criteria complicates data ordering. If, for example, we wanted to order activities by their finish dates, we would have to accommodate the possibility that more than one activity in a zone would share a single scheduled finish. This raises a need for a secondary ordering criterion. The tabular report writer could sequence records with the same value in their primary order field relative to one another by the value in the secondary order field. In our example with finish dates as the primary sorting criterion, we might specify start dates as the secondary sort criterion. Activities within each zone that shared a finish date would be ordered relative to one another by their start dates.

In some cases, the values in both the primary and secondary criteria might match, raising a need for third or even higher-level criteria to ensure that every record has a unique sequence discriminator. If we do not specify adequate ordering criteria, the tabular report writer defaults to a field with a unique value on each record like either the sequence of the records in the modelfile, or their node identifiers.

We need not specify page break and zone criteria for every report. We can group data into regional zones without page breaks, and specify page breaks without zones; the particular combination of these grouping techniques depends on the requirements of our report. Without data ordering criteria, however, the tabular report writer will sequence the entire report by its default mechanism. In whatever combination, these three mechanisms enable us to organize our data in ways that maximize the effectiveness of the reports.

Other Format Considerations

Several other features of the tabular report writer control information that borders the report, such as titles, legends, comments, and other textual descriptors. A standard report can have any number of appearances, and the tabular report writer can achieve the desired appearance in any number of ways. It is sufficient to state that the tabular report writer must accommodate our need to place additional textual information around the borders of our reports.

Special Reports

Some common project modeling reports have specialized attributes that require individual description. Notable special cases include the Soon-to-Come-Due and Predecessor-Successor reports.

As we track the progress of a project, a report listing all activities and events due to occur in the near term proves very useful. This list, commonly called a *Soon-to-Come-Due report,* imposes very complex data selection requirements, some of which we have mentioned. The Soon-to-Come-Due report should display all activities scheduled to occur during a specified time window defined by a start date and a finish date. This sounds simple, but some variations complicate the process.

Figure 15-1 shows four activities that occur during a specified time window (April 1 through June 30). Activity B finishes, activity C starts, and activity D both starts and finishes in this time window. These four events (two starts and two finishes) must without question appear in a Soon-to-Come-Due report. Conversely, Tasks A and F start and finish either prior to or after the window, excluding them from the report. These five nodes illustrate the obvious choice of events for a Soon-to-Come-Due report.

An activity like E, however, creates a conflict. This task starts prior to the time window and finishes after it; it is on-going during the entire specified window. Should it be listed on the report? More generally, should we report all activities on-going during the window or just those that start and/or finish in the window? The answer depends on the user's requirements, so the user should be able to choose whether the report lists these activities. This choice significantly affects the combination of select statements.

To define Soon-to-Come-Due reports, we must also consider whether to judge activities' inclusion in the time window only by their baseline dates (BES

Figure 15-1

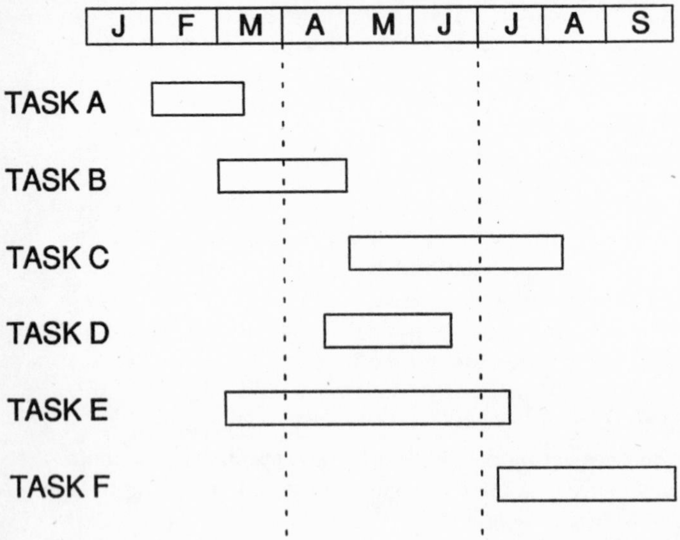

and BEF), or by both baseline and projected dates (PES and PEF), or by projected dates alone? Tasks quite commonly have projected dates in the time window with baseline dates outside the window, or vice versa. Again, the answer should match the requirements of the user, although either way, it adds further complication to the combination of selection statements.

Choices for these two options create four possible Soon-to-Come-Due report variations. We could generate either a standard report with variations, or a series of four reports. Each variation would have its own set selection statement, but all of the other options (field control, page break, zoning, and ordering) need to be established by the user. The uniqueness and complexity of this multiplicity of select criteria for the Soon-to-Come-Due report requires that we segregate it from the general record listing in the tabular report writer.

Another report that displays activities along with their predecessors and/or successors significantly benefits model analysis. This *Predecessor-Successor report* requires distinct and very complex ordering criteria. The tabular report writer must order activities by different criteria than it applies to the activity's predecessors and/or successors. This special report format employs all of the normal tabular report writer features as discussed, but they can affect only the activity records in the modelfile. The user should choose whether to include each activity's predecessors, or successors, or both in the report, but the tabular report writer controls the format of these records (fields, order, etc.) for simplicity. This unique requirement suggests that the Predecessor-Successor report be segregated from the general record listing in the tabular report writer.

Tabular reports can be the most valuable means of communication between the database and the user. They show only data without any graphics and are relatively fast to produce. They require, however, a very powerful tabular report writer, that is still simple to use. This has a major influence on the selection of a software application.

MODEL PLOTS

Basic Layout or Format

Model or logic plots facilitate model analysis by graphically displaying the relationships between activities and their interfaces or constraints. This picture of all three model elements together shows clearly how they relate to one another. Activities appear as boxes or rectangles, subdivided into small compartments that show specified field values of each activity's attributes. Constraints appear as lines between the node ends, with labels indicating constraint types and lag values. (Finish-to-start constraints, the most common type, generally carry no labels, but all other types are indicated.)

Figure 15-2

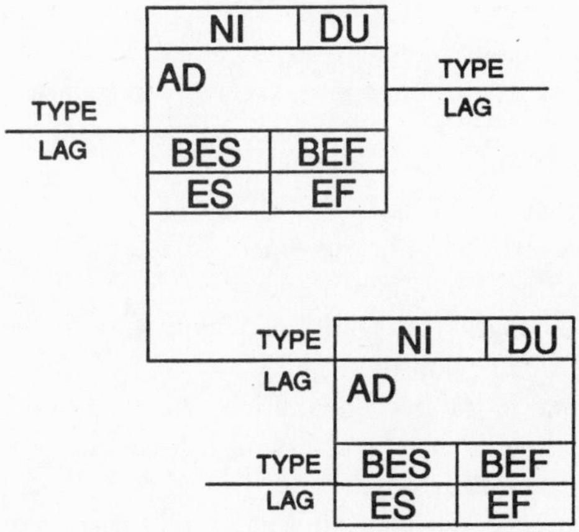

Many model plots appear throughout this text to depict the modeling processes. A general sample is illustrated in Figure 15-2. The user can control many aspects of the model plot's appearance through the software interface called the *model plot writer* (MPW).

Node Format and Control

The node or activity box portrays the data pertaining to all activities and events. Much like the tabular report, we need to control what information appears here and how it appears. Also, as with tabular reports, no plot has enough space, nor does the user need to show all the fields associated with each activity at any one time. We need the ability to:

1. Designate the fields to appear in each node
2. Format the location of these fields within the node
3. Format the length of each field

Model plot writers accomplish these three requirements in different ways, but again it is only important that they do it easily by whatever method. This could be done through a graphic interface in which the model plot writer provides a blank node where the user specifies the fields to depict in their relative positions, with definitions of their lengths.

Figure 15-3 shows such an interface. In the first line, we specified four fields: NI, DU, CAL, and S2, in that order, which defines the fields and their positions within the node. The figure also shows how we specified the length of each field: NI has eight characters, DU has four characters, CAL has two characters, and S2 has six characters. Note that we have elected to show only part of the S2 field. The line on which each field appears and its sequence within that line establish its position relative to all other fields within the node. We enter a delimiter character, in this case a slash (/), to separate the fields. In the actual graphic node, this becomes a vertical line which cordons off a separate compartment for each field.

In Figure 15-3, we could have placed any of the record fields beneath or above the node. This reflects an enhancement of the model plot writer that allows us to place specified fields above and/or below the node box.

Specifying field lengths does not adequately format them. Consider, for example, the activity description field (AD). Summing the number of characters in every field on a line in the node determines the total line length. This gives a length for the first line of 20 characters, 60 for the second, 26 for the third, etc. To lines with more than a single field (i.e., lines 1, 3, 4, and 5), we would like to add blank spaces before and after each field to improve the visual presentation of the field values. These spaces increase the lengths of these node lines, as do the vertical box delimiters, boosting the lengths of line 1 to 31 characters, of lines 3 and 4 to 34 characters each, and of line 5 to 28 characters. Plotting this node as formatted would create a very long, narrow node with a great deal of empty space (i.e., the last 30 characters of lines 1, 3, 4, and 5). The 60-character length of the activity description field causes this problem.

Figure 15-3

NODE FORMAT - USER CONTROLS

```
NI - 8 / DU  -  4 / CAL  -  2 / S2  -  6
AD - 60 - 30 WW
BES - 9 / S3  -  8 / BEF  -  9
ES - 9 / S4  -  8 / EF  -  9
SPTR - 6 / ASF  -  2 / AOS  -  3 / FPTR  -  6
```

Some additional field formatting, however, to break the AD field into two 30-character lines solves this problem. These lines employ word wrapping, that is, the text breaks after the last word that fits within the specified length. This would place the activity description on lines 2 and 3 in the node, each of them about 30 characters long (30WW), making each node one line wider. The total length of the node would correspond to the longest remaining line, or only 34 characters, giving a better shape both aesthetically and for efficient space use.

To finish formatting the fields, we need to select a date convention with the delimiters (i.e., dash or slash) to separate days from months from years. We must add these few additional node format requirements beyond those addressed in the graphic interface (Figure 15-3) to the capabilities of the model plot writer. Once we satisfy all of these needs, the node format would appear as in Figure 15-4. This example shows only the field names, whereas the actual plot would include each node's parameters. This information allows us to perform our analysis.

Figure 15-4

<table>
<tr><td>NI</td><td>DU</td><td>CAL</td><td>S2</td></tr>
<tr><td colspan="4">AD</td></tr>
<tr><td>BES</td><td colspan="2">S3</td><td>BEF</td></tr>
<tr><td>ES</td><td colspan="2">S4</td><td>EF</td></tr>
<tr><td>SPTR</td><td>ASF</td><td>AOS</td><td>FPTR</td></tr>
</table>

We have supplied all the fields necessary for the activities, but not the parameters that describe the events (start and finish) associated with each activity and events node. A simple switch in the software interface turns the display of event data on or off. When the switch is turned off, only the activity information appears, as in Figure 15-4. When this switch is turned on, any event indicated by a value in either the SEF field or the FEF field will appear in the proper milestone shape above the appropriate start end or finish end of the node. The text value in either the SED or FED field appears next to the milestone. The result is illustrated in Figure 15-5. If this event information appears, we must ensure that no activity fields are formatted to appear above the node to prevent these values from overwriting the event information.

Figure 15-5

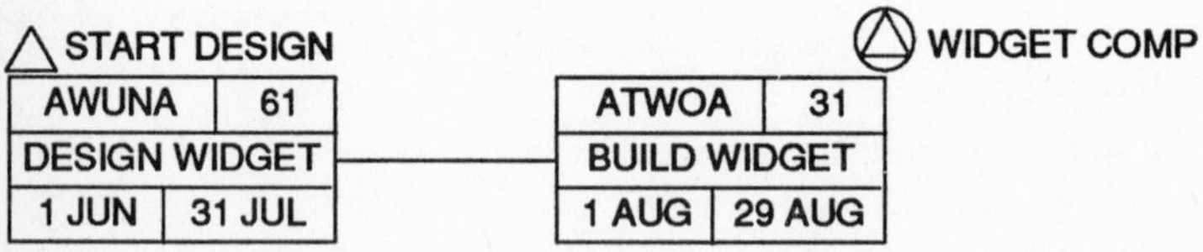

Besides the parameters of the activities and events, some other key characteristics of the project need emphasis. Almost any model plot features some especially important nodes or some parameter or other facet that needs special attention. To meet this need, we can vary these nodes' shapes to indicate their special characteristics along the lines of the examples in Figure 15-6.

Figure 15-6

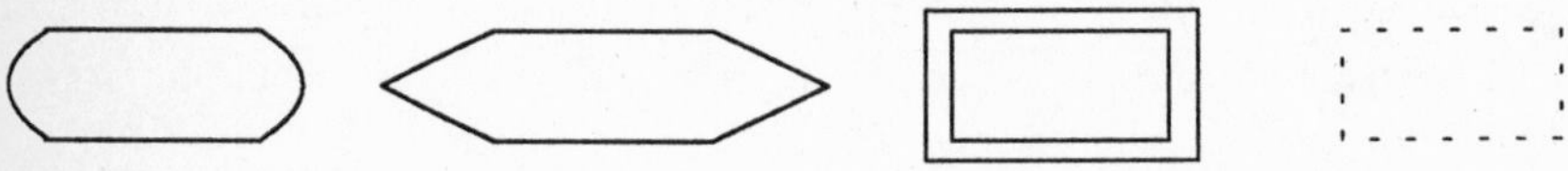

We could identify several model parameter conditions by alternative node shapes, including:

1. Nodes on paths below certain threshold values of time reserve
2. Start and finish nodes
3. Horizontal interface nodes
4. Events nodes (DU = 0)
5. Nodes with specified values in structure fields

Somehow, the model plot writer must provide the ability to link a criterion to a special node shape. For example, we might instruct the model plot writer to draw all logical start and finish nodes as ovals, or all nodes with zero or negative path time reserve with pointed ends, etc. Such a process would resemble a select statement, since presence of a specified criterion (IF), invokes a special node shape (THEN). This method visually highlights certain data in the model very effectively.

Horizontal interface nodes present a special problem. These nodes can be one of two types, cross-model interfaces, and intramodel interfaces. As described in Chapter 14, *cross-model interfaces* are special events nodes that link modelfiles together by simulating interfaces or constraints between them. Certainly these important nodes should stand out in any model plot, and this is relatively easy to achieve through a simple select statement specifying discriminating fields already present on these nodes (INI, INT, and IFS). For instance, any node with a value in the INI field is a cross-model interface, so it should have a specified node shape.

The second type of integration node, *intra-model interface nodes,* arise any time we define a portion of a particular modelfile to plot through a select statement. A designated node may have predecessors or successors that do not meet the selection criterion, but we need to plot them for proper analysis along with the selected nodes. A special node shape can flag these nodes, saving us the difficult job of identifying them with a criterion statement. This feature will be more thoroughly discussed later in this chapter.

We need a model plot writer with all of these features to portray activity and event data adequately. Without any one of them, the analytical effectiveness of our model plot diminishes significantly. For instance, to focus our analysis, we need to control the fields portrayed, their relative positions, and their formats within the nodes. We must display essential event data suitably to make it visible to our analysis. Finally, we need to spot other unique model characteristics, as well, and alternative node shapes make this fairly easy. These features are essential requirements of our model plot writer.

Constraint Control

In addition, we need a flexible process to tackle the difficult work of drawing the constraints correctly according to the left-to-right convention, passing from predecessors to successors listing types and lag values. Otherwise, model plots would not be complete.

In addition, constraint records in the modelfile can have many attributes beyond the five required parameters (PNI, SNI, TYP, DU, and CAL). Displaying extra information like titles, descriptions, structure codes, and other bits of pertinent data would enhance the analysis greatly. For example, the textual description of the constraint can identify the work effort represented by the type and lag value. Printing this additional data on or below the interface lines would help an observer to understand a particular constraint.

Since considerable analysis focuses on paths, we also need a means of highlighting a particular sequence of work. Vivid flags on paths that meet certain criteria, such as minimal values of time reserve or particular values in a structure code (e.g., a series of tasks performed by a subcontractor), aid easy recognition. Just as we flag nodes with alternative node shapes, we can flag constraints with variant line styles such as the line types illustrated in Figure 15-7.

Figure 15-7

ALTERNATE LINE STYLES

To accomplish this, the model plot writer needs a facility to link certain constraint conditions to particular line types. Again, it does this by accessing the values maintained in the record fields. By such a method, we can highlight entire paths, assign the nodes on a particular path a special shape, and draw the constraints on the same path as special line types. This makes paths with time reserve problems or other work sequences related by some common criterion stand out boldly on the plot.

Vertical Position Control

Besides properly formatting the shapes of the nodes and the styles of the constraint lines, we also need to control their positions in relation to one another. Of the many schemes devised to portray these model elements in linear sequences, the most satisfactory place them either by rank or by a set time scale.

Rank order can define the sequence of a group of nodes by the sequence number of its predecessors. The process assigns a rank to each node by adding one unit to the largest rank among its predecessors. It then arranges the nodes in sequence by rank. To better understand this, examine Figure 15-8, which shows rank order sequence for a sample model.

Node B1 and E1 are rank 1, nodes A1, C1 and D1 are rank 2, as each has only one predecessor rank (1). Nodes A2, B2, D2 and E2 are rank 3 as each has two predecessor ranks.

Figure 15-8

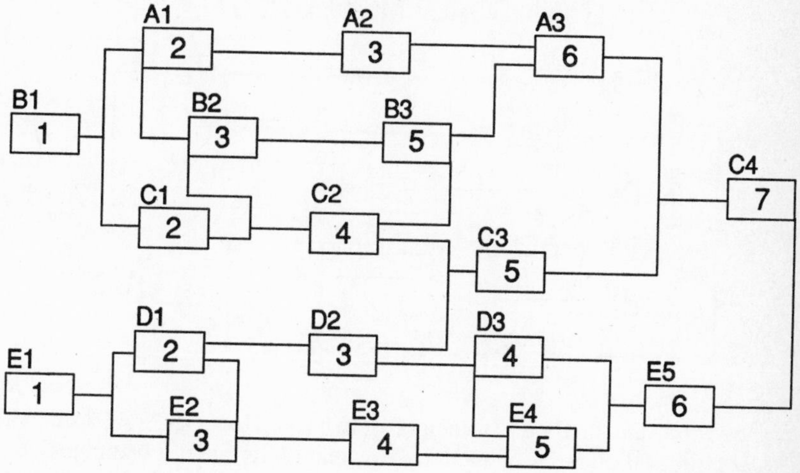

Node A3 has two predecessors, A2 and B3, with ranks of three and five, respectively. A3 takes a rank of six, one unit higher than its highest-ranked predecessor. Node B3 also has two predecessors, B2 and C2, with a maximum rank of four, making the rank of node B3 five. The nodes lie in order from left to right by rank as determined from their predecessors.

The model plot writer generates a plot in rank order by establishing a column for each rank spaced across the page. Rank one nodes appear in the first column, rank two nodes in the second column, etc. Adequate distance between the node columns allows for the proper depiction of the constraint lines, which flow smoothly from predecessors in the left-hand columns to successors, which always lie to the right in higher-rank columns, automatically satisfying the left-to-right date convention. Figure 15-9 illustrates the model from Figure 15-8 plotted in rank order.

Rank 1 nodes appear in the first column, rank 2 in the second, rank 3 in the third, and so on. Adding Early Start and Early Finish dates has no effect on the nodes' positions. Rank five, for instance, shows three tasks: B3 finishes on August 2, C3 finishes on August 23, and E4 finishes on September 7. These tasks spread across more than a month, a considerable amount of time in this simple example which spans only about five months overall. In larger models, nodes in the same rank column can show calculated and/or schedule dates years apart. Their position in the model plot depends solely on how many predecessors they have, and ignores their schedules. This can distort the analytical per-

Figure 15-9

RANK SCALE MODEL PLOT

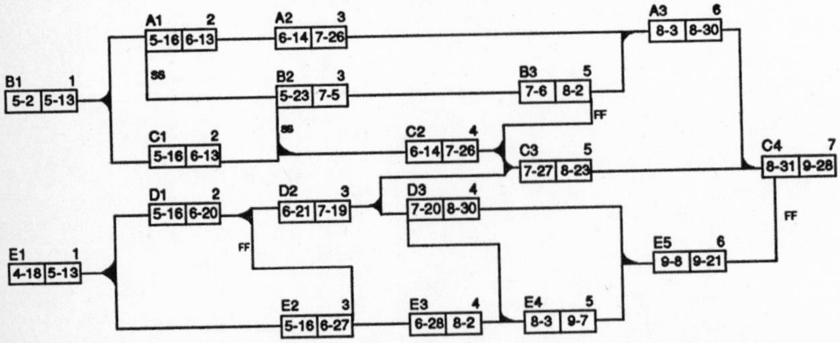

spective of time, the method's principal drawback. However, rank order also tends to compress the length of model plots, one of its main advantages.

Note that graphically plotted models typically show a single line where two or more constraint lines merge. Small flow indicator ticks at intersections aid in understanding the flows of the interfaces.

The main alternative to rank-order plotting, *set time scale plotting,* defines columns to indicate specified periods (months, two months, quarters, etc.) and spaces them across the page. The process then places each node in the column that corresponds to its value in a date field common to all of the nodes and specified by the user, like the baseline start (BES), or the baseline finish (BEF), or the Early Start (ES), or the Early Finish (EF), etc. Each column runs from the start date of one period to the start date of the next period, and all nodes with dates in the chosen field between these two values appear in the column.

Figure 15-10 illustrates the example from Figure 15-8 plotted in a set time scale by the nodes' Early Finish (EF) dates. Nodes A2, B2, C2, and D2 all have Early Finishes in July, so they appear in the column for that month. Nodes A3, B3, C3, D3, and E3 all have Early Finish dates in August, placing them in that month's column.

The nodes' positions within each column vary to accommodate various predecessor and successor conditions. B3, for instance, appears to the far left of the August column because it has a successor, A3, that also finishes in August.

Because all nodes can have different durations and almost an infinite combination of constraints can bind the various nodes, this format can create some unusual constraint patterns. The September column illustrates a simple example of this. Even if we shift node E4 to the far left of the column, the Finish-to-Start constraint to E5 violates the left-to-right convention because node E5 also has a successor, C4, in the September column, and available space simply will not fit all three nodes and their constraints in the proper flow sequence. In this case, it seems a simple matter to add a little more width to each column and accommodate the September situation. In larger models, however, this is not always feasible. Typically a set time scale plot might show some constraints in violation of the left-to-right convention. This is the only serious drawback of the technique.

Figure 15-10

SET TIME SCALE MODEL PLOT

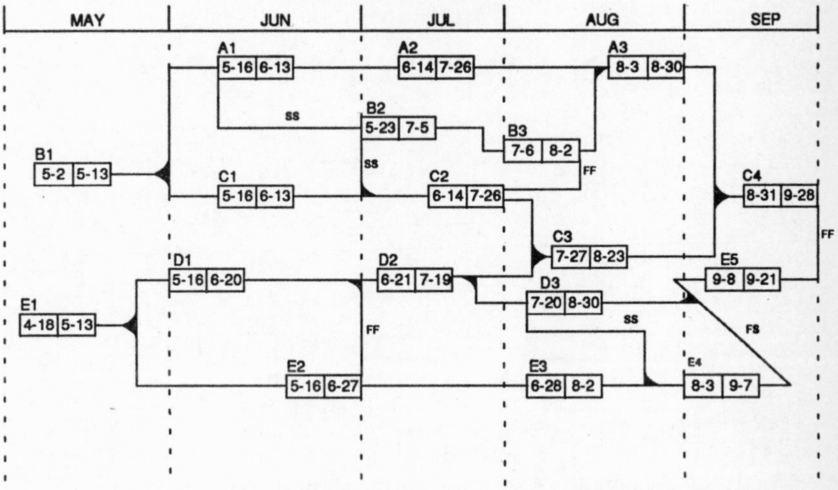

Either rank order or set time scale positioning provide the necessary controls to place nodes from left to right. Either format creates a disciplined, systematic diagram of minimal length to assist us in our analysis.

Horizontal Position Control

Having learned to position nodes from left to right along the length of the plot, we must also control the nodes' positions from the top to the bottom with the height of the plot (sometimes called its width). We call this *horizontal position control.*

This process recalls the techniques of regional zoning in tabular reports, so it is also called *regional banding* or *zoning*. To initiate the process, we specify a field or subfield to dictate placement of each node in the plot. The process divides all nodes in the output product into groups based on their values in this zone field. It then plots the groups in order from top to bottom by the alphanumeric sequence of the values in the zone field.

Figure 15-11 illustrates the effects of regional zoning. This example specifies the first character of the node identifier as the zoning subfield, dividing the diagram into A, B, C, D, and E zones. Nodes A1, A2, and A3 all appear in the top-most zone as the *A* in their zoning subfields is the smallest alphanumeric value of the five. Likewise nodes C1, C2, C3, and C4 all appear in the *C* zone in the middle of the plot. The titles that identify each zone come from a file that the user establishes in the model plot writer linking the zone values to the textual titles.

Figure 15-11

REGIONALLY ZONED MODEL PLOT

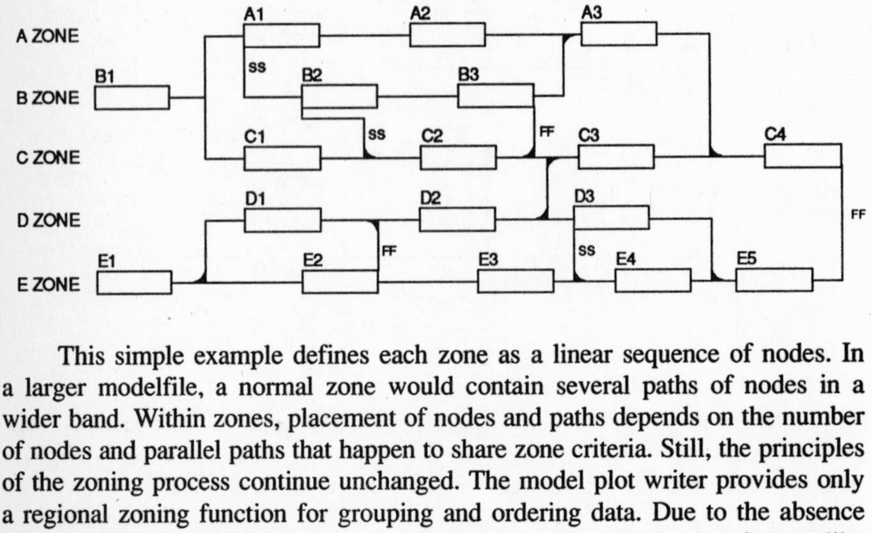

This simple example defines each zone as a linear sequence of nodes. In a larger modelfile, a normal zone would contain several paths of nodes in a wider band. Within zones, placement of nodes and paths depends on the number of nodes and parallel paths that happen to share zone criteria. Still, the principles of the zoning process continue unchanged. The model plot writer provides only a regional zoning function for grouping and ordering data. Due to the absence of any practical need, it provides no data order or data pagination feature like those in the tabular report writer.

Combining the vertical and horizontal positioning processes allows us to control graphic displays as though they were matrixes of zones and columns. The model plot that results gives invaluable assistance for analysis, as the nodes appear at specific locations dictated by their project structures and schedules.

Other Display Controls

Of the several other features that enhance model plots, only one is essential to our needs. In practice, large modelfile plots can hide specific problems. Therefore, the model report writer must help us isolate portions of the data in a particular model. This allows us to focus our attention on only the pertinent data. As in the tabular report writer, this process is called *data selection* or *filtering*.

In tabular reports, this process segregates records (activity and events nodes and constraints) from the modelfile that match designated criteria in the appropriate data fields. Just as in the tabular report writer, we designate a field or subfield, an operative symbol (e.g., =, > , <, etc.), and acceptable values. We can specify multiple criteria by linking individual criteria with AND and OR operators. In the model plot writer, however, we only designate the records for activity and events nodes we want to see on the plot; the process automatically includes the appropriate constraint records.

The model plot writer should provide an additional option as an integral part of the data selection process. It should allow the user to include or ignore the intramodel interfaces along with the appropriate nodes. This gives the option of showing how the model segment interfaces with the rest of the model. Predecessors of the selected nodes in intramodel interfaces provide inputs to the selected segment, and successors receive the output of the segment. Such information can be quite useful in the analysis of the segment.

Figure 15-12 is an example of this process. Through data selection, we have identified all nodes that contain the value *XC1A* in a certain structural field, identifying the nodes DECA, MEGA, GIGA, and TERA. Ignoring the intramodel interfaces, we would see only these nodes along with their connecting constraints (DECA-GIGA, MEGA-GIGA, and GIGA-TERA) on the plot. Displaying the intramodel interfaces would add every predecessor and successor of any of these nodes, along with the appropriate constraints. The model distinguishes intramodel interface nodes, which do not satisfy the data selection criteria, by alternative node shapes. The example highlights these interface nodes with double lines.

Figure 15-12

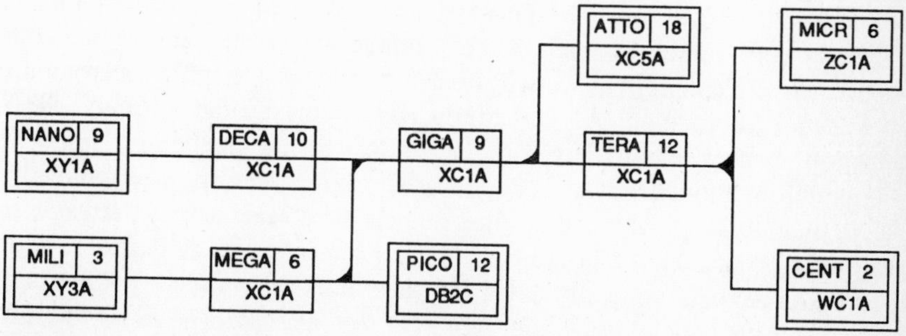

Notice, also, that the intramodel interface nodes display the same fields as the selected nodes. Our analysis requires the same information on the intramodel interface nodes that we have on the selected nodes.

Even adding the intramodel interface nodes, the data selection process produces a relatively small model plot which allows us to narrow our focus to a particular segment of the project. This helps us direct our analysis toward a specific purpose instead of becoming muddled or confused by a great deal of extraneous data.

Other components of the model plot writer help the user format the model's graphic appearance. We can specify titles and subtitles for the plots and control their placement, size, and font of the characters. We should also be able to define the character size of all text data displayed in the model plot, and to add a legend. This should briefly describe the fields displayed in the nodes, as well as their locations within the node. It could also provide format information about constraints. We should be able to place comments or other messages in specified sections of the plot. All of these border features enhance the understanding of the displayed data.

A model plot writer must be very comprehensive in order to satisfy our need for a powerful, but flexible analysis tool. We have discussed the essential features, including proper node and constraint formats, horizontal and vertical position controls, data selection, intramodel interface depiction, and format and legend controls. These enhancements to the basic model plot give the user the best tools with which to dissect and examine the model.

GANTT BARCHARTS

Basic Format

The most common type of graphic output product that we generate for project models is called the *Gantt chart*. Often, it is also called a *barchart*, which is fine as long as we distinguish this diagram from histograms, which are also sometimes called *barcharts*. The Gantt chart displays model information in a way that is very simple to read. Some promote the Gantt chart as an entire methodology of planning, but we will treat it as simply a graphic illustration of project model information. As powerfully and simply as they display project information, Gantt barcharts have nothing to do with the method by which we derive the information. This fact is key to understanding their proper application.

We measure displayed data against a linear time scale across the page, the distinguishing characteristic of the Gantt chart. We draw bars to represent activities and arrange them vertically down the page, measuring their lengths against the time scale to indicate schedule. Each bar begins at the point on the time scale when the activity starts and completes at the point when the activity finishes. Events appear as geometric shapes at the point on the time scale when they occur. Unlike the set time scale model plot, the lengths of the bars on a Gantt barchart indicate the durations of the corresponding activities.

A simple Gantt barchart is illustrated in Figure 15-13. It groups the tasks and events together to show how they relate to one another. For instance, we can see the schedule of Task A2 along with its relative sequence to Tasks A1 and A3. We can also see selected events like the start of A1 and B2, and the finishes of A2, A3, and B3. Events depiction is controlled by the values in the SEF and FEF event fields, as will be explained later. This basic relationship of a time scale to activity information can be used in an infinite variety of ways. Therefore, we need a very powerful facility for varying formats and customizing barcharts to meet our needs, which we call a *Gantt plot writer* (GPW).

Figure 15-13

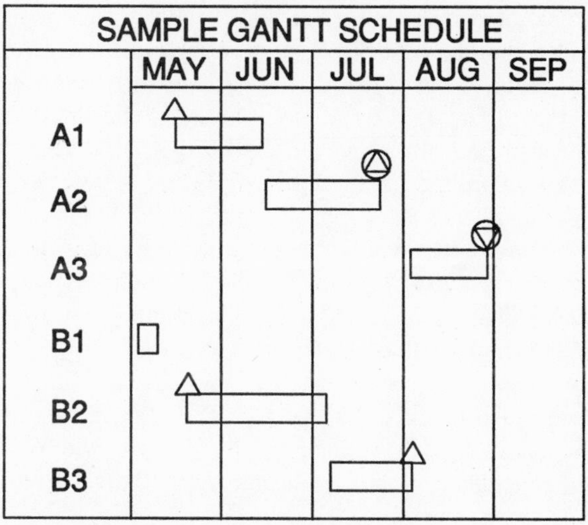

As its most critical function, the Gantt plot writer lets us control the way the bars and milestone shapes portray model data relative to activities and events. Each activity or node record has several start dates (BES, PES, AES, ES, and LS) and several finish dates (BEF, PEF, AEF, EF, and LF). Sometimes we may want to display only one type of start and finish dates (e.g., ES and EF dates). At other times, we may desire to display two or more date types for each record (e.g., baseline versus projected and actual dates). Figure 15-13 illustrates a simple Gantt chart where the bars reflect the events' and activities' Early Start and Early Finish dates.

In a more complex chart, however, we might portray two or even three pairs of date types, by one of two graphic techniques. The first technique, called *overlaid bars,* differentiates date types by superimposing various bar styles (solid-line bars, dotted-line bars, half-width shaded bars, etc.) to form a single, complex bar. Figure 15-14 shows an example of this format.

The other principle technique places similarly distinguished bars one above the other. In this method, called the *over/under technique*, the upper, or over, bar depicts one set of information and the lower, or under, bar depicts another. The purpose of the output product, the user's preference, and the specific information shown dictate both the technique and the way it is used. Figure 15-16 illustrates this bar type format.

Chapter 12 discussed the importance of comparing the various date types (early to late, baseline to status, etc.). Gantt charts do this extremely well. For example, to compare schedule information to status, we need to portray all of the baseline, actual, and projected dates that pertain to each activity and event. The overlaid bar technique best displays these three sets of dates. To be effective, this format requires bar styles that appear distinct from one another even when overlaid. For Figure 15-14, we have selected solid-line bars for the baseline data, dotted-line bars for projected dates, and half-width shaded bars for actual dates. Even superimposed, these bar styles still allow us to distinguish the start and finish date of each bar type.

At the left margin (RAW), this Gantt chart lists baseline, projected, and actual dates for each activity. This unnecessarily duplicates the bars' depiction of this data, but it enhances the example by illustrating the relationships between the RAW data and the bars that display it.

Activity A1's baseline start of May 16 and baseline finish of June 13 define the start and finish of the solid-line bar. This task actually started on May 10 and finished on June 10, as illustrated by the half-width shaded bar. The task's projected start date is made irrelevant by the actual start, so it does not appear.

Task B1 has also actually finished, but its actual start date is identical to the baseline start date, which the chart shows clearly even though the bars are superimposed. The half-width shaded bar allows us to discern both the actual and the baseline start dates.

A solid-line bar also shows the baseline dates of Task A3. This task's projected start and finish dates define the dotted-line bar. Task B3 has neither actual nor projected start dates. Therefore, assuming the projected start matches the baseline start, the dotted-line bar begins at that point. This bar style appears as only a solid line when the solid-line and dotted-line bar styles coincide. Note that the constraints of this model make the status of Task B3 inconsistent with that of its predecessor, Task B2. We know this because we remember the con-

Figure 15-14

NI	BES	AES	PES	BEF	AEF	PEF
A1	5-16	5-10	5-15	6-13	6-10	-
A2	6-14	6-13	-	7-26	-	7-15
A3	8-3	-	8-15	8-30	-	9-9
B1	5-2	5-2		5-13	5-25	
B2	5-23	5-26		7-5	-	7-15
B3	7-6			8-2		8-12

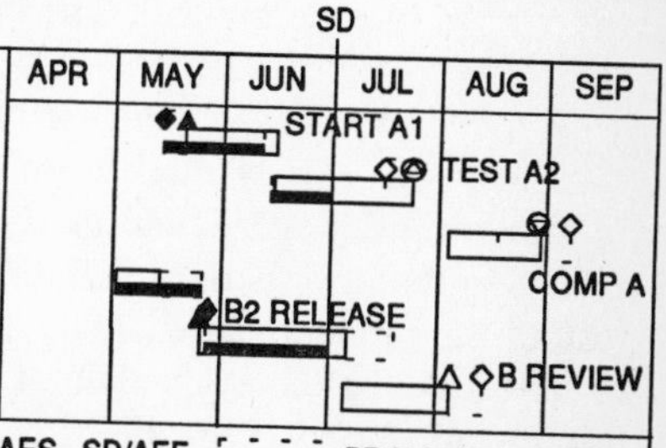

LEGEND: B/L: BES - BEF ACTUAL: AES - SD/AEF PROM.: PES - PEF

stant relationship between B2 and B3 (FS), but the Gantt chart does not clearly indicate this information, which is the principal drawback of the format.

Activities A2 and B2 are also similar in that both have actually started, but neither has actually finished. These on-going tasks create a dilemma of sorts in our graphic display. To portray these activities' actual start dates, a half-width shaded bar begins at that point. The task has no actual finish, however, to define the end of the bar. We cannot assume an actual finish date, yet we must stop the bar at some point.

The simple answer is to end the bar on the status date, as illustrated, but then the finish end of the half shaded bar indicates no more than the project's status date (SD). In order to mark some important data with this end of the bar, we can use it to indicate relative progress. For instance, if we assess an activity as two weeks behind schedule, then we end the half-width shaded bar two weeks before the status date; for a task that is three weeks ahead of schedule, we shade to the right of the status date by three weeks. The difference between the projected finish of the task and its baseline finish provides the same information, but this shaded bar method does have some merit and should be available. It does, however, require the user to gather and maintain additional information on each activity and at least one additional field.

As explained, events appear in Gantt barcharts as geometric shapes at certain points along the bars. The values in the SEF and FEF fields determine which events appear on a particular chart. Typically, these fields designate, for one thing, a schedule hierarchy, defining the highest schedule level at which a particular event will appear. This is used in conjunction with the summarization process such that when we generate a chart for a particular schedule level, the GPW displays the appropriate events.

The events field also defines the milestone shape. A table matches characters that might appear in the SEF or FEF fields with geometric symbols. In

Figure 15-14, the code in the SEF fields of nodes A1 and B2 corresponds to an upright triangle, whereas the value in node A2's FEF field corresponds to the circled upright triangle symbol. The codes that define these shapes correspond, in turn, to some information about the events, like their schedule levels or who performs the work. These geometric shapes must be symmetrical and come to a point so they can clearly mark a point on the time scale. The event's textual description from the event field usually appears beside each event.

This symbology should interpret the date data of the record to serve some purpose of the product. To chart the schedule information, the milestone shape might reflect the baseline schedule of the event (as in Figure 15-13). When status information for an event yields an actual or projected date that differs from the baseline date, a second symbol, usually a diamond, tracks the current schedule. A line joins the pair of symbols (as in Figure 15-15). If the status information comes from an actual date, the two event shapes are filled in or shaded (Figure 15-15b), but if the status information comes from a projected date, both symbols are open (Figure 15-15a). If the actual date of an event matches the baseline date, no second shape appears, just the baseline shape, filled or shaded (Figure 15-15c). These graphic conventions hold true for any event, whether the start or finish of an activity or an events node.

Figure 15-15

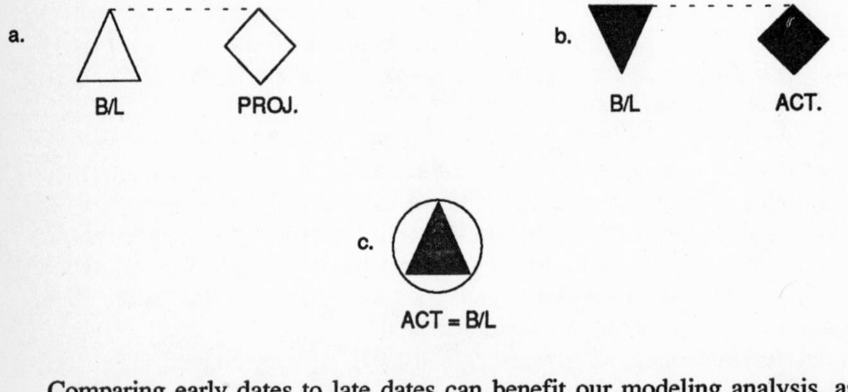

Comparing early dates to late dates can benefit our modeling analysis, as well. The over/under barchart style does this well. Figure 15-16 illustrates this format for tasks A1, A2, A3, and C1. The solid-line bar displays the Early Start and Early Finish dates, and the short arrows attached by lines indicate the late dates. Space between the arrows and the starts and finishes of the bars measures positive or negative time reserve. Activities A1, A2, and A3 show positive reserve, while C1 has negative reserve. Remember, this time reserve does not belong to these activities, but rather to the worst-case paths on which they lie.

In this format, event markers usually appear above the upper bar. The event markers indicate baseline and early dates, as is typical in this kind of comparison.

Gantt charts can mix or alter the various graphic formats as necessary for various effects. A chart could, for example, compare baseline dates to late dates with overlaid bars. Another could compare schedule data (baseline, projected, and actual dates) in an over/under chart. We could overlay two bars, each reflecting different information, and add an underbar to show a third set of information. The Gantt chart can be as complicated as the information we need to relate. The Gantt plot writer must adapt flexibly to our requirements for displaying information.

Figure 15-16

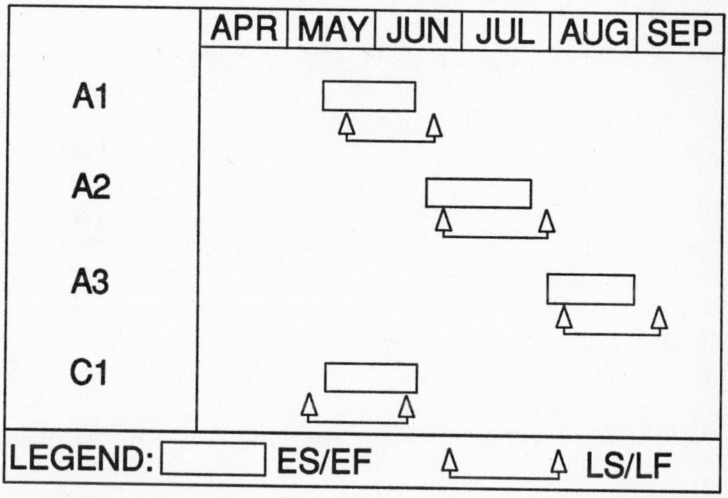

Record Attributes Window

At their left margins, Gantt barcharts normally feature a *record attributes window* (RAW). This section displays specified fields or parameters regarding each activity in line with the activity's bar to enhance the information in the chart. Aesthetics limit this section to about a third of the product width, the barchart section filling the remainder. This section usually resembles a fairly narrow tabular report listing certain parameters of activities.

There are many requirements of the GPW relative to the record attributes window. The user needs to specify the fields in this section, their sequence or placement, their headings or titles, and some aspects of their field formats. For simplicity, the Gantt report writer should perform all of these functions in the same way the tabular report writer performs similar functions, as explained earlier.

Paging, Zoning, Ordering, and Selecting Data

The Gantt report writer must provide several other features to format the data in the chart. Since we are displaying the same activity records as in tabular reports, but in a different way, all of the functions for grouping data (paging, zoning, and ordering) and for selecting data provide equal benefits. Even so, it is appropriate to review each of these functions as they affect the Gantt chart.

Often, the user needs to break a Gantt chart into segments to disseminate to various persons. This requires control of page breaks. The Gantt plot writer groups the data by the values in a user-specified field, then advances to the next page every time this value changes. As an example, suppose we must produce Gantt charts for our sample model, breaking pages on the first character subfield of the node identifier. Thus, activities A1, A2, and A3 would appear on page 1; B1, B2, and B3 on page 2; C1, C2, and C3 on page 3; etc. In practice, a paging group may fill several pages before the page breaks.

The user can also segregate data into groups or regional bands within a page by specifying a field or subfield. After the GPW separates the paged section by the appropriate criterion, it groups the data by the values in the zoning field, orders it in alphanumeric sequence, and then plots it. Horizontal lines, blank lines, and/or zone titles separate and identify the zones.

Figure 15-17 shows a zoned Gantt chart. The zone criterion is the first character subfield of the node identifier. The GPW breaks the plot into five zones, one each for nodes beginning with *A*, *B*, *C*, *D*, and *E*. Figure 15-17 illustrates only bands A and B with titles and a dashed line to separate the zones. This produces a graphic effect of a series of groups of activities, related by common characteristics.

Figure 15-17

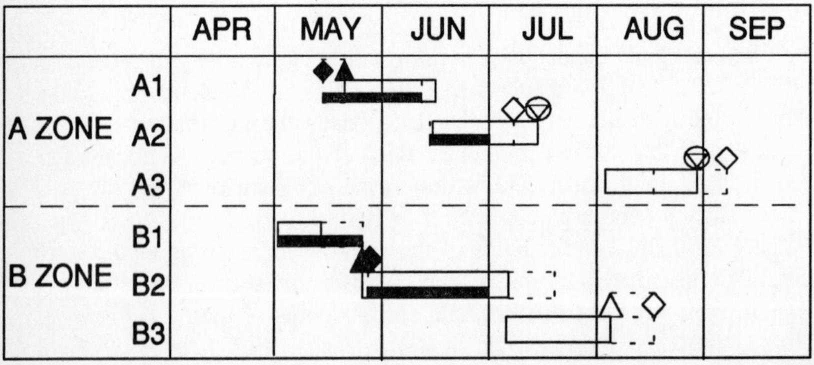

Within a zone, a specified ordering criterion controls the sequence of the data. We identify primary, secondary, tertiary, etc. ordering fields or subfields. Within each zone, the GPW sorts the records by the values in the primary order field (subfield) in alphanumeric, date, or numeric sequence depending on the type of field. It orders any records with identical values in the primary field relative to one another by the values in their secondary fields (subfields) and so on through tertiary, etc., until it discriminates the sequence of each record with respect to all others. A default criterion, like record sequence, built into the GPW distinguishes those records for which the user specified inadequate sort criteria. The records in Figures 15-13, 15-14, and 15-16 have all been ordered by the node identifier field.

In addition to ordering and grouping data, the GPW must allow us to select a portion of the records for display. It does this just as the other report writers did for both model plots and tabular reports. Like model plots, however, it only pertains to node records since Gantt barcharts do not display constraints.

From-Until Dates

We also need to limit the data for a Gantt chart according to the calendar. Projects can span many years, but Gantt barcharts commonly cover relatively short periods of time. The From-Until feature of the GPW selects data for the period between two user-specified dates to produce these partial charts. The user specifies the earliest date, called the *From date*, first to establish the beginning of the time period, followed by the later date, called the *Until date*, to establish the end of the period. The chart then shows only the records that contain data falling between these two dates. This function can either isolate a specific time frame within a project or simply frame the chart calendar. Sometimes this feature is called *windowing*, as it defines a time slice or window on a project.

Summarized Gantt Barcharts

Chapter 13 discussed the concept of summarized Gantt barcharts. These extremely powerful reports can represent massive amounts of information in very small spaces. Refer to Chapter 13 for detailed discussion of summary output products.

Figure 15-18 plots our example model in a summary chart, again summarizing data by the first character of the node identifier field. Compare this to the zoned chart in Figure 15-17. Regional zoning groups detail activities by some criterion, whereas summarization combines the information of detail activity groups into a single activity based on a similarly specified criterion.

Figure 15-18 lists the summarized schedule data in the record attributes window. The barchart plots this information as for a detail barchart. Summarization is actually a process, rather than simply a different kind of chart. Note that this example does not show any milestones, though it easily could have by following the process explained in Chapter 13.

Figure 15-18

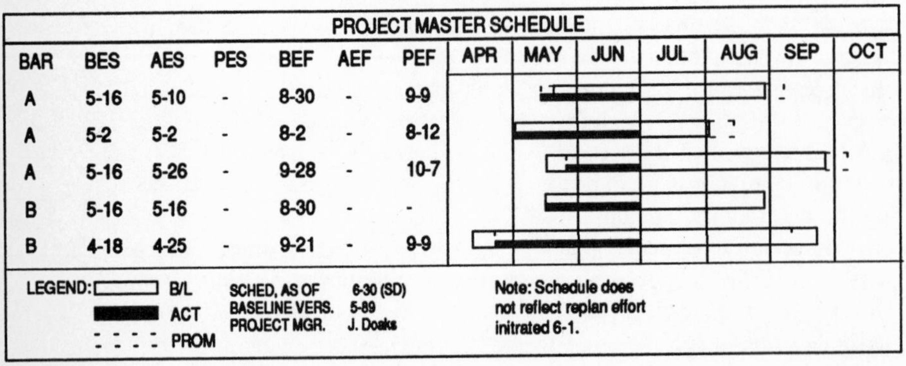
PROJECT MASTER SCHEDULE
BAR BES AES PES BEF AEF PEF
APR MAY JUN JUL AUG SEP OCT
A 5-16 5-10 - 8-30 - 9-9
A 5-2 5-2 - 8-2 - 8-12
A 5-16 5-26 - 9-28 - 10-7
B 5-16 5-16 - 8-30 - -
B 4-18 4-25 - 9-21 - 9-9
LEGEND:
B/L
ACT
PROM
SCHED, AS OF 6-30 (SD)
BASELINE VERS. 5-89
PROJECT MGR. J. Doaks
Note: Schedule does
not reflect replan effort
initrated 6-1.

Legends

Adding legend or border information to Gantt barcharts requires a very flexible GPW because virtually no standards guide us in this area. We can, however, discuss some features in general terms. Most charts require a title, which could be two or three lines long. It can appear at the top or the bottom of the chart and in larger type than the other text.

Some Gantt charts need symbology legends, some do not. Legends should define the date fields the activity bars and event shapes portray, along with other graphic conventions such as shading in the activity bars and milestones to indicate status information.

Most of the other information needed to explain Gantt barcharts takes the form of various textual blocks. This includes signature blocks, product comments, and schedule configuration control terminology (status date, schedule revision number, etc.). The user needs the ability to set aside and define sections of the chart for this information.

Summary of Gantt Barcharts

The Gantt barchart is the most flexible, and therefore the most complicated, of all the output products. Many believe its importance is grossly exaggerated, but we must not underestimate its value either. We need to exploit Gantt charts' easy and flexible communication of information. With this understanding and a powerful GPW, we can disseminate a broad spectrum of information regarding our project model to a large community.

HISTOGRAMS, X-Y PLOTS, AND PIE CHARTS

The last family of output products, including the histogram, the X-Y plot, and the pie chart, is often called *management graphics*. Many database software packages provide these as fundamental output products, but their usefulness in project modeling is very limited. A pie chart could map the distribution of the values in structure fields, but such a map would have questionable value. Therefore, we do not need a true report writer for management graphics. Instead, a way to compile specific charts with some flexibility can satisfy the requirements of this product family.

For one example of a useful management graphic, an X-Y plot could portray time reserve erosion. Chapter 12 discussed such a plot and an example appeared in Figure 12-3. This allows us to track the rate of time reserve erosion (ROE) and compare it to the rate of progress (ROP) of the paths passing through any event. In order to accomplish this, we need a process that periodically measures the time reserve at specified events and puts out a routine graphic report for each.

This report tracks time reserve at several milestones in a project. The user must list the node identifiers and events within the nodes (starts or finishes) for the process to track.

The user should also specify an acceptable value of time reserve for each reported milestone. It may be zero, a positive value, or even an increase from the original level. The plot shows a straight line between this point and the time reserve measured for the event at the report's origination. This allows us to measure the acceptable rate of reserve erosion. We must try to establish and/or maintain time reserves on or above this line.

Periodically (e.g., weekly, monthly, etc.) at a designated time, the process captures the current time reserve for each specified event and stores these values for reporting on the current and all subsequent periods' reports. It plots the new and previous points, and then extrapolates the line by the least squares method to project the trend of the tracked points to demonstrate the effects of continuing the present course. This line represents the actual rate of time reserve erosion.

As an optional feature, this output product might provide a From-Until windowing capability to limit the plot to a specific time frame. It might also plot the rate of erosion lines from a specified number of previous periods to display trends more dramatically. Lines lying very close together indicate a constant trend. If the line of each period lies above that of the preceding period, the trend reflects improvement; if preceding periods' lines lie above the line of the current period, the trend is declining.

The plot function may also allow the user to include or leave out a tabular listing of the product data. In addition, it may provide the ability to add border text such as a title, subtitle, legend, and axis labels.

This output product provides nowhere near the extensive flexibility of the other products, but these few options more than adequately support our requirements.

Another useful X-Y plot tracks schedule performance of a specified list of project milestones, as in Figure 15-19. The chart combines three curves: the first tracks the cumulative baseline schedule dates, the second tracks the cumulative actual dates up to the status date, and the third starts at the end of the actual curve and tracks the cumulation of all projected dates. If the actual date curve lies below the baseline curve, the gap between them indicates the cumulative number of missed milestones. Downstream, this curve should merge with the baseline curve at some point. The point at which the projected curve merges with the baseline curve indicates when performance will meet the schedule.

Figure 15-19

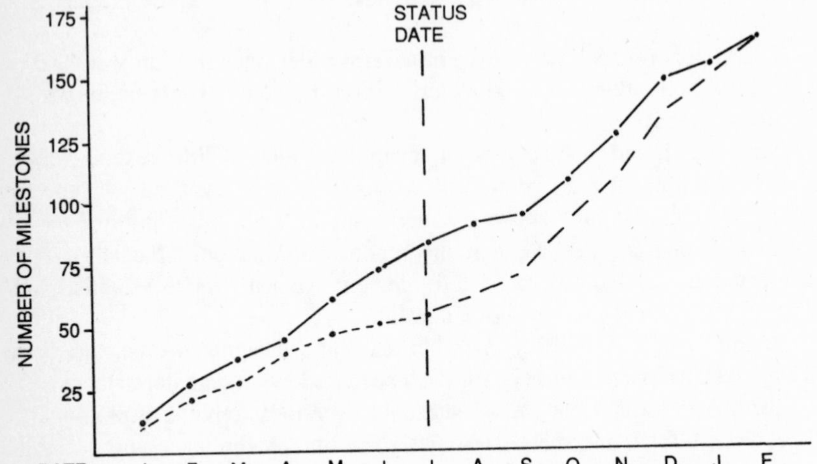

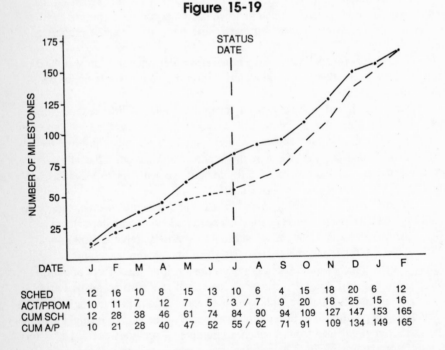

DATE	J	F	M	A	M	J	J	A	S	O	N	D	J	F
SCHED	12	16	10	8	15	13	10	6	4	15	18	20	6	12
ACT/PROM	10	11	7	12	7	5	3 /	7	9	20	18	25	15	16
CUM SCH	12	28	38	46	61	74	84	90	94	109	127	147	153	165
CUM A/P	10	21	28	40	47	52	55 /	62	71	91	109	134	149	165

This kind of X-Y plot should allow From-Until windowing to limit data to a particular time frame. It should allow the user to include or leave out a tabular listing of the data as shown at the bottom of Figure 15-19. Finally, it should provide the ability to add border text such as a title, subtitle, legend, and axis labels.

The discussion in Chapter 16 of resource modeling employs histograms and X-Y plots more extensively to illustrate resource requirements and resource pools.

SUMMARY

Each of the four principal families of output products, tabular reports, model plots, Gantt plots, and management graphics, has its own requirements and capabilities. Software applications called report writers generate each of these kinds of charts.

The tabular report writer (TRW) creates a textual listing of data. It should provide the user the opportunity to specify the fields in each report and their sequence, and to define field column headings and field formats.

It should also group data as directed by the user by breaking between pages, creating regional zones within pages, and ordering data. It should allow the selection of data from the modelfile through simple and complex selection statements and provides for report titles, legends, and user comments. The TRW should also produce two special types of tabular reports: Soon-to-Come-Due and Predecessor-Successor reports.

The model plot writer (MPW) creates model plots in which nodes appear as compartmentalized boxes with lines representing constraints joining the appropriate ends of predecessors to the appropriate ends of successors. The MPW should allow the user to specify the fields to appear, their placement, and format in each node. It should depict event information as geometric shapes at the appropriate node ends and labeled with the events' descriptions.

The MPW should further allow matching of special node shapes to certain conditions through values in specified node record fields. It should also match various line styles to constraints through values in a specified constraint record field.

The MPW should allow the placement of nodes horizontally in the plot either in rank order or according to a set time scale. It should also provide for the placement of nodes vertically in the plot through regional zoning, which groups the data into horizontal bands according to values in a specified node record field.

The MPW should allow for the selection of nodes and their associated constraints for a plot by user-specified criteria. The user needs to choose whether to display or suppress intramodel interface nodes not included by the select statement, but which are immediate predecessors or successors of the selected nodes. It should also allow users to add titles, legends, and comments.

The Gantt plot writer (GPW) produces Gantt barcharts, which represent activities as bars with start and finish dates measured against a time scale.

Events should appear as pointed graphic shapes along the bars measured against the same time scale. The GPW should provide the user the ability to control orientation of the bars according to any of the start and finish dates (early, late, baseline, projected, or actual).

The Gantt chart should allow linking of particular date types to particular bar types. The GPW needs to superimpose the bars that indicate different types of dates for a single record (producing overlaid bars) or place one above the other (producing over/under bars). It should compare two dates associated with an event by a second attached graphic shape (usually a diamond).

The user should be able to specify which fields appear, their sequence, and their formats in the record attributes window. The GPW should provide for the grouping of data in the chart by breaking between pages, defining regional zones, and ordering data, all according to values in user-specified fields.

The user needs to select or filter data for the chart by entering simple or complex selection statements. The user should also be able to specify from and until dates to limit data on a chart to a particular time frame.

The Gantt chart should also generate a product from summarized data according to a user-specified field or subfield and an event selection mechanism. It should allow the user to add titles, legends, and comments.

Mangement graphics, including pie charts, histograms, and X-Y plots, do not add much to project modeling, but some specific reports are useful. An X-Y plot can show the rate of time reserve erosion versus the rate of project progress for selected events in the modelfile. A set of X-Y curves can also show schedule performance, comparing baseline dates to actual and projected dates. Almost all other management graphics that contribute anything useful deal with resource modeling, which will be discussed in Chapter 16.

CHAPTER 16

Resource Modeling

Up to this point, the project modeling process has measured task duration only in time units. In reality, the span of work depends on the application of resources (labor, materials, equipment, facilities, etc.) over a period of time. In other words, the accomplishment of work depends principally upon two variables, time and resources. To model a project solely as a function of time is to look at it with one eye closed.

We must expand our analysis to include resources before we can hope to understand our project fully and manage it proactively. We have not discussed resource analysis earlier because its prerequisite is time modeling. Resource modeling, in fact, complements CPM time modeling. Unless we first structure the model to calculate time reserve properly, adding resource data is just more useless information. We can also erect a house, but without a solid foundation, it soon falls into ruin.

Now that we have thoroughly examined the time modeling process, we can proceed with resource modeling. Like the process itself, the objectives or results of this modeling technique complement those of time modeling:

1. To understand the resource requirements over time of both the project as a whole and each task within the project.
2. To compare cumulative resource requirements to the projected or actual resource levels.
3. To smooth the peaks and valleys in the requirements for certain resources such as specific labor skills, tools, or facilities.
4. To track, measure, and control resource costs.

In order to understand how to achieve these objectives, we first need to explore the seven steps of the resource modeling process:

1. Establish a resource profile for each task.
2. Establish a pool for each resource.
3. Sum the requirement for each resource throughout the project relative to a specified period.
4. Compare cumulative resource requirements to the projected resource availability.
5. Determine the allocation of resources adequate to meet the requirements of the project within the time constraints, but do so in the most efficient manner.
6. Convert resource requirements to dollar equivalents.
7. Track and control these parameters.

Ideally, our modeling will yield a mathematical relationship between the duration of a task and the resources necessary to accomplish it. This relationship would tell us how much time would be necessary to accomplish a task applying a given amount of resource, or how much resource would be necessary to accomplish the task in a specified amount of time. From this relationship, we could truly optimize time and resource use for our project.

In discussing time reserve analysis, we stated the principle that an increase in resources can, in general, reduce the duration of a task. Unfortunately, no fixed or simple relationship defines these proportions, either in general or for a specific task. Even a large-scale manufacturer that repeats tasks many times over predicts their spans by standards that are at best averages. Too many variables can affect the outcome.

Although we do not understand how to achieve the ultimate result, resource modeling still provides tremendous benefits. Just as we found with time modeling, any improvement in our understanding and control of a project's future helps us toward a successful outcome.

RESOURCE LOADING OF TASKS

As the first step in resource modeling, we must load the resource requirements for each activity relative to its span. In other words, we must determine how much resource is needed to accomplish a task within a specified duration. We can then extend this information about the resource requirements of each task to determine the resource requirements of the entire project. We find four general categories of resources, each with its own peculiarities:

LABOR—the work effort of people necessary to complete a specific task

MATERIAL—the physical objects consumed or transformed in the accomplishment of a specific task

TOOLS—the fixtures or other equipment that acts on the materials to accomplish a specific task

FACILITIES—buildings or other work space that house work on a specific task.

Any task could require inputs from of any or all of the various resource categories. A manufacturing task could require specific craftsmen, tools, and fixtures to assemble and process the materials in a facility with adequate room, power, light, etc. An engineering task could require specific engineers, design and analysis equipment, and an appropriate workspace. Computer time and the paper on which to produce the product could be considered the consumed material of the task. Virtually every effort requires application of some or many classes of resources.

We cannot practically, and we need not, measure and control every resource through the modeling process. The effort is labor intensive and therefore must be weighed against value returned. We may limit our analysis to one category such as labor, or to labor and a few critical tools and facilities. Regardless of how much or how many of the project's resources we incorporate into our model, the principles generally remain the same.

To determine the resource requirements of a task, we need certain information about each resource we want to track. We store this information as a repetitive series of data items for each task, which closely resembles a series of records within each activity record. For lack of a better term, we will call these *resource subrecords*. Basically, each resource subrecord must specify the amount of a specific resource needed and define the period within the task when it is needed. Subrecords for each of the categories of resource have their own requirements.

We define requirements for material, for instance, on a list or bill of materials. One field (called *RESID)* of the subrecord must identify the specific material item with a unique identifier code. Another field (called *RESQTY)* registers the quantity of the material necessary to accomplish the task in units pertinent to the item (e.g., number, pounds, inches, liters, etc.).

The third (OFFSET) and fourth (RESPAN) fields define when we need the material during the task. Since material is almost always added at a point in time (event), from then on becoming part of the task's effort, we specify the need for material either over the span of the task or at a point offset from the task's start. We might, for example, need the material after completion of 25 percent of the task or three weeks after it starts. Typically we specify offset as time units (days) from the task's initiation. We need not specify a span for material requirements because they become part of the task or are consumed and cannot be made available for other tasks.

This method of defining resource requirements produces a *step resource profile* (SRP). A simple example could look like:

TASKA - Assemble Model Catapult - 2 Hours

RESID	*RESQTY*	*OFFSET*	*RESPAN*
PLANKA	4	0.0	-
PLANKB	2	0.6	-
PLANKC	2	1.0	-
DOWLA	1	1.0	-
DOWLB	2	1.6	-

This task, to assemble the base structure for a Cub Scout project, is estimated to take two hours. We have determined that we need five specific material items: three sizes of small boards and two different dowels. Five subrecords store the requirements for these items. We designated the first 8" x 1/2" x 1/4" board as PLANKA. We need four of these (RESQTY = 4) to perform the task at the beginning of the effort (OFFSET = 0). We also need two 3" x 1/2" x 1/4" boards labeled PLANKB. We need them when we have progressed 6/10 of an hour into the task (OFFSET). Since all these boards become part of the task's output, no duration (RESPAN) is specified. Similarly, we can define the requirements for all other materials.

We define facilities and tooling in a similar fashion. We would first set up a list of available fixtures and tools, similar to a material list, assigning each item a unique identifier. We would then enter the tool list identifiers in resource subrecords specifying quantity (RESQTY) and timing (OFFSET). Unlike materials, however, tools do not become part of the product, so we must specify how long the task needs the tool or in some other way define when it becomes available for another task (RESPAN). We can address these questions simply with a step resource profile (SRP) specifying how long after the task's start the need for the tooling arises (in OFFSET) and how long the need lasts (in RESPAN). Suppose, as an example, we define the need for a specific tool as arising two days after the task's start by entering a 2 in the OFFSET field, and we state that we need it for five days by entering a 5 in the RESPAN field. This leaves it available for another task during the first two days of our task's duration and seven days after its start.

In modeling facilities, we generally deal not with multipurpose floor space, but rather with specialized work areas such as a unique laboratory. As we did for tools, we list these special facilities and assign each a unique code. In the task subrecord we designate the specific facility that is needed (in RESID), the time after the initiation of the task when the need arises (in OFFSET), and the duration of the need (in RESPAN). We could also enter a 1 for the quantity of the resource needed in RESQTY.

Labor, the most important category for us to track, tends to complicate resource loading more than the other categories. In some ways, the procedures remain identical to those for the other resource types. We begin, for example, by listing the various labor types and assigning each a specific code. The difference comes in how we best specify the quantity and time requirements for each task. For one alternative, we can utilize the step resource profile technique in virtually the same way as for the other resource categories: specify the quantity of a specific labor resource needed (by a head count) along with offset and duration values. However, labor requirements tend to vary over the span of a task, and we measure labor in hours rather than days.

We can overcome the first problem by specifying labor requirements in steps by repeating the same resource identifier in several subrecords. As an example, suppose we need one unit of a specific labor resource (ENGA) for the first 5 days of the task, then we need two units of the resource for the next 10 days and four units of the resource for the remaining 10 days. To model this situation, we define a single resource (ENGA) in three subrecords, one for each variation. These subrecords would look like:

RESID	*RESQTY*	*OFFSET*	*RESPAN*
ENGA	1	0	5
ENGA	2	5	10
ENGA	4	15	10

An alternative step profile technique adds these resource subrecords together. We would specify a need for one unit of the resource for the entire duration of the task. After 5 days, we would add a second unit, also for the balance of the task. Finally, 15 days after the task's initiation, we add two more units, again for the balance of the task. This structures the resource profile in the same way that we actually apply the resources. The following subrecords would represent this situation:

RESID	*RESQTY*	*OFFSET*	*RESPAN*
RESA	1	0	25
RESA	1	5	20
RESA	2	15	10

Regardless of our method, to model this situation requires three subrecords because we change the resource level three times. A plot of both of these resource profiles would graphically illustrate the quantity of resource required versus the duration of the task, as in Figure 16-1.

Figure 16-1

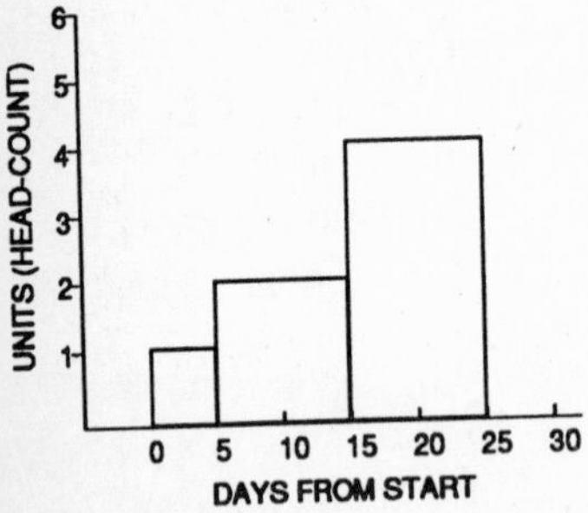

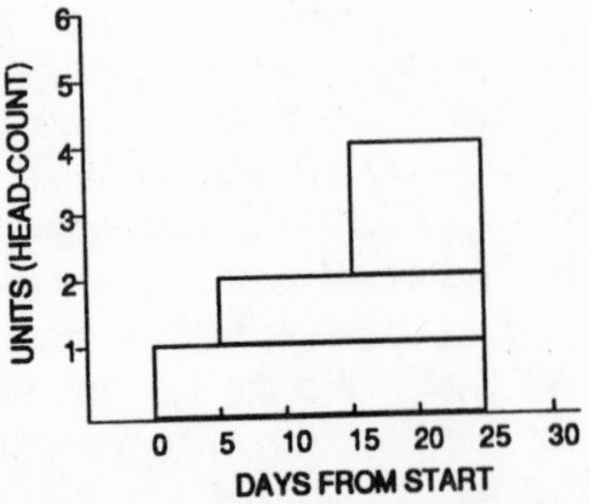

These methods do model a changing resource, but they require a considerable number of parameters (12) to do it. In a more practical method, called a *periodic resource profile* (PRP), we would establish a fixed period for measuring resources (e.g., weeks, months, etc.) and then specify the number of units (e.g., labor-hours, labor-weeks, labor-months, etc.) needed to accomplish the task for each period. We would measure the previous requirement for engineers in hours per week. A need for one person would add 40 hours per week, a need for two people would add 80 hours per week, etc. The subrecord would look like:

RESID	*PER1*	*PER2*	*PER3*	*PER4*	*PER5*
ENGA	40	80	80	160	160

This would achieve the same purpose as the resource profiles illustrated earlier by only a single subrecord, requiring us to input only 6 parameters instead of 12. Also, it specifies resource requirements directly in the units (labor-hours) in which we wish to measure and control the resource, which could also be accomplished in an SRP. Figure 16-2 diagrams this resource profile.

Comparing this graph to those in Figure 16-1 shows that both of these methods specify identical labor resource requirements. The principle difference is that step resource profiles require that we generate a set of parameters every time the level of the resource changes, while periodic resource profiles require that we generate a single parameter only when the specified period changes.

Figure 16-2

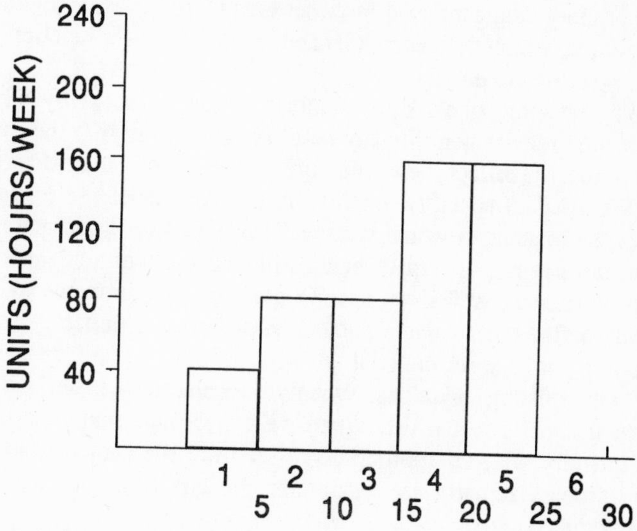

This is a very important distinction, and it is best to have both methods available.

Again we must emphasize that adding resource records to a model increases the need for maintenance of the modelfile considerably regardless of which method (SRP or PRP) we use. Tracking and controlling every resource item would probably not repay this effort, but tracking and controlling critical resource items, especially labor, would. We must make this important decision before we enter the first resource parameter. After loading the resource requirements we have chosen to manage for each task, we can move on to the next step.

RESOURCE POOLS AND SUMMARY ANALYSIS/LEVELING

Resource Pools

Based on some criteria, we also establish the resource pools for the project, which define the projected availability of any resource throughout the project at any point in time. This defines availability of the various individual materials, tools, facilities, and labor that we need to track and control, maintaining each as an individual resource pool. We establish each resource type as a separate record or series of records in a project structure called *resource files*.

Besides the resource code names, fields in the records store textual descriptions of each resource along with project structural information for sorting and selecting. The primary data contained in these records, however, is the time profile of each resource, which defines its planned availability as a function of time throughout the span of the project.

We specify resource pools by virtually the same process by which we specify the resource requirement for any task. They differ in that the quantities in resource pool records reflect the availability of a resource rather than requirements for it. We need to specify quantities in the records of the resource file consistent with the manner in which we specified requirements in activity subrecords. When we begin to compare availability versus requirement, our processes will not work very well if we specify the labor requirements of tasks as head counts and define the available pool in labor-hours or dollars.

The resource pool file for material must list all of the materials used in the project along with existing quantities. When we receive new materials, we add the appropriate quantities to the list. When we expect new materials to arrive, we enter the quantity with an appropriate offset from the project start. As the manufacturing process actually uses materials, the quantities in each resource record will be reduced.

We maintain the list to track the amount of material currently available or expected to arrive for the remaining tasks of the project. We could derive this list from a material ordering system that processes and controls the purchases from vendors.

We track the same key facilities and tooling that we tracked in the model as pools in the resource file. The file lists each facility or tool with quantities. If the availability of either a facility or tool changes, as when a new lab begins operation six months into the project or we lose access to a tool for maintenance or refurbishment, then we must define an appropriate adjustment in the time profile for resource availability. As for a resource requirement, a subrecord of the resource record specifies the quantity, offset from project initiation, and duration of the availability of each tool and facility.

As for resource requirements, labor pools present the most complexity of all the resource types. We define them by the same two methods by which we specified resource requirements. In a step resource profile (SRP) pool, we specify the amount of available labor (in units such as labor-hours, head counts, etc.), the offset after project initiation when it becomes available, and the duration over which it is available. Any time these values change, we must add another resource pool subrecord to define the new parameters. The process is so similar that we can define the same fields as for task requirement definition.

For example, suppose a project had a pool of electrical engineers. For the first six months, it had 10 of these engineers available, after which the pool increased to 15 for the next six months, finally peaking at 20 for the remaining year of the project. The following labor pool record and subrecords would model this situation:

(Record) Electrical Engineers: ELENG

RESQTY	*OFFSET*	*RESPAN*
10	0	181
15	181	181
20	362	362

The RESQTY field in each subrecord states the available amount of resource. The number of days after the initiation of the project when the labor resource becomes available appears in the OFFSET field. The duration over which the resource is available appears in the RESPAN field. This cumulative data defines the overall availability of this labor classification, as illustrated in Figure 16-3.

Figure 16-3

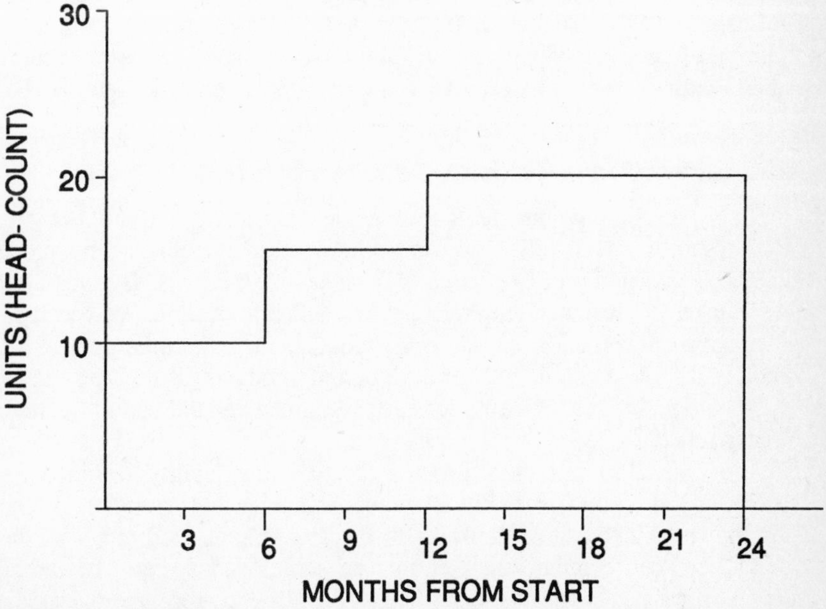

As for resource requirements, we can define available resources by another method called the *periodic resource profile (PRP) pool.* This requires no subrecords, but only a sufficient number of fields within each resource pool record to specify a resource level for every planning period (usually a month). Each field stores the available amount of resource for a specified period starting with the first (PER1) in hours per period (months). Our previous example would look like:

(Record) Electrical Engineers: ELENG

PER1	PER2	. . .	PER6	PER7	PER8	. . .	PER12	PER13	PER14
1520	1600		1520	2880	2400		2400	2800	3200

In this record, the value in the the first field (1520) reflects the number of available hours in the first month (152) multiplied by the resource level for the month (10). Each succeeding field shows each subsequent month's available working hours multiplied by the resource level (found by a headcount). This establishes the total available hours of the engineering resource for each month. This gives the same results as the step profile, although it requires more parameters. As a rule, step resource profiles require more entries for an activity's requirements than does a periodic resource profile. The reverse is true for pools. It is not inconsistent, however, to define requirements using a PRP and specify the pool using an SRP. This is, in fact, the preferred method.

Through these methods, we specify the total availability of each resource. Our model now shows the resource requirements of each activity and the amount of resources available to meet those requirements.

Periodic Summation of Resource Requirements (PSRR)

After defining resource requirements for each activity and specifying pools that state each resource's availability, we can begin resource modeling with a process called *periodic summation of resource requirements* (PSRR). This gives a total profile of all project resource requirements as a function of time. We specify the period of summation (i.e., in days, weeks, months, etc.), then the process adds up the total resource requirements for each period, producing a time-phased profile of the cumulative requirements for each resource over the period, assuming a particular schedule.

Figure 16-4 illustrates this process for two key facility resources: the Cramapaniola Lab (Lab 1) and the Pseudophandicular Lab (Lab 2). A Gantt barchart shows bars for all of the tasks in the project that need either of these labs so that we can compare their current schedules and spans. The periodic summation process calculates the cumulative need for each resource during each period (in this case, weeks). The results appear at the bottom of the example.

Figure 16-4

LAB RESOURCE REQUIREMENTS

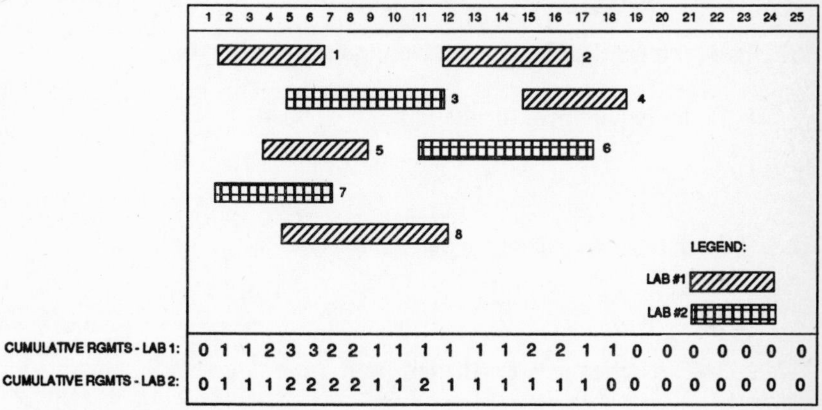

We can list the cumulative requirements for Lab 1 as follows:

Week 1—no requirement for the lab
Week 2—1 requirement for the lab from Task 1

* * *

Week 4—2 requirements for the lab from Tasks 1 and 5
Week 5—3 requirements for the lab from Tasks 1, 5, and 8
Week 6—3 requirements for the lab from Tasks 1, 5, and 8
Week 7—2 requirements for the lab from Tasks 5 and 8

* * *

Week 9—1 requirement for the lab from Task 8

* * *

Week 12—1 requirement for the lab from Task 2

* * *

Week 15—2 requirements for the lab from Tasks 2 and 4

* * *

Week 18—1 requirement for the lab from Task 4

* * *

Week 20—no requirement for the lab

Similarly, we can list the cumulative requirements for Lab 2:

Week 1—no requirement for the lab

* * *

Week 3—1 requirement for the lab from Task 7

* * *

Week 5—2 requirements for the lab from Tasks 3 and 7

* * *

Week 8—2 requirements for the lab from Tasks 3 and 7

* * *

Week 11—2 requirements for the lab from Tasks 3 and 6

* * *

Week 14—1 requirement for the lab from Task 6

* * *

Week 17—1 requirement for the lab from Task 6

* * *

Week 19—no requirement for the lab

Similarly we can create periodic summations (PSRRs) for every other resource type that is distinguished by an identification code relative to any specified period (weeks, months, etc.) and regardless of their units (as long as they are consistent.

Resource Leveling—Materials, Facilities, and Tools

Figure 16-4 shows that our original schedule does not match the cumulative resource requirements very well. The demand for Lab 1 goes from 0 to 3 to 1 to 2 and back to 0, and the demand for Lab 2 bounces back and forth between 0, 1, and 2 as well. We generally prefer to avoid this variation as we would like to use critical facilities optimally, keeping them working as much as possible and shutting them down only for maintenance or refurbishment. The cumulative requirement of a resource gives valuable information for our pursuit of this goal. It may guide us in rearranging our schedule to improve the use of key resources.

Cumulative requirements might also exceed capacity. If only one of these labs were operating, we could not meet the schedule displayed in Figure 16-4. We would have to delay most of these tasks pending the availability of a suitable facility.

To discover such conditions, we compare the cumulative requirement for a given resource to its availability as indicated by our resource pools. Especially for the inert resource categories (material, facilities, and tooling), we may have to rearrange the schedule to avoid any conflict.

The process by which we optimize the use of each resource and adjust our schedule for the availability of a given resource is called *resource leveling* or *resource allocation*. Through it, we develop viable, yet attainable schedules. We will continue with the problem from Figure 16-4 to explore this process.

The resource pools provide the following information:

LAB1

RESQTY	*OFFSET*	*RESPAN*
1	0	25
1	5	20

LAB2

RESQTY	*OFFSET*	*RESPAN*
1	1	24

Lab 1 is available for the duration of the project, and a second, equivalent facility becomes available, as well, after the first five weeks for the rest of the project. Lab 2 is available after the first week for the rest of the project.

Now we can assign the available resources to scheduled tasks. At the scheduled start of each task, we examine the resource pool for adequate resources to initiate the task. We then assign this resource until task completion, when it returns to the pool and becomes available for other work. If the pool shows no resource available at the start of a task, we delay its start until a sufficient resource returns to the pool or in some other way becomes available. In effect, we delay our peak requirements into the valleys.

Figure 16-5 illustrates this process, matching the scheduled tasks from Figure 16-4 with the appropriate pools. Task1 uses Lab 1 first. It can begin at the beginning of the second week on schedule as a Lab 1 remains available in the pool. This task continues using the resource until its completion on the sixth week, then the lab returns to the pool. When we try to initiate Task5 at the beginning of the fourth week, we find no Lab 1 in the pool, so we delay its start. At the beginning of the sixth week, a second Lab 1 enters the pool and we can assign it to Task5. The task occupies this resource until its completion at the end of the 10th week.

Figure 16-5

LAB RESOURCE LEVELED

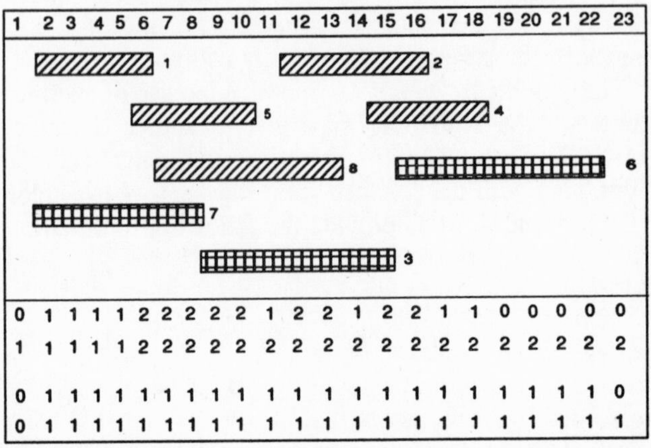

CUMULATIVE RQMTS - LAB 1:	0	1	1	1	1	2	2	2	2	2	1	2	2	1	2	2	1	1	0	0	0	0	0
AVAIL - LAB 1:	1	1	1	1	1	2	2	2	2	2	2	2	2	2	2	2	2	2	2	2	2	2	2
CUMULATIVE RQMTS - LAB 2:	0	1	1	1	1	1	1	1	1	1	1	1	1	1	1	1	1	1	1	1	1	1	0
AVAIL - LAB 2:	0	1	1	1	1	1	1	1	1	1	1	1	1	1	1	1	1	1	1	1	1	1	1

When we attempt to initiate Task8 at the beginning of the fifth week, again we find no resource in the pool. We delay its start until the lab used on Task1 returns to the pool when the task finishes at the end of sixth week. It is then reassigned to Task8 which occupies this lab until its completion seven weeks later at the end of the 13th week. Task2 needs this lab next for its start at the 12th week. The lab used in the accomplishment of Task5 waits in the pool, and Task2 holds the resource until it finishes at the end of the 16th week. As we attempt to begin Task4 at the 15th week, we find the resource used by Task8 has been returned to the pool. We can begin Task4 as scheduled and complete it after four weeks.

By this process, we have leveled the schedule to correspond to resource availability, but we cannot say that we have a totally optimal schedule. Lab 1 sits idle during weeks 11 and 14. We would have to pull in our schedule, however, in order to improve resource use. If we could start Task2 and Task4 one week earlier, we would level the use of Lab 1 throughout the project. Otherwise, we could use this down time for refurbishment and routine maintenance, if necessary, or as a cushion if any of the previous tasks overrun their schedules.

To level use of the second lab, we have to delay the starts of Task3 and Task6 until a suitable lab becomes available. The single Lab 2 facility forces us to schedule our tasks serially even though no constraint relationships may require it.

Other considerations may affect resource use and leveling, as well. For instance, at one point both Task5 and Task8 waited for Lab 1. When one became available in the sixth week, we assigned it to Task5 because that task was scheduled to start earlier. We might have given the resource to Task8 instead, however, if it were more important to our project.

When two or more tasks compete for the next available resource, we need a mechanism to decide which activity should get it. We do this through a process that resembles a data ordering statement, specifying primary, secondary, tertiary, etc. priority fields to break ties. We could decide based on values in a schedule start field (BES), or a time reserve (SPTR) field, or even a structure field (S1) containing some kind of a priority code. Comparing the chronological or alphanumeric sequence of the values in the primary field distinguishes which activity has priority over another for a resource. Any tasks that have the same value in the primary field fall in order based on the secondary field, and so on. This should enable us to assign resources to meet our project's needs better.

Often, we can substitute an alternative resource during times when the pool shows inadequate primary resources available. To do this, we would identify the alternative in the resource subrecords of the individual tasks as for any other resource requirement, except that we add a flag field or fields to indicate that the resource is an alternative and to specify which resource it can replace. During resource leveling, if we find no primary resource available, we can check for alternative resources. If the task specifies one that is available in the pool, we initiate the task using it.

These two processes share a significant flaw in that both work only when two or more tasks are competing for a resource prior to their initiation. If one task is slated to begin later than the other, then neither the process to assign resource priorities or that to identify alternative resources works properly. Thorough planning for critical resources requires, however, that we look ahead at least some distance to apply these principles. We need to add an offset defining a time period by which to look ahead prior to the assignment of resource priorities and/or alternative resources.

Normally, when a task uses a resource from the pool, it holds that resource until its completion. The accomplishment of sufficiently critical tasks can, however, justify taking a resource away from another on-going task. This interrupts one task's accomplishment but it is sometimes necessary. This requires another flag for tasks critical enough to take resources from other tasks. To determine which on-going task surrenders its resource to this critical task, we could use either the priority mechanism discussed earlier or establish a unique set of criteria for this purpose. This is an example of two sub-processes required to enhance our basic process to provide us better control over resource allocation.

All of these processes together do not perfectly reproduce a human decision process. Resource leveling cannot economically accommodate every contingency that might arise. When we assign resources to tasks, we cannot simply apply a set of rules; logical relationships and rationale are not always consistent in assessing which tasks proceed and which must sit and wait. For instance, we may judge one situation by one set of priority criteria and another by different priority criteria. We may elect to use an alternative resource to avoid waiting five weeks for a primary resource, but choose to delay the start of the task to get a primary resource if we had to wait only three weeks. For another problem, in some cases an alternative resource could cause the task to take longer, and the duration of the task would have to reflect this impact.

Several other situations can become more complex than our processes can control. In general, however, resource leveling can at least assist us in the efficient assignment of material, facility, and tool resources to specific tasks. The process is not perfect, nor an end in itself, and its limitations may force us to go beyond it to reach a decision. Still, the cumulative or summed resource requirements (PSRR) alone provide considerable benefit even if we choose to personally assign priorities and level utilization.

We must also remember that this process is not a one-time effort. As we progress through a project, the schedule changes due to many factors. A task may take longer than expected or begin late because of many factors. Many things can change the cumulative requirements for a resource and our plan must evolve in order to maintain relative consistency. We must periodically review our analysis to reassess current resource assignments, and this takes considerable time and effort. We cannot simply push a button to resolve all our problems through a computerized algorithm.

Labor Requirement Summation and Leveling

Labor is probably the most important resource to sum and level because it varies the most and usually costs the most. The principles basically repeat those for the other resource categories, but some unique problems arise. Let us examine a simple resource problem to illustrate this. The Gantt chart in Figure 16-6 shows nine tasks that require the same two labor groups: EEs and MEs (electrical and mechanical engineers). The schedule of each task corresponds to the time line of weeks across the top. For each weekly period, the total resource requirement (PSRR) appears along the bottom of the example.

Figure 16-6

SAMPLE RESOURCE SCHEDULE

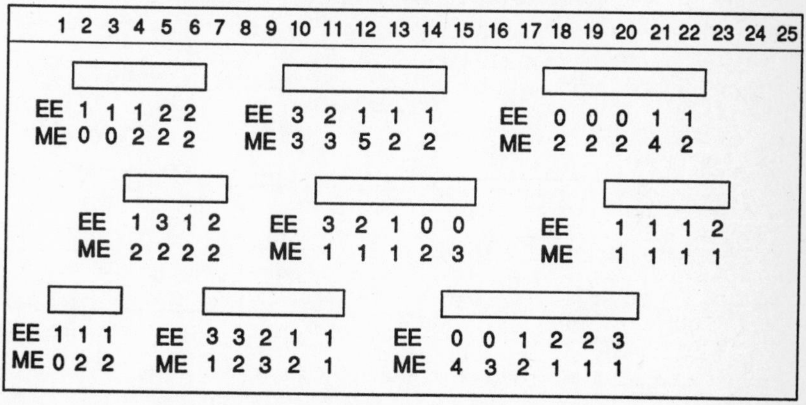

TOTAL - EE: 1 2 2 2 5 3 5 3 2 4 6 3 2 1 0 0 1 2 2 4 2 2 2
TOTAL - ME: 0 2 2 4 4 4 3 2 3 5 5 6 3 4 7 3 2 3 3 4 5 3 1

Examining each task shows how the resource requirements can vary through the span of a task. Some require the most labor at the start end (front loaded tasks), some require the most at the finish end (end loaded tasks), some are level loaded, and the requirements of some seem almost random. No simple equation can define the labor necessary to accomplish a task because each has its unique pattern of resource requirements. When we compare the requirements of each task with those of the other tasks and summarize the total requirements, the result almost always exhibits peaks and valleys. In the example, our requirement for EEs ranges from 1 to 6, back to zero, up again to 4, then ends at 2; the need for MEs likewise increases and decreases. We usually cannot hire and then dismiss personnel so rapidly. We must maintain resource availability at relatively stable levels. We need to bring personnel on board as the work becomes available and build to a level that will accomplish all of the work on the project's schedule. We must do this without an excess of resources as this would be very inefficient. To achieve the optimum requires adequate resources to offset the peaks with the valleys in a reasonable time frame relative to our schedule.

If we take the conventional approach to resource leveling or allocation, we examine each task as it is scheduled to occur, determine if the resources are available in the pool to accomplish the task and schedule it if they are. If any resource is short for any portion of the task, the entire task is delayed until such time as adequate resource is available. This process worked well when we dealt with material, facilities, and tooling, but labor is a different matter. The fallacy of this method can be illustrated through the problem in Figure 16-6. If we assume that our pool contains 3 EEs and 4 MEs, the result of resource allocation using our current process is illustrated in Figure 16-7. In our original schedule the last task was completed by the end of the 23rd week, but our resource plan indicates that we will not finish until the 37th week (15 weeks late—a 65 percent overrun to our original plan).

Figure 16-7

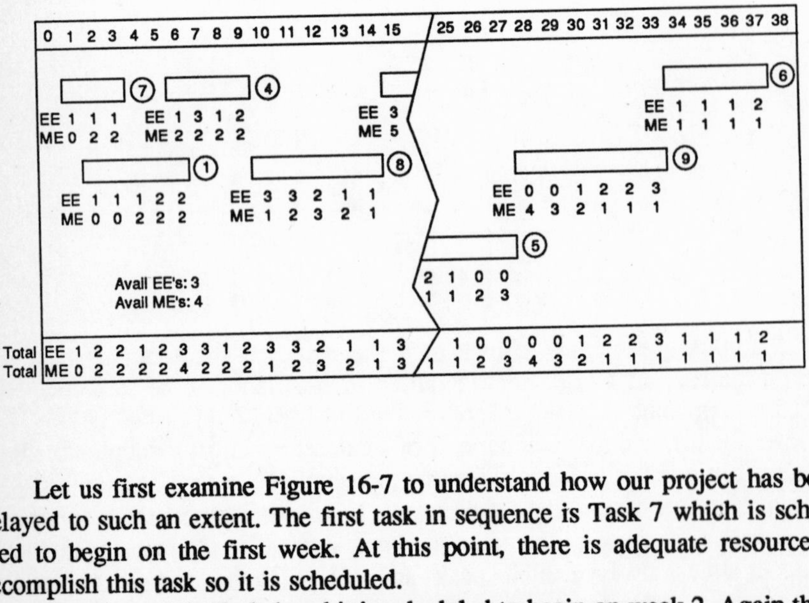

Let us first examine Figure 16-7 to understand how our project has been delayed to such an extent. The first task in sequence is Task 7 which is scheduled to begin on the first week. At this point, there is adequate resource to accomplish this task so it is scheduled.

The next task is Task 1 and it is scheduled to begin on week 2. Again there is adequate total resource and we schedule this task as originally planned.

Task 4 is scheduled to start on the 4th week, but this will cause an overload in week 5 of EEs. Therefore we must delay the task's start until week 6 where it can be accomplished without causing an overload.

Task 8 is scheduled to start on week 7, but this will cause an immediate overload of EEs. The start of this task will have to be delayed until week 10 in order for it to have adequate resource.

At this point we have already slipped the finish of our project by 3 weeks, and this delay continues to grow as we proceed down the path. It culminates in a 15 week total delay. In addition to the delay, we are also not getting efficient utilization of our resources. There are only two weeks in the schedule where all four MES are busy.

However, let us take a different approach to our resource allocation. If we plot the total resource requirement versus the pool for EEs, we end up with Figure 16-8. It shows graphically the peaks and valleys of the resource requirements. We need to match against this a reasonably stable resource pool, yet still finish the total work within a reasonable span of time. Two EEs could eventually accomplish all of the work of these tasks, but it would take through the 31st week to do it.

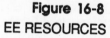

Figure 16-8

EE RESOURCES

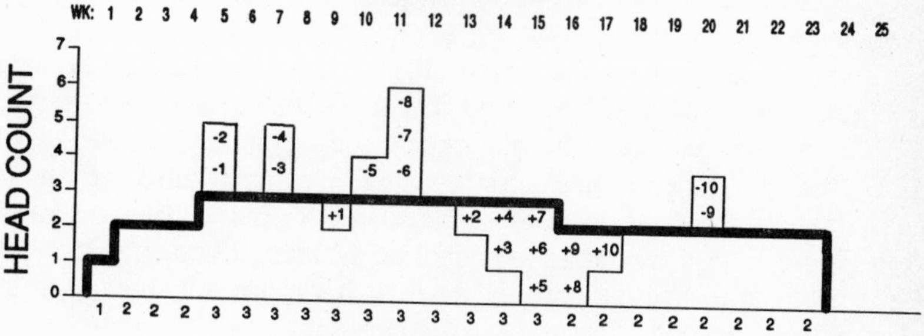

We will need three EEs for about three months in order to come anywhere close to our schedule which was the level used in Figure 16-7. We correlate a buildup of 3 EE's to our schedule such that all three are working by the end of the fourth week (illustrated by the bold line). For a period of time, this defines the level that we maintain.

We can determine how long we need the resource by counting each requirement represented by a grid square that occurs above our availability line (numbered as negative values). These are called *overloads*. We must offset every overload with a comparable availability (undedicated grid box) below the line, or an *underload* (numbered as positive values). Once the pool offsets all of the overloads (–10) with a comparable amount of underloads (+10), we have accommodated our planned work load. We could also state this mathematically, as equality between the area under the resource availability (bold) line and the area under the cumulative requirements (PSRR) line.

Since we generally cannot do any work early, we must time the underloads to follow their corresponding overloads (+1 in the 9th week offsets –1 in the 5th week, +2 in the 13th week offsets –2 in the 5th week, etc.). Our example includes an exception to this where we offset the –9 and –10 overload on the 20th week with underloads in the 16th (+9) and 17th (+10) weeks. This would require that we accomplish some portions of these tasks earlier than planned. If this is not feasible, we would have to maintain our resource level at two through the 24th week and release our third EE two weeks earlier. Either way, we produce a resource plan that satisfies our work load. In this case, we only used 3 EEs for 11 weeks, and 2 EEs for another 11 weeks, but this represents almost perfect efficiency.

A similar analysis on the MEs appears in Figure 16-9. To meet our requirements, we must build up to a level of four by the 10th week. If we hold this level through the 24th week, then we should accomplish all of the work. Our plan offsets all overload requirements (–1, –2, –3, etc.) with comparable, properly timed underloads. Our total overload (–11) matches the total underload (+11) leaving the area beneath the two lines equal. Again this represents almost perfect efficiency.

The question that we must ask is, is reality closer to the situation in Figure 16-7 or Figures 16-8 and 16-9? We find that if we fail to plan our resource utilization in advance, the situation will more closely equate to 16-7. We will inefficiently use our resources and miss schedule. We should also see that we used approximately the same level of resource in both plans. The cost of the overrun in Figure 16-7 is enormous and could be avoided. It certainly should be all that one needs to realize the potential savings that can result from good disciplined resource allocation.

We must be careful in that we cannot plan with perfect efficiency because we cannot estimate resource requirements perfectly. As difficult as it is to estimate the duration of a task, it is even more difficult to estimate the task's resource needs. However, even with these inaccuracies, the potential savings are enormous.

We can also see that the method of labor allocation should be different than the simple methods used to allocate material, facilities, and tooling. Our

Figure 16-9

ME RESOURCES

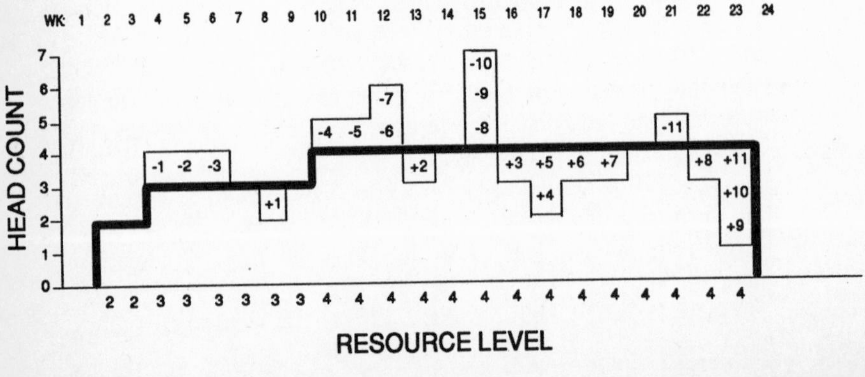

equal area method establishes a reasonable level of personnel, one that we can practically achieve and one that will still satisfy our schedule. To level facility and tool resources, we compared available resource levels to cumulative requirements. For labor resources, however, we have determined a resource level that satisfies our requirements, a somewhat different process for the peculiar traits of the resource. We can then compare our leveled requirements to our available resources to ensure that our resources will actually meet our needs.

This analysis, among other flaws, has not considered the impact of the absence of one type of resource on the other type. For instance, we plan to have only three MEs available on the fourth week. Just one of the two tasks needing two MEs will get them, the other will initially have to make do with one. Yet the two needed EEs will be available. Thus, one of these tasks will have all of its resource needs of one type, but only half of the other. Does this cause a problem? We don't know because our analysis assumed that each resource type is independent of the other in the performance of each task. Our process only ensures that each task receives its total requirement reasonably close to its schedule. It is not a perfect method, but it certainly provides us a basis for intelligent planning of our labor.

We cannot stop the process after devising initial resource plans in the beginning of a project, because change and status affect them as well. Failure to accomplish tasks according to plan can change the resource requirement significantly. Figure 16-10 shows a very simple project of four tasks scheduled to be accomplished between the 2nd and the 17th weeks. These tasks have FF constraints and they share common resources. The cumulative requirement (PSRR) of each resource type (*A* and *B*) appears at the bottom of the plan. Below the plan, resource plots illustrate the personnel loading required to meet our schedule. The *A* labor skill builds up to a level of four by the 5th week and holds this level through the 16th week. The *B* labor skill builds up to a level of five by the 6th week, holds this level through the 15th week and then declines to three for the final two weeks. This is our initial plan.

A status assessment after five weeks reveals a delay in the finish of the first task of two weeks. Figure 16-11 shows the impact of this slip. The late completion of the first task will also delay the completion of the second task through the finish-to-finish constraint, but only by one week. The project absorbs the slip before it affects the other two tasks.

A more serious problem arises with regard to the resources. The compilation of the cumulative resource requirements after the slip shows that resource type *A* now peaks at 7 and 8, up from 6, and resource *B*, which hovered around 5, is now between 7 and 8, as well. The slip created a near-term "bow wave" of resource requirements. If we do not alter our plan, this overload will travel downstream like a flash flood, ultimately delaying completion of the project.

Figure 16-10

INITIAL CONDITION

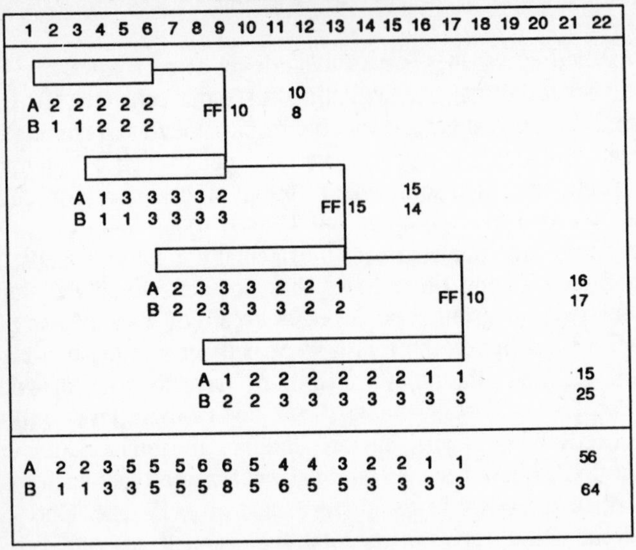

A. RESOURCE

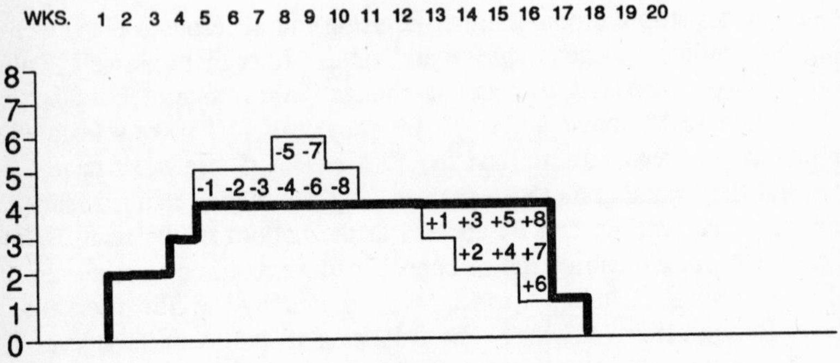

B. RESOURCE

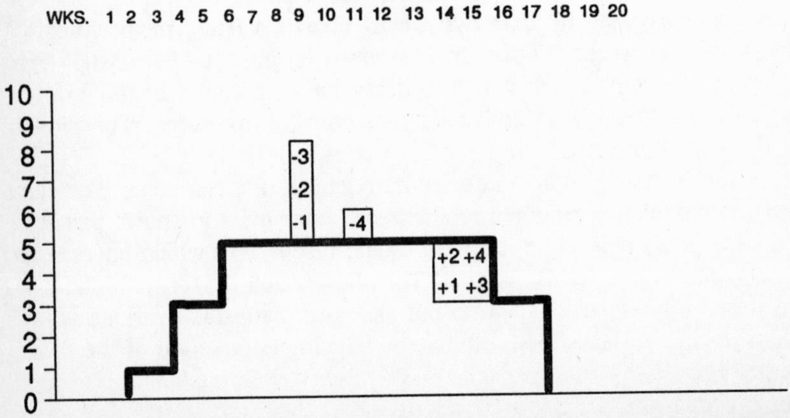

At the lower half of Figure 16-11 is a resource profile of both *A* and *B* which illustrates our problem. We must offset our overloads in the 7th through the 10th weeks with underloads below the available resource line. To maintain current resource levels, we would have to extend the completion of the project one week to offset the overloads. We simply lack enough people to complete the tasks as scheduled. (The impact is not greater because of the minute size of our project and the short spans of time.) In order to correct this situation, we must replan the resources of the project.

Figure 16-12 illustrates such a replan. The setback caused by the delay of the first task's completion creates the need for an additional person of each resource type (*A* and *B*) for a period of time. If we can acquire them according to the replan, the project should finish on schedule because we can offset the overload more quickly. This is illustrated in resource plots at the bottom of Figure 16-12.

If we can not acquire the additional resources, we would have to look at working overtime to make up our deficit. Working four people a reasonable amount of overtime would in time do the job of an additional resource and correct our problem.

This very simple exercise illustrates why we must continue to update our resource analysis just as we do our time analysis. A project is not static. The external forces of change and the internal forces of status constantly alter the resource requirements. We must adjust our plans to maintain them in line with the project at hand. This makes the process of managing the project resources every bit as challenging as time management.

THE COST OF RESOURCES

Once we understand the detailed and overall resource requirements of our project, the next logical step is to compute their cost. This very simple process quickly becomes very complicated because of standard accounting practices. Such are the realities of modern business. Once again the category of resource (material, facilities, tools, and labor) affects the costing process.

Each item of material, for instance, has a current cost which may also include taxes, handling charges, and overhead. We could list this total amount as a parameter of the item in the resource pool. As the production process uses each item of material, we can charge its equivalent dollar amount to the task and subtract it from the pool. This resembles a ledger as we add costs to the project and then subtract them from the company inventory. In addition, our model specifies the time phase of each task and its associated material, telling us how much we need and when. From this information, we could develop a time phased budget plan for material which directly affects our project's cash flow.

Figure 16-11

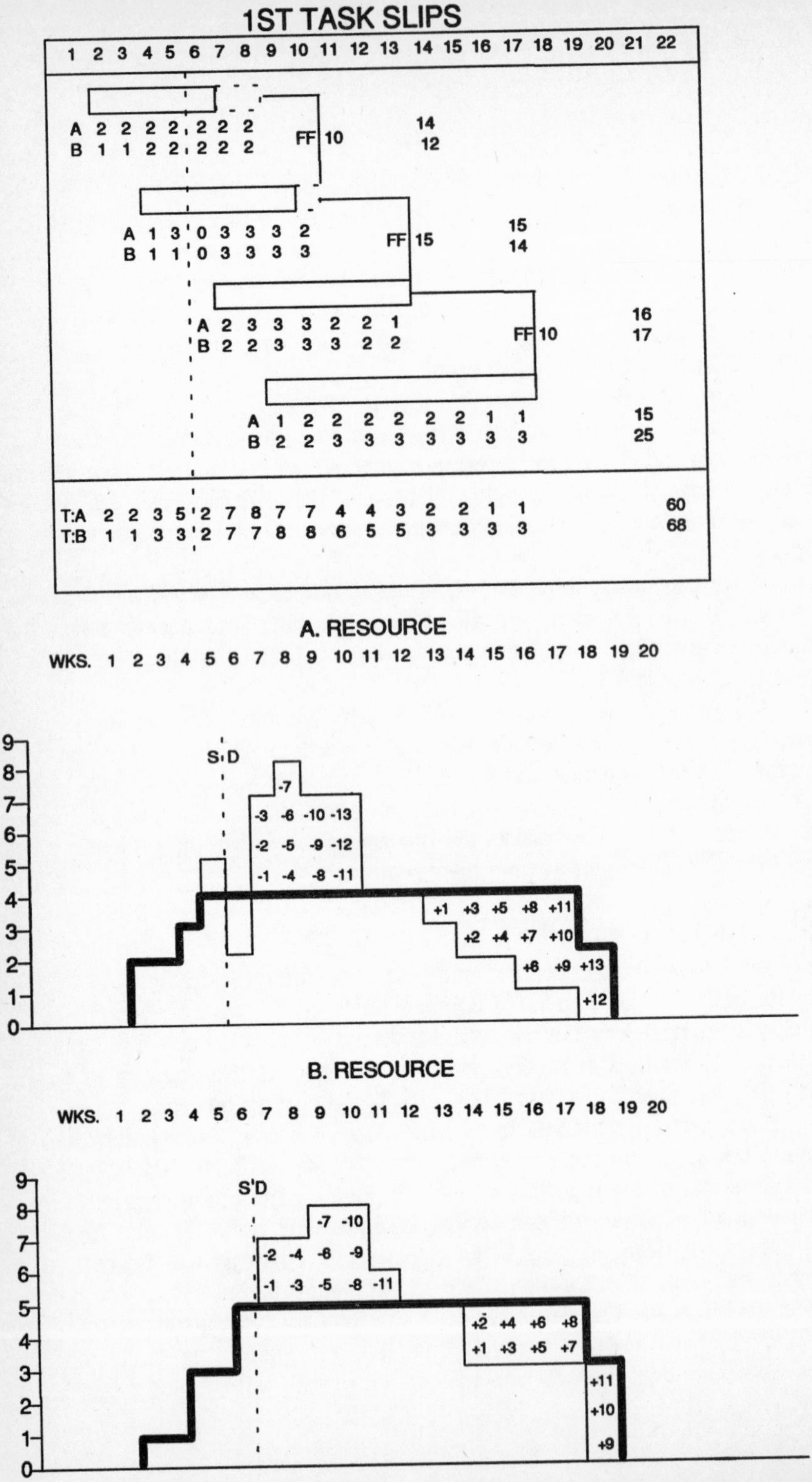

1ST TASK SLIPS
1 2 3 4 5 6 7 8 9 10 11 12 13 14 15 16 17 18 19 20 21 22
A 2 2 2 2 2 2 2
B 1 1 2 2 2 2 2
FF 10
14
12
A 1 3 0 3 3 3 2
B 1 1 0 3 3 3 3
FF 15
15
14
A 2 3 3 3 2 2 1
B 2 2 3 3 3 2 2
FF 10
16
17
A 1 2 2 2 2 2 2 1 1
B 2 2 3 3 3 3 3 3 3
15
25
T:A 2 2 3 5 2 7 8 7 7 4 4 3 2 2 1 1
T:B 1 1 3 3 2 7 7 8 8 6 5 5 3 3 3 3
60
68
A. RESOURCE
WKS. 1 2 3 4 5 6 7 8 9 10 11 12 13 14 15 16 17 18 19 20
S D
-7
-3 -6 -10 -13
-2 -5 -9 -12
-1 -4 -8 -11
+1 +3 +5 +8 +11
+2 +4 +7 +10
+6 +9 +13
+12
B. RESOURCE
WKS. 1 2 3 4 5 6 7 8 9 10 11 12 13 14 15 16 17 18 19 20
S'D
-7 -10
-2 -4 -6 -9
-1 -3 -5 -8 -11
+2 +4 +6 +8
+1 +3 +5 +7
+11
+10
+9

Figure 16-12

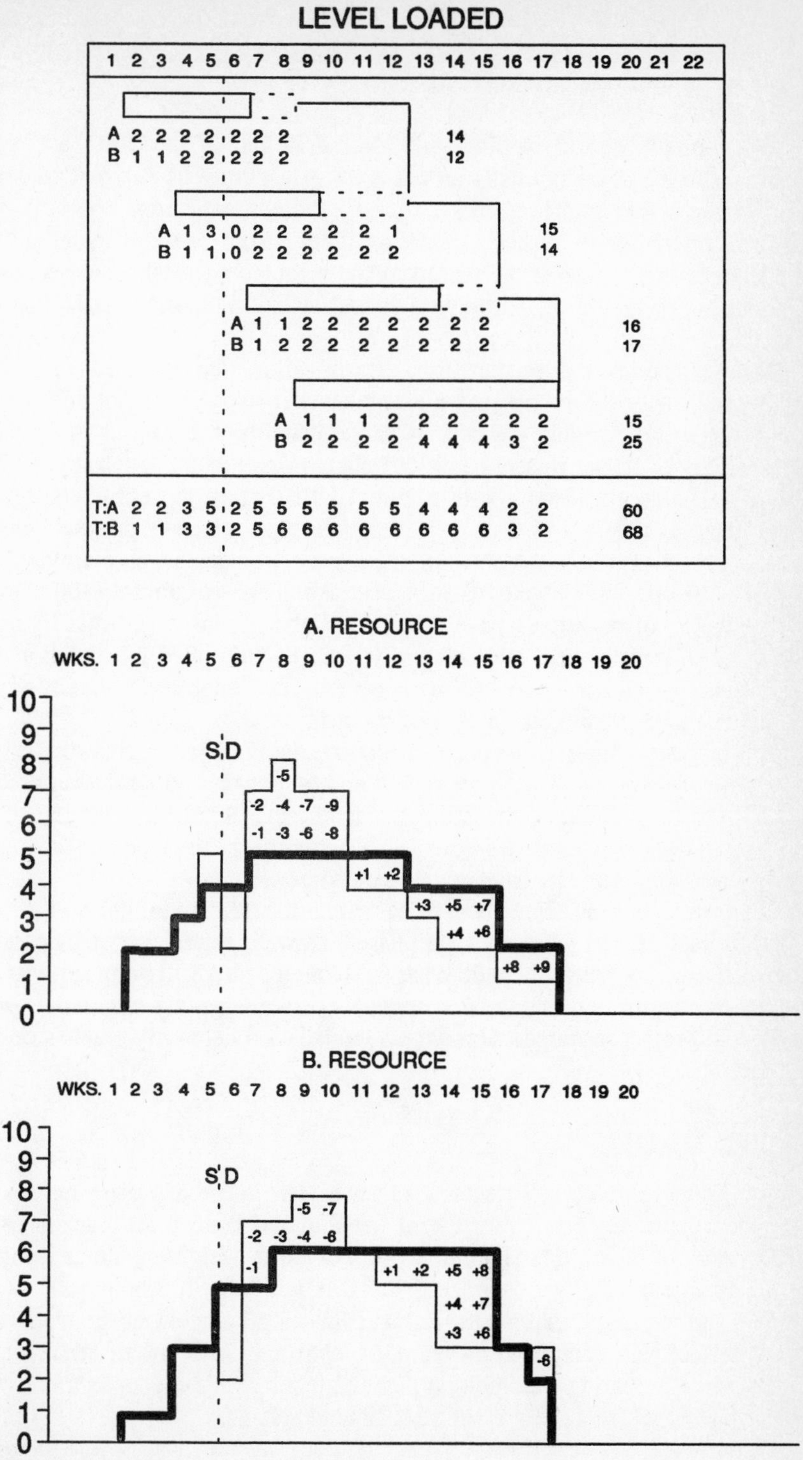
LEVEL LOADED
1 2 3 4 5 6 7 8 9 10 11 12 13 14 15 16 17 18 19 20 21 22
A 2 2 2 2 2 2 2 14
B 1 1 2 2 2 2 2 12
A 1 3 0 2 2 2 2 2 1 15
B 1 1 0 2 2 2 2 2 2 14
A 1 1 2 2 2 2 2 2 2 16
B 1 2 2 2 2 2 2 2 2 17
A 1 1 1 2 2 2 2 2 2 15
B 2 2 2 2 4 4 4 3 2 25
T:A 2 2 3 5 2 5 5 5 5 5 5 4 4 4 2 2 60
T:B 1 1 3 3 2 5 6 6 6 6 6 6 6 6 3 2 68
A. RESOURCE
WKS. 1 2 3 4 5 6 7 8 9 10 11 12 13 14 15 16 17 18 19 20
10 9 8 7 6 5 4 3 2 1 0
S.D
-5
-2 -4 -7 -9
-1 -3 -6 -8
+1 +2
+3 +5 +7
+4 +6
+8 +9
B. RESOURCE
WKS. 1 2 3 4 5 6 7 8 9 10 11 12 13 14 15 16 17 18 19 20
10 9 8 7 6 5 4 3 2 1 0
S.D
-5 -7
-2 -3 -4 -6
-1
+1 +2 +5 +8
+4 +7
+3 +6
-6

We normally charge facilities and tooling as part of the overhead of the company. Therefore, we normally do not convert resources of this type to dollar amounts in resource modeling. In some cases, though, we charge special facilities on an hourly or daily basis, handling them just as we did material. We append the facility's rate to the record in the resource pool, and then compute our planned expenditure by multiplying these rates by the spans of tasks that use the facilities.

Different companies account for labor in different ways. Basically, however, we cost labor by multiplying the number of personnel by the appropriate labor rate. To this, we add the cost of fringe benefits and various and sundry overhead figures. All of this provides the planned cost of labor resources.

For instance, the first task in Figure 16-10 (before the slip) requires 10 personweeks (2 + 2 + 2 + 2 + 2) of resource type *A* and 8 personweeks of resource type *B* (1 + 1 + 2 + 2 + 2). A personweek equals one individual for one week. To find the budgeted cost of this task, then, we multiply the number of personweeks of resource type *A* (10) by its labor rate and add that figure to the number of personweeks of resource type *B* (8) multiplied by its labor rate. To this sum we add the cost of fringe benefits and appropriate overheads to calculate the total planned labor cost of the task.

By this same process, we could determine the cumulative cost of each resource (56 personweeks of resource type *A* and 64 personweeks of resource type *B*) for the entire project, or the project cost in total. Since our cumulative resource plan is time phased, we could also determine a time phased cost plan against which we could measure actual expenditure.

Determining rates, fringe benefit costs, and overheads requires an involved process in itself. Often these are time phased, as well, since labor rates escalate periodically as do fringe benefit costs and overheads. Software applications every bit as complicated as our CPM applications are generally required to properly cost all project resources. Our simple explanation here only touches on this subject.

EARNED VALUE

In loading the resource requirements of each task into the project model, we create the opportunity for a very useful tracking and control mechanism called *earned value*. In effect, this is an effective method of weighting status information. If, as an example, we estimate that a task will take 100 hours of labor to complete and we assess that we have accomplished 50 percent of the work, then we have effectively completed 50 hours of labor and 50 hours of labor remain to complete. Comparing time-phased planned labor requirements to this earned value indicates whether we are ahead or behind schedule.

Figure 16-13 illustrates three detail tasks (A, B, and C) and one summary task (SMRY). Each detail task will take 200 hours and four weeks to accomplish, but we have spread the hours differently over the four weeks of each task to correspond to the effort needed to accomplish each of them. Task A takes 50 hours each week, so it is a level- or even-loaded task. Task B is end loaded, as it requires that we concentrate effort in the fourth (85) and third (75) weeks as compared to the first (15) and second (25) weeks. Task C is front loaded, requiring more effort in the first (75) and second (50) weeks than in the fourth (25) and third (50) weeks. Our status assessment indicates that we have completed 50 percent of each of these tasks. Because of the spread of labor, this would indicate a different level of accomplishment for each task.

Suppose that at the end of the first week we have completed 25 percent of each task, and at the end of the second week we have completed an additional 25 percent. In total, we have accomplished half of the work of each task. Comparing this to the plan, however, gives a different schedule status for each effort.

We have completed 25 percent of Task A, the even-loaded task, the first week and 25 percent the second. At the end of the second week, we planned to have accomplished 100 hours (50 + 50) or 50 percent of the task. The estimate of 100 hours of the work completed means that we are on schedule. If we had completed 75 hours, we would be behind schedule whereas if we had completed 125 hours, we would be ahead of schedule. We can portray this situation graphically through the shading of the bar. To indicate the on-schedule status of the first task, we shade the bar up to the end of the second week, the status date.

Figure 16-13

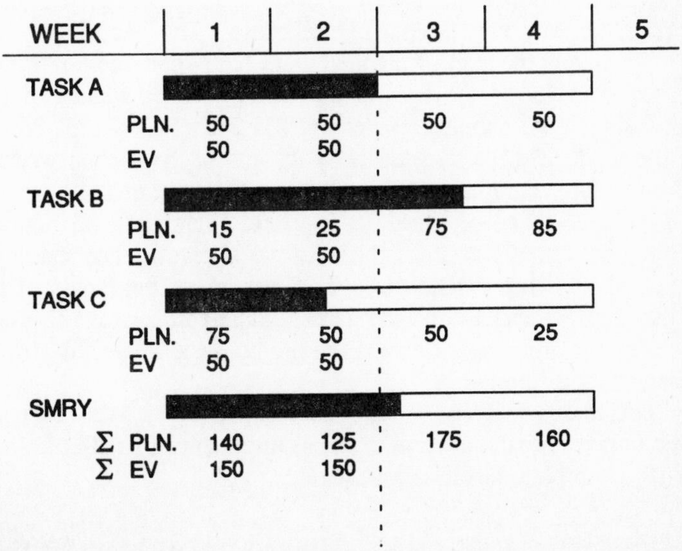

The schedule for Task B, however, called for completion of only 40 hours (15 + 25) or 20 percent of the task at the end of the second week. In fact, we estimate that we have accomplished 100 hours, or 50 percent of it, so we are ahead of schedule. We have completed 60 hours (100 – 40) of the work scheduled for the third week. Our original estimate called for completion of 75 hours of work in the third week, and we have completed 80 percent of that work (60/75 = .8) by the end of the second week. To illustrate this, the shading in the Gantt chart extends to cover 80 percent of the third weeks' length.

Task C is behind schedule, though. We should have completed 125 hours (75 + 50) or 62.5 percent of the task by the end of the second week, but we have completed only 100 hours, or 50 percent. This corresponds to 50 hours of accomplished effort per week. Half of our effort in the second week (25 hours) contributed to completion of the work scheduled for the first week (50 + 25 = 75). The other 25 hours of the second week contributed to work scheduled for that week, completing half of that 50 hours of planned work. In the Gantt barchart, the shading of the Task C bar covers through only half of the second week.

At a glance, the shading in the schedule shows that Task A is on schedule, task B is almost a week ahead of schedule, and Task C is about three days, or one-half week, behind schedule. This effectively allows us to visually measure and reflect the schedule performance of on-going tasks.

This is a very powerful process, as, more importantly, it provides us a common medium on which to base all of our status. This means, among other things, that we can summarize our data. The fourth bar, SMRY, represents a cumulation of all three of the detail tasks. It starts with the earliest start date and finishes with the latest finish. In practice, summary tasks normally correspond to various levels of a project hierarchy. This structure is portrayed through field codes in the same way we summarized schedules. The bar for the summary task's schedule covers the sum of planned work hours per week of Tasks A, B, and C. Its earned value is the sum of the weekly earned values of the three tasks. Cumulatively, we should have completed 265 (140 + 125) hours or 44 percent of the work of these three tasks by the end of the second week, yet we estimate that we have completed 300 hours or 50 percent of the work. We are, therefore, 6 percent ahead of schedule, in effect completing 35 hours (300 – 265) or 20 percent (35/175 = .2) of the third week's work. The shading on the summary task should then extend through 20 percent of the length of the third week. In this way, we can extend the earned value comparisons to assess performance for major groupings, or even the project as a whole, to provide very useful management information.

In addition to this analysis, we can also compare actual labor costs to either earned value or schedule completion to assess the financial health of our project. This gives us a complete business assessment.

Even this very brief explanation should make clear that earned value provides a very objective assessment of the progress of on-going tasks. It tells us where we are in the accomplishment of each task by estimating how much of the task we have completed. Resource loading allows us to understand the relationship between this estimate of work completed and our plan. We can extend this measurement up through our schedule hierarchy by summarizing the information. This enables us to measure total on-going performance relative to plan very effectively to determine if our efforts have us ahead or behind.

SUMMARY

The accomplishment of work is a function of both time and resources. The resource modeling processes extend the techniques of project time modeling to achieve a more complete assessment of work accomplishment.

Resource modeling has four objectives:

1. To understand the resource requirements of both the project and each task within it relative to time.
2. To compare cumulative resource requirements to projected or actual resource levels.
3. To level or smooth the peaks and valleys in requirements for certain resources such as specific labor skills, tools, or facilities.
4. To track and control resource costs.

The resource modeling processes consist of seven basic steps:

1. Establish a resource profile for each task.
2. Establish a pool for each task.
3. Sum the requirements of each resource throughout the project relative to a specified period.
4. Compare the cumulative resource requirements to the projected resource availability.
5. Determine the allocation of resources adequate to meet the requirements of the project within the time constraints, but do so in the most efficient manner.
6. Convert resource consumption to dollar equivalents.
7. Track and control these parameters.

Resources fall into four basic categories: labor, materials, tools, and facilities.

We attach resource requirement information to activity records through fields in resource subrecords. We model material, tool, and facility requirements through a step resource profile (SRP). In a subrecord, we specify the amount of necessary resource (RESQTY), the time of its need after the task's start (OFFSET), and the duration of its need (RESPAN). If the resource requirements of an activity change, we enter a new resource subrecord.

We model labor through either a step resource profile or a periodic resource profile (PRP). In this second method, we enter the requirements of a resource type for specified periods in a single subrecord of the activity record.

We store resource pools in a separate datafile containing a record for each distinct resource. Fields in these records define each resource in the same manner as activity subrecords define its requirements.

Once the user defines resource requirements for each activity, the process then sums the cumulative resource requirements for each distinct type for each specified period (weeks, months, etc.) to compile a periodic summation of resource requirements (PSRR). Comparing this measure of cumulative resource requirements for a specific resource type to the pool of the same resource reveals any shortfalls (overloads) or surpluses (underloads).

Through various means we allocated the available resources to the tasks, in a fashion that allows us to best meet schedule requirements, yet in as efficient a manner as possible. Two of these processes were discussed:

1. Matching available resources to each task.
2. Equal area method of requirement—availability allocation.

Other more refined techniques are necessary to accurately model true resource allocation scenarios.

Once we have allocated resources, we can determine a cost equivalent for each using various financial parameters and the schedule: rates, burdens (overhead), escalation, etc.

An earned value method is an extremely powerful mechanism to overlay on resource loaded tasks. It provides a common medium for detail and summarized performance measurement (cost and schedule).

Epilogue

Successful project modeling has its price. In making any investment, we must evaluate it knowledgeably and prepare well to spend assets wisely. Even then, risks remain. Many pitfalls block the path to success, and it is best to map them out up front.

Many problems arise from misconceptions regarding the modeling process. In the most common mistake, users underestimate the complexity and effort required to achieve worthwhile results. Project modeling is very labor intensive; it requires sharp skills and long, hard work. This is the principal investment in project modeling: the time and expense to train people, and then exertion of effort necessary to complete the job properly.

In another common mistake, users place an exaggerated trust in computer applications, believing that the right software will solve their problems. Even the most expensive, upper-end computer tools only complement the process; they do not themselves develop or maintain viable project models, analyze the processed data, develop practical and effective workaround solutions, or take action to implement these solutions. These jobs are all part of the workload of the practitioner. Computers also add another level of complexity and processes encompassed by the software application are anything but "push-button."

Understanding the user's burden, however, does not in any way diminish the importance of the computer. We simply must thoroughly understand its role. The computer, like the CPM process itself, is a tool, and tools do only what the knowledge, skill, and effort of the user make them do. The finest hammer can no more build a house, than the finest pen can write a novel. Neither can the best CPM tool plan and manage a project itself. The magic of any tool is in the user. In the hands of a master it accentuates that person's skills and efforts; in the hands of a novice, it is nothing but an expensive ornament.

Some believe that modeling processes arise naturally from simple intuition. This mindset leads directly to disaster. If these processes were so intuitive, then cost overruns and delinquent schedules would not be nearly so commonplace. The model of a project is every bit as complicated as the project itself for three reasons. First, there is virtually an infinite variety of ways to combine the model elements and parameters to simulate a myriad of practical situations. Second, most projects require management of a large number of activities and interfaces, and this sheer volume adds complexity. Finally, usually before we have even finished the first iteration, we must begin to alter the model to keep pace with the project's dynamics. All these things complicate the modeling process immensely.

We can conclude from this that the key to modeling success is the skilled professional who not only takes the time to run the process correctly, but knows what to do with the process' output. Such knowledge and competence do not come in an instant. We must be willing to learn and grow according to our abilities and efforts. We must develop and progress from the rudimentary to the sophisticated.

An understanding of the sequence and flow of a project generates a better schedule than mere intuition. A set-back schedule from the backward pass further substantiates the planning process as it helps us determine valid schedule requirements. Forward-pass assessments of the schedule impacts of slips and changes on successor tasks adds still further clarity. Eventually, we can advance to time reserve management and extend our project management efforts into the future by proactive methods. These are only a few of the interim steps on the evolutionary path that we can follow.

This all requires extensive learning. The project modeling method employs specific expertise in many areas, and ignorance in any one of them can be fatal. Together the facets of the process work like the legs on a table—take away any one of them and the structure collapses. Only on a solid, complete foundation can we build a stable process.

Elements of Successful Modeling

1. The Project Modeling Processes. Project modeling mathematically simulates reality. Valid project models require thorough understanding of model elements and parameters as well as the relationship between the real situation and the model that represents it.

2. The Intricacies of the Project. Understanding the project at the model's level of detail requires an enormous effort. Further, no one can model in isolation. The information for model parameters and elements must come from those who are responsible for accomplishing the project. Every facet of the process requires coordination and consensus of all involved.

3. The Relationship between the Model and the Computer Database. Even though a model may generate a single set of parameters and elements, no two software applications treat them the same. Semantic and syntactical variations can bring both dramatic differences and subtle nuances. We must understand how to represent each modeling parameter in the format of our software. If our software application lacks certain features, we must derive suitable alternatives or workaround solutions. We must also understand how to configure, input, and maintain our data.

4. The Processes and Output Products of the Software Application. To maximize our efforts, we must understand how our application processes the data, and then how to format this data in suitable output products. This facilitates our analysis and our communication of the results.

5. The Response of Project Management. The project management must, a) understand what to expect from the modeling processes, b) know how to interpret the model data and isolate problems, c) understand what options are available to resolve the problems, and d) monitor progress, analyze, and adjust.

Mastery of these skills could very well take years. Without knowledgeable users who can develop and analyze project models and knowledgeable managers who can develop plans and implement solutions, the process will never fulfill its expectations. These people must make and fulfill a commitment or the effort will always fall short. The promise of proactive management cannot be kept without competence in all five aspects.

Part of the growth of both user and manager toward enlightened understanding includes comprehension and acceptance of many fundamental principles of project modeling:

> In project management, it is not as important to manage the tasks and events as it is to manage the interfaces of effort. (Corollary: a project is not an independent collection of activities and events, but rather a coordinated, interdependent sequence of effort.)
>
> Time is a key resource to any effective problem resolution. More time to react to a problem provides not only a greater choice of solutions, but a better chance of success for any course of action.
>
> PROACTIVE MANAGEMENT—In order to influence the outcome of a project positively, it is far better to direct its

future course based on an imperfect understanding than it is to control only the known present.

To fail to plan a project initially is to admit ignorance of the requirements for accomplishing the project's objectives. To fail to replan the project's effort continually as it progresses is to admit ignorance of how little one actually knew in the first place.

The dynamics of a project caused by change and status assessments are an integral part of the reality of project management. Our tools, our methods, and we ourselves must adapt to these dynamics to succeed.

Any utility of computer applications depends on a thorough understanding, first of the processing they perform, and second, of all of the application's features.

Path time reserve is a property of a path, and not the individual activities along the path.

Attempting to control a project by monitoring schedule performance is a reactive methodology; managing time reserve is a proactive methodology.

Attempting to control a project by monitoring actual expenditure is a reactive methodology; managing the resource requirements of future effort is a proactive methodology.

Schedule performance and cost performance are integrally related. Very few factors can affect only one of these. We must learn to manage them and their associated reserves together.

Even if we develop the skills, maintain the motivation, and spend the time, we still have nothing if we do these things without discipline. *Discipline* must be an integral part of the process; it is by far the most important aspect. We must maintain discipline in gathering model information, our modeling methods, our status methods, our baseline control methods, our problem resolution processes, our database structure, our data functions, and our computer operations. Every facet of the modeling process depends on discipline.

Yet, the discipline must be practical. All too often we tend to confuse discipline with bureaucracy. Discipline comes not from filling out forms, sequestering printers and plotters behind locked doors, or establishing different groups to handle the various facets of the process; it comes from defining the fundamental process and then taking the necessary steps to properly accomplish each of them in a judicious but thorough manner.

Project modeling then is a technology, a philosophy, a discipline, a methodology. Above all, it is a major endeavor, and one that should be undertaken only with the proper commitment. The potential rewards are enormous. Planning and careful management maximize financial and schedule success.

> Nothing else we can do offers a greater promise of return for the assets invested.

Thorough and continuous planning, more than any other effort, offers the greatest opportunity for successful achievement of project management's schedule and financial objectives.

APPENDIX A

Arrow (AOA) versus Precedence (AON) Modeling Techniques

Two of the many contrasting methods of project modeling have the names Arrow and Precedence. This book has employed the Precedence method, also sometimes called *Activity on Node* (AON) modeling or *Precedence Diagramming Method* (PDM), ignoring the Arrow method, also called *Activity on Arrow* (AOA) *modeling, Arrow Diagramming Method* (ADM), or *IJ*. This choice does not imply any fundamental deficiency in the Arrow method. It is simply very confusing to examine project modeling in Arrow and Precedence simultaneously. Now that we have explored the Precedence process thoroughly, however, we can learn to translate the concepts in the alternative technique. This appendix will briefly examine the Arrow method, especially noting its differences to Precedence.

The similarities between these two methods far exceed the differences. Both methods model projects by simulating the three basic elements: activities, events, and constraints. Both follow identical processes, first performing a forward pass, then a backward pass, and finally calculating Path Time Reserves.

The only major difference lies in the fundamental element of the model, the definition of the *node*. Precedence defines nodes as activities, each with a start and a finish. Arrow nodes represent events, however, which define an instant of time. This fundamental variation creates a very different model as it defines the project elements completely differently. Since events appear as nodes, activities lie between start nodes and finish nodes. They take the form of node-to-node relationships, records of which contain the task's span. Constraints are a little more complicated in Arrow, occurring in two forms. Examples simplify the explanation.

Figure A-1 portrays a simple model of three activities bound by two constraints. The top diagram illustrates the model in Precedence, the bottom diagram illustrates it in Arrow. The Precedence model consists of three nodes and two constraints while the Arrow model consists of four nodes and three node-to-node relationships. Arrow node 1 corresponds to the start of Precedence node A; Arrow node 2 corresponds to the finish of the Precedence node A. Precedence activity A appears in the Arrow model as nodes 1-to-2.

The Arrow node 2 serves double duty; it is the finish of the first task (1-to-2), and also the start of the second task (2-to-3). Because of this, it also models the relationship between the first and second activities. In the Arrow method, an event (node) can indicate both the finish of a predecessor and the start of a successor.

Figure A-1

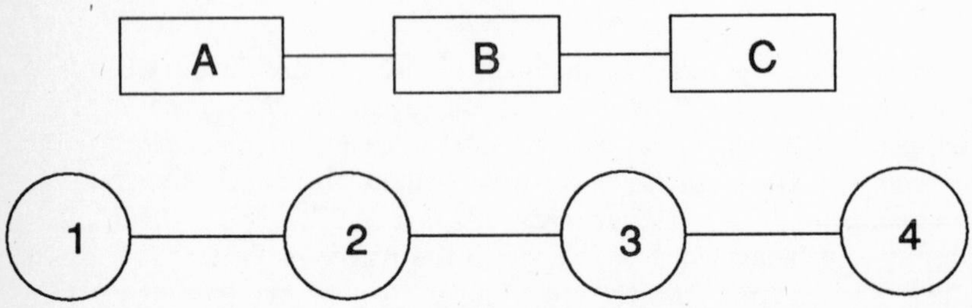

Figure A-1 shows that Precedence models these tasks in five records (three nodes and two constraints), while Arrow takes seven records (four nodes and three node-to-node relationships). However, the basic record in Arrow is not the node record (representing events), but rather the node-to-node record (representing activities). We could create separate event records only for the significant milestones, defining all other nodes in activity (node-to-node) records. This would trim this Arrow model to only three activity (node-to-node) records. Only when we wish to emphasize an event should we define it as a separate record. In our example, we may wish to define the project start and finish events, again increasing the record count to five. We may elect to highlight just the finish giving four records. The number of records necessary to model in Arrow versus Precedence is classic "apples and oranges." It provides no basis to prefer one method over the other; it is virtually a wash.

The second type of Arrow constraint is more complicated than the first, where a node then defines the relationship between two tasks in terms of the finish of one and the start of the other. However, tasks can be interrelated in another way, as we will see through our second example. The Precedence model in Figure A-2 contains seven activities and the relationships between them. Two paths join nodes A and C, the first through node B and the second through node D. Likewise, one path between nodes E and G goes through node F and the other through node D. Activity D becomes a funnel for several paths, complicating this model considerably.

Figure A-2

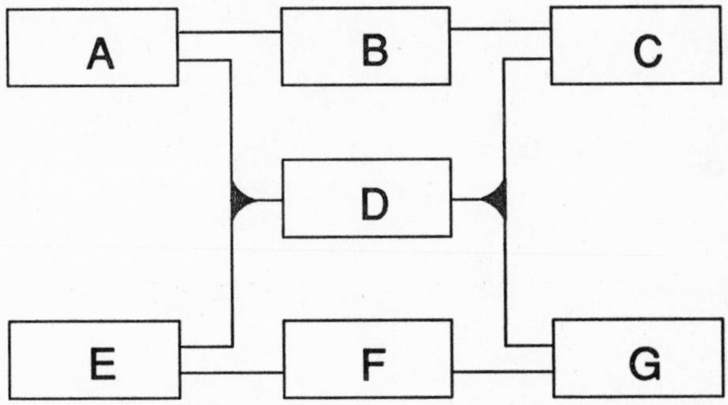

To model this relationship in Arrow, however, nodes can initiate more than a single task, creating a Y-like branch at the common node. An accommodating software application would allow us to create more than one activity between the same two nodes to model different activities. Neither of these techniques adequately models the situation, though. Instead, we would normally create a separate activity, adding a pair of nodes, to model activity D. Figure A-3 illustrates an Arrow model that represents task D with the node-to-node relationship 5-to-6.

In such a case, we must initiate this task and terminate it properly. The start of this task would have to depend on the finish of both activities A and E. Activity A is modeled by the node-to-node relationship 1-to-2, and activity E is represented by the node-to-node relationship 7-to-8. Attaching the finish nodes of these two activities (nodes 2 and 8) to node 5 models the relationship correctly. Node-to-node relationships with no duration (2-to-5 and 8-to-5) serve the purposes of our constraint relationships. This is the second method of modeling task relationships in Arrow.

Likewise, we need to attach the finish of node 6 to the start of tasks 3-to-4 and 9-to-10 with two durationless node-to-node relationships (6-to-3 and 6-to-9) to completely model the proper relationships. A durationless node-to-node relationship in Arrow is called a *dummy*, an unfortunate choice of words with its connotation of stupidity. In fact, it is simply an alternate method of constraining activities together.

Figure A-3

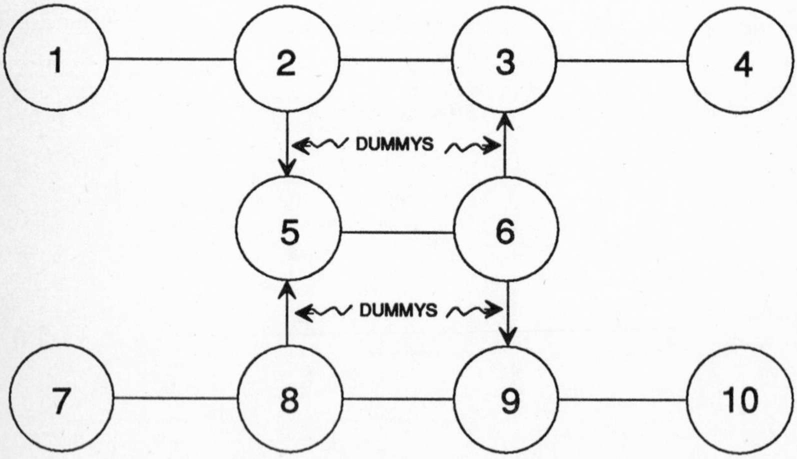

This very brief description of the Arrow methodology reveals fundamental similarities to the Precedence methodology. In practice, each has advantages and disadvantages, but we can model any project in either Arrow or Precedence. Each creates an entirely different model structure, but each defines every activity, every event, and every constraint relationship. Both yield identical calculated data.

The choice between them breaks down simply to personal preference. Like the Lilliputian war in *Gulliver's Travels,* any debate centers on nothing more than the choice of which end of the egg to open. It is far too important to get on with modeling our projects than to waste time haggling over which technique to use. Besides, most software applications have made the choice for us.

APPENDIX B

PERT versus CPM

Just as Precedence and Arrow accomplish similar goals by different methods, the Critical Path Method (CPM), the modeling process used in this text, contrasts with PERT (the Project Evaluation and Review Technique). Many think of them as separate processes, but each duplicates almost every facet of the other. They are even more alike than Precedence and Arrow. Their single variation does slant each toward a different purpose, though.

PERT preceded the Critical Path Method (CPM), but being second does not automatically imply improvement. The two processes simply provide us more than one way to accomplish modeling objectives. Both combine the same elements (activities, events, and effort relationships or constraints), and both process data through a forward pass and a backward pass, followed by time reserve calculations. The methods differ only in their techniques for determining activity durations.

CPM defines duration as a single value, the most current estimate of the work units (shifts, days, weeks, etc.) necessary to accomplish a task. PERT also employs single-value activity durations, but the user does not input it directly. Instead, the process derives each duration from three user-specified time estimates (spans). That is, instead of inputting a single value, we load three values for the duration of each task and the process calculates a final value for the duration used for processing.

The first of these user-input values, called the *most likely* (m) *span* or duration, virtually equates to the value we would input in CPM. PERT requires input for two other durations, called the *most pessimistic* (b) and the *most optimistic* (a). An optimistic duration estimates the length of time necessary to accomplish the task if everything goes right; a pessimistic duration estimates the length of time necessary to accomplish the task if everything goes wrong. These definitions depend a great deal on interpretation. Everything going right could include divine intervention, just as everything going wrong could include nuclear holocaust. We must refine our definitions to provide guidelines to realistically determine activity spans.

More specifically, we define the optimistic span as the shortest length of time in which we would accomplish the task if we repeated it 100 times (the 1-percent chance). The pessimistic span is the longest length of time in which we would accomplish the task if we repeated it 100 times (the 99-percent chance). By defining these two values in this manner, we exclude the most absurd conditions. Statistically, the 1-percent chance and the 99-percent chance of task occurrence are three standard deviations, or three sigma, from the mean, appropriate end points in a statistical analysis.

Once we have defined the three duration estimates for each task, the PERT process resolves them into a single value for each activity, on which it then bases its time analysis. This results in a weighted average of the three spans:

$$DU = (a + 4m + b) / 6$$

This equation weights the most likely duration heavily (four times). The result will also be skewed toward any large difference between the intervals a-to-m versus m-to-b. In most cases, this skews the result to the right of (greater than) the most likely (m) duration because, first, users tend toward optimistic estimates of how long it will take to accomplish something. Second, more things can generally go wrong with greater impact than can go better than expected. Therefore, the determination of a most likely duration (m) is generally independent of risks. The PERT process, by skewing each task's duration to reflect its risks, takes these factors into account.

When PERT calculates the durations of tasks in this manner and then processes them through a forward pass, the resulting end date (EF) of each path reflects the cumulative skewing of the tasks along the path, giving an assessment of the length of time necessary to accomplish the paths in light of some potential problems.

This tends to slant the PERT process in a rather different direction as a planning tool. Instead of planning based solely on expected durations, the PERT process considers some of a project's potential problems. It is one of the original risk-analysis techniques. At the same time, however, this does not exempt PERT as a time modeling tool because virtually any process or method discussed in this book could apply to it, either directly or with some adaptation. We cannot escape the conclusion that the PERT and CPM methods do not differ significantly, raising the logical question, why has PERT lost the status of preferred methodology?

The reasons are: 1) better risk-analysis tools do exist, and 2) the three time estimates tend to muddle time reserve analysis. A more sophisticated statistical problem evaluation process called *Quantitative Risk Analysis* also uses three time estimates, treating these values as a range or distribution of time rather than as a single value. Subsequent data processing results in a distribution of information as well giving an assessment of the risk of each individual task and the resulting impact to the paths on which each lies. From this information, we can determine a level of confidence (50-percent chance, 80-percent chance, etc.) of attaining a certain required date, or conversely we can set a date based on a certain level of confidence.

Time reserve analysis works best with key parameters (durations, lags, constraint types, date targets, etc.) set directly by the user. This analysis does not try so much to determine when downstream events might happen, as to assess the potential impacts to those events of near-term perturbations. This is commonly referred to a *cause-and-effect evaluation*. Setting the parameters directly gives a clearer picture of the relationship than deriving them indirectly through a calculation.

For these reasons, PERT is not used very much any more. It remains available, however, and could serve an enterprising user, who only needs to understand how it works and what it can do.

APPENDIX C

Basic Requirements of a CPM Computer Processor

I. ANALYTICAL PROCESSES (Chapters 1-12)

A. GENERAL REQUIREMENTS

1. Support the presence of multiple logical starts and finishes, flagging each of these nodes. It must also indicate whether or not these nodes are properly date targeted (with a date predecessor, actual date, or time reserve parameter on a start node, and a date successor or time reserve parameter on a finish node).
2. Provide a date convention that reflects start dates as the beginning of the work period and finish dates as the finish of the work period.
3. Provide for certain data validation and set flags, including:
   a. Total number of activity records
   b. Total number of milestones (nodes with event parameters and DU = 0 nodes)
   c. Total number of constraints of each type (FS, FF, SS, and SF)
   d. Ratio of events nodes to total nodes
   e. Ratio of activities to constraints
   f. Identification of logical start nodes (nodes without predecessors of their starts)
   g. Identification of hanging start nodes (start nodes without date predecessors)
   h. Identification of logical finish nodes (nodes without successors of their finishes)

i. Identification of hanging finish nodes (finish nodes without date successors)

j. Detection of true loops (in which date calculations require their own output as input)

k. Identification of constraint node incompatibility (incompatibility between the nodes identified as node records and the nodes identified in constraints)

l. Identification of updating out-of-sequence conditions (any status data that contradicts model logic)

m. Identification of any dates input by the user that conflict with the calendar

n. Identification of the critical path (nodes and constraints) from a user-specified terminus node back to a start node

B. NODE RECORD PARAMETERS

1. The application should manage detail models of at least 1,000 activities and summary or integrated models of 10,000 activities.
2. Each node should have at a minimum all parameters identified in Appendix G or a mated relational database.
3. The application must allow for date targeting. At a minimum, it must provide a direct date interface for:
   a. Date predecessors, start-no-earlier-than (SNE) dates, that affect a node's ES date calculation
   b. Date successors, finish-no-later-than (FNL) dates, that affect a node's LF date calculation
4. The application must allow the user to assess status of a project using both actual and projected information. The method should use actual (AES and AEF) and projected (PES and PEF) dates (see Section II).
5. The application must allow the user to designate events or milestones properly as node ends. Appendix G lists the appropriate event parameters.
6. The application must properly process the model through the forward and backward passes and calculate the four date fields [Early Start (ES), Early Finish (EF), Late Start (LS), and Late Finish (LF) dates]. It should calculate these dates for completed and on-going activities, as well.
7. The application must calculate Path Time Reserves properly using the following equations:

a. SPTR = LS – ES

b. FPTR = LF – EF

8. The application must allow for and properly identify external constraint stretching:

   a. In the forward pass, it should identify (source and amount) any finish-end predecessor (FF constraint) that overrides the EF calculation of a node without changing the ES date calculation. In this case, the difference between the task's ES and EF dates is greater than the task's duration

   b. In the backward pass, it should identify (source and amount) any start-end successor (SS constraint) that overrides the LS calculation of a node without changing the LF date calculation. In this case, the difference between the task's LS and LF dates is greater than the task's duration

C. CONSTRAINT RECORD PARAMETERS

1. The application should manage detail models of at least 3,000 constraints and summary or integrated models of 30,000 constraints.

2. Each constraint should have at a minimum all of the parameters identified in Appendix G.

3. The application must support the principal constraint types:

   a. Finish-to-start (FS)

   b. Start-to-start (SS)

   c. Finish-to-finish (FF)

4. The application must allow the user to attach a lag value (an integer value that forces a delay across the constraint) to any constraint. It must associate this value with an appropriate calendar.

5. The application must allow for unlimited multiple predecessors or successors of a node and process them properly:

   a. In the forward pass, by the latest early date rule

   b. In the backward pass, by the earliest late date rule

II. SCHEDULE TRACKING AND CONTROL PROCESSES (Chapter 12)

A. BASELINE—The application should:

1. Retain the current baseline dates of nodes in user-input fields.

2. Allow the user to set the baseline dates of all, several, or a single task(s) from any other date field in the record.

3. Allow the user to update the baseline dates of all, several, or a single task(s).

B. STATUS (ACTUAL AND PROJECTED)—The application should provide the ability to:

1. Input and maintain actual start and finish dates on each activity/event.
2. Input and maintain projected start and finish dates on each activity/event.
3. Maintain baseline and current schedule (status) separate from analytical dates (early and late dates).
4. Simulate status (actual and/or projected) in the model parameters (SNE and FNE dates) for impact assessment.

III. CALENDAR UTILITIES PROCESSES (Chapter 10)

A. ESTABLISHING CALENDARS—The application should support multiple calendars, allowing the user to specify different base units (hours, shifts, days, weeks, etc.), define different workday (or nonworkday) patterns (5-day week, 6-day week, etc.), and define holiday tables that can eliminate specific dates for work in any workday pattern.

B. DURATION AND DATE CALCULATION—The application should provide a facility to utilize the basic date math equation:

$$FIN = ST + DU - 1$$

After the user provides two of these values and the appropriate calendar, the application should compute the third.

C. OVERTIME TIME RESERVE CALCULATION—The application should provide a facility to calculate the number of workdays (for extended workday calculations), nonworkdays, and holidays between two user-specified dates.

IV. CODE FIELD SUMMARIZATION PROCESSES (Chapter 13)

The application must summarize data based on user inputs to code fields in the following records:

A. ACTIVITY RECORDS—It must summarize each group of detail activities by a user-designated field or subfield, generating a summary activity record with the earliest start date and the latest finish date in each field (start: BES, PES, AES, ES and LS; finish: BEF, PEF, AEF, EF and LF).

B. EVENT RECORDS—It must select summary events or milestones by a user-designated method, attaching the appropriate data to the correct summary activity record defined by the event's summarization code.

C. CONSTRAINT RECORDS—It must select summary constraints for which the two nodes (predecessor and successor) have different summarization field (subfield) values, attaching the data to the appropriate summary activity with either a predecessor or successor bias.

V. HORIZONTAL INTEGRATION PROCESSES (Chapter 14)

A. DATA STRUCTURE—The application must allow the user to designate any events node as an interface event through an entry in a special field (INI). The application designates each interface node as either a predecessor or successor (INT) depending upon the model interfaces.

1. INI flagged events node with no successors: INT = P.
2. INI flagged events node with no predecessors: INT = S.

B. PROCESS—The application must merge copies of all designated detail models together, complete a total CPM process, and insert parameters into the interface nodes that simulate the total model effects, as well as label each interface node with its interfaces' modelfile names in the original detail models.

VI. WHAT-IF MODELING (Chapter 12)

A. The application must allow the user to perform a what-if exercise with the following features:

1. Creating a temporary copy or clone of the model on which to perform the analysis. A portion of any model may also be selected.
2. A database editor by which the user can make the changes necessary to model the what-if scenario. A textual section is also useful where configurations, ground rules, assumptions, etc., can be described.
3. Reports of any disparity in any calculated data between the what-if model and the original model.
4. An option of saving the what-if model, otherwise disposing of it properly.

VII. USER INTERFACE (Chapter 10)

A. DATABASE EDITOR—The application must feature a range of user interfaces with the modelfile ranging from full screen edit (FSE) to user friendly (UF) to accommodate a broad range of user ability. The database editor will accommodate:

1. Data record entry.
2. Data record change or edit.
3. Data record deletion.

B. FUNCTIONS—The application must provide the user several functions to aid in the maintenance and control of the database, including:
   1. Control over the records selected from the database for edit.
   2. Control over the record fields displayed for either data entry or edit and the sequence in which they appear.
   3. The ability to link two or more data files if using RDBMS for simultaneous access for data entry and edit.
   4. The ability to establish fixed values in specified fields for data entry.
   5. The ability to make mass changes to the database, including:
      a. Selecting records for data changes through a user-defined criterion
      b. Setting specified values meeting that criterion, including removing or deleting a value in certain fields
      c. Deleting blocks of records by similar criteria
      d. Reordering the sequence of data in the modelfile

VIII. OUTPUT PRODUCTS—TABULAR REPORTS (Chapter 15)

Tabular reports are the principal interface between the modelfile and the user. In order to properly display data, the user must control certain format details through a tabular report writer (TRW) with the following features or characteristics:

A. REPORT LENGTH AND WIDTH—The user must control the length of the lines (i.e., number of characters across the page) and the number of records displayed per page.

B. REPORT FIELDS—The user needs a way to:
   1. Specify the fields in each report and their sequence across the page.
   2. Specify and format the headings over each field.
   3. Format the fields displayed, including length and word wrap for text fields.

C. DATA SELECTION OR FILTERING CRITERIA—The user needs a way to select a subset of records from the modelfile for display by:
   1. Specifying any field or subfield in a record.
   2. Specifying a wide range of criteria statements (e.g., $=$, $>$, $<$, $\neq$, etc.).

3. Specifying acceptable values in the field or subfield.
4. Linking more than one selection criterion statement together with AND and/or OR statements.

D. DATA GROUPING CONTROL—The user must have the ability to group data in several ways through values established by either the user or the application in fields or subfields to achieve:

1. Page break control.
2. Regional zoning or banding.
3. Data order through primary, secondary, tertiary, etc., fields or subfields.

E. TITLES, SUBTITLES, USER COMMENTS—The application must allow the user to add other necessary information to each report page such as titles and comments.

F. SPECIAL REPORTS—SOON-TO-COME-DUE AND PREDECESSOR-SUCCESSOR REPORTS

1. To meet special requirements, the user needs a special format for soon-to-come-due reports which includes:
   a. User-specified time windows
   b. User-specified date types (baseline, projected, early, late, etc.)
   c. Choice between inclusion of all records occurring in the window or just those which either start or finish
2. To meet special requirements, the user needs a special format for predecessor-successor reports which includes the ability to select each task's predecessors, or successors, or both.

IX. OUTPUT PRODUCTS—MODEL PLOTS (Chapter 15)

The model plot is the most complete tool for project model analysis as it graphically depicts the activities, events, and constraints. It requires a very powerful and flexible model plot writer (MPW) to control the details in the plots.

A. NODE FORMAT AND CONTROL—The model plot shows node (activity and event) information in a rectangular box subdivided into field compartments. The MPW must allow the user to:

1. Specify fields and control their placement in each node box.
2. Define the format of each field, including length and word wrap for text fields.
3. Relate special node shapes to highlight certain activity conditions (e.g., start and finish nodes, nodes with certain levels of time re-

serve, nodes with specified values in a field, interface nodes, intramodel interface nodes, etc.).

4. Portray events (node ends) as graphic attributes of nodes with proper graphic and textual annotation.

B. CONSTRAINT (LINE) CONTROL—The constraint information appears on lines that connect nodes. The MPW must:

1. Depict constraints accurately (e.g., from the start end of the predecessor to the start end of the successor for an SS constraint, etc.).
2. Define constraint type, lag value, and any user specified descriptive text on or near each constraint line.
3. Allow the user to specify alternative line styles for specific purposes (e.g., specified levels of time reserve along a path, constraints with specified project structure information, etc.).

C. NODE PLACEMENT CONTROL (SEQUENCE)—The MPW must allow the user to control the placement of nodes across the horizontal axis by certain criteria, including:

1. Rank ordering.
2. Set time scale ordering by values in a specified date field relative to a fixed width time scale.

D. NODE PLACEMENT CONTROL (ZONING)—The MPW must allow the user to control the placement of nodes along the vertical axis. Just like regional zoning or banding in tabular reports, the user specifies:

1. Zone criteria.
2. Zone order from top of page to bottom.
3. Graphic indications of zone breaks (skipped lines, horizontal lines, titles, etc.).
4. Zone titles corresponding to the value of the zoning field in each group.

E. DATA SELECTION OR FILTERING CONTROL—The MPW must allow the user to select a subset of records from the modelfile for display by:

1. Specifying any field or subfield in a record.
2. Specifying a wide range of criteria statements (e.g., = , > , < , ≠, etc.).
3. Specifying acceptable values in the field or subfield.
4. Linking two or more selection criterion statements together with AND and/or OR statements.

5. Intramodel interface node depiction in an alternative node shape of the user's choice, containing all of the information specified in section A (node format control).

F. OTHER CONTROLS—The MPW must let the user enter page size, titles, subtitles, legends, user comments, and other page format controls.

X. OUTPUT PRODUCTS—GANTT MULTIPLE MILESTONE BARCHARTS (Chapter 15)

Gantt barcharts serve an enormous variety of purposes in many formats, so they require a utility with tremendous flexibility called a Gantt plot writer (GPW). Unlike the other product types, Gantt barcharts have specific sections with their own peculiar requirements, including the body and Record Attributes Window.

A. BASIC FORMAT—A Gantt barchart displays data for activities as bars with start and finish dates that correspond to a time scale. Events appear as pointed geometric shapes that correspond to the same time scale. Any of the start and finish dates (early, late, baseline, projected, and actual) can appear.

B. BARCHART BODY CONVENTIONS

1. Relate any date type to a particular bar style, either superimposing bars depicting different information or drawing them one above the other.
2. Milestones can relate two date types for an event (e.g., baseline to early, baseline to status, early to late, etc.) with two geometric shapes attached by a horizontal line.
3. Milestone symbols are properly annotated with descriptive text and dates.
4. When more than one milestone appears above a bar, stack them as necessary to prevent overwriting.
5. The data portrayed in the barchart body must align horizontally with the information in the Record Attributes Window.
6. The user controls the size of graphic shapes and size and placement of text.
7. The user controls the page size and the number of records displayed per page.

C. DATE STRIP

1. The user specifies the units (days, weeks, months, quarters, years, etc.).

2. The user specifies the date window with from and until dates.
3. Calendar strips for parts of the chart can vary (e.g., first third of the product divided into months, the middle third into quarters, and the last third into years).

D. RECORD ATTRIBUTES WINDOW
1. The user specifies fields and controls their sequence across the window.
2. The user adds field headings and controls their format.
3. The user specifies field format, including length and word wrap for text fields.

E. DATA SELECTION OR FILTERING CONTROL—The GPW must allow the user to select a subset of records from the modelfile for display in both the barchart body and the Record Attributes Window by:
1. Specifying any field or subfield in a record.
2. Specifying a wide range of criteria statements (e.g., $=$, $>$, $<$, $\neq$, etc.).
3. Specifying acceptable values in the field or subfield.
4. Linking several selection criterion statements together with AND and/or OR statements.

F. NODE PLACEMENT CONTROL (ZONING)—The GPW must allow the user to control the placement of nodes along the vertical axis of the product, just like regional zoning or banding in tabular reports. The user places the data records in both the barchart body and the Record Attributes Window, specifying:
1. Zone criteria (field or subfield) to determine in which zone or band each node appears.
2. Zone order from top of page to bottom.
3. Graphic indications of breaks between zones (skipped lines, horizontal lines, titles, etc.).
4. Zone titles corresponding to the value of the zoning field in each group.

G. DATA ORDER—The user specifies data order through primary, secondary, tertiary, etc., fields or subfields for data records in both the barchart body and the Record Attributes Window.

H. DATA SUMMARIZATION—The GPW must generate a chart that accurately and adequately depicts the information derived from the processes of section IV for the data records in both the barchart body and the Record Attributes Window.

1. The user specifies the summarization criterion as a field or subfield.
2. The user specifies the milestone selection mechanism.
3. The GPW displays constraints depending on activity summarization criteria and predecessor or successor bias.

I. OTHER PAGE FORMAT CONTROLS—The GPW must allow the user to enter titles, subtitles, legends, user comments, and other page format controls.

XI. OUTPUT PRODUCTS—MANAGEMENT GRAPHICS (Chapter 15)

Management graphics, including pie charts, histograms, and X-Y plots, contribute little to project modeling, with certain exceptions.

A. TIME RESERVE EROSION PLOTS measure the rate of project progress relative to rate of Path Time Reserve erosion for selected events in the model.

B. SCHEDULE PERFORMANCE PLOTS measure the cumulative baseline schedule requirements versus actual and projected performance.

C. RESOURCE AND QUANTITATIVE RISK ANALYSIS creates the most practical use for management graphics.

XII. APPLICATION SYSTEM REQUIREMENTS

Several other practical considerations influence the choice of a computer application, including:

A. PERFORMANCE CONSIDERATIONS

1. Processing speed for both analytical functions and output product generation.
2. Hardware and software requirements: CPU, storage, terminals, plotters, communication, etc.
3. Multiple user capability.

B. PRODUCT SUPPORT

1. Quality and cost of training.
2. Quality and cost of user documentation.
3. Quality and cost of technical support.
4. Quality and cost of software maintenance.

Note: This appendix does not cover the requirements for resource and risk modeling.

APPENDIX D

Qualitative Evaluation of Project Models

To supplement the project modeling process, we need a way to evaluate project models overall, both as analytical tools to identify time reserve, and as schedule tracking and control tools. Such a process could not judge the truth of details like whether a task's duration should be 30 or 45 days, or whether a constraint of a certain type exists between two tasks. We must assess the validity of these parameters through less objective means and do so continuously throughout the project's performance.

Yet there are many characteristics of a model that reflect clues about its quality as an analytical tool. This appendix outlines these mechanisms. We must understand that these checks are not absolute rules, but guidelines that indicate trends or tendencies of a model. We must apply proper understanding to this evaluation.

These evaluations can be conducted by single users as a self-audit or they may be expanded into a process routinely performed on each modelfile.

I. PROJECT MODEL MECHANICS

A. MODEL COMPOSITION

DETERMINE THE RATIO BETWEEN THE TOTAL NUMBER OF ACTIVITIES TO THE TOTAL NUMBER OF CONSTRAINTS: A low ratio indicates undefined constraints. A ratio of 1.00 means that each activity has an average of one input constraint and one output constraint, which would seem low for most projects. Though the exact proportion varies from project to project, ratios in excess of 2.00 should be normal.

B. ACTIVITY NODE CHARACTERISTICS

1. ASSESS ACTIVITY DURATIONS: Although we have no quantitative method of assessing the validity of any one task's duration, we can measure the project as a whole for adequate rolling wave decomposition. As a rule of thumb, durations of tasks in the first third of a project should not exceed 5 percent of the project's total span, durations of tasks in the middle third should not exceed 10 percent, and durations of tasks in the last third should not exceed 15 percent. These values are guidelines, not hard rules.
2. CHECK FOR EXTERNAL CONSTRAINT STRETCHING: Each task that experiences stretching in either the forward pass (creating an EPSA) or backward pass (creating an ESSA) should be identified. In each case the cause of the stretching should be determined and a response should be established. We would respond differently to an EPSA than to an ESSA.

C. START AND FINISH NODE DISCIPLINE

1. DETERMINE THE RATIO OF START AND FINISH NODES TO TOTAL NODES: A large ratio (in excess of 0.10) would indicate a fragmented model and a probability of many missing constraints.
2. CHECK FOR PROPER DATE TARGETING ON START AND FINISH NODES: We must ensure that each pass begins properly with appropriate target dates on the start and finish nodes. Each start node should have an SNE, PES, AES date or time reserve parameter (SSTR) and each finish node should have an FNL date or time reserve parameter (FSTR).

D. CONSTRAINT/INTERFACE DISCIPLINE

1. CHECK FOR FINISH-TO-START (FS) CONSTRAINTS WITH LAG VALUES: Lags on finish-to-start constraints usually indicate date rigging. All lag values must represent time consumed by effort for valid time reserve calculations. To be valid, a lag on an FS constraint could be replaced with a node that represents definable project effort—something that is legitimate path length.
2. CHECK FOR START-TO-START (SS) CONSTRAINTS WITHOUT LAG VALUES: SS constraint types should have lag values representing a portion of the predecessor task. The absence of a lag value on an SS constraint normally indicates improper modeling of task overlap. The value of the lag should be less than the predecessor task's duration.

3. CHECK FOR FINISH-TO-FINISH (FF) CONSTRAINTS WITHOUT LAG VALUES: FF constraint types should have lag values representing a portion of the successor task. The absence of a lag value on an FF constraint normally indicates improper modeling of task overlap. The value of the lag should also be less than the successor task's duration.
4. CHECK FOR START-TO-FINISH (SF) CONSTRAINTS: This is an impractical constraint type and its use is discouraged.
5. CHECK FOR NEGATIVE LAG VALUES: Negative lag values (or lead) is an improper method of modeling task overlap and is discouraged.

E. EVENT CONVENTIONS

1. CHECK FOR EVENTS NODES WITH ONE-UNIT DURATIONS: Events nodes with durations of one unit almost always indicate date rigging stemming from confusion about date conventions (start date after finish date). This is an improper method of modeling events nodes. Activities can have one-unit durations, however.
2. DETERMINE THE RATIO OF EVENTS NODES TO TOTAL NODES: A large ratio (in excess of 0.10) indicates an inflated activity count and, more than likely, some unnecessary events nodes. It could also indicate date rigging. The main problem here is that the model has insufficient activities (tasks with duration) to define the work of the project.
3. CHECK FOR PROPER STRUCTURE OF EVENTS NODES: Since nodes with no span are established solely to report an event of the project, they must contain appropriate event structure coding. This would include values in the SEF and SED fields (for an event measuring a start) or the FEF and FED fields (for an event measuring a finish).

F. EXCESSIVE OR IMPROPER DATE TARGETING

1. CHECK FOR EXCESSIVE DATE TARGETING: Determine the ratio of date predecessors (SNE/PES dates and FNE/PEF dates) plus date successors (FNL and SNL dates) to the total number of nodes in the model. A large ratio indicates date rigging (or a model in its latter stages). Although date targets are essential, too many begin to influence the calculated dates of the model more than the constraints, lags, and task durations.

2. CHECK IMPROPER DATE TARGETING: Start-on (SON) or finish-on (FON) date targets are discouraged because these parameters are unrealistic.

G. OTHER ANALYTICAL DEFICIENCIES OR ERRORS

1. CHECK FOR LOOPS: Loops are fatal analytical errors that cause a node's date calculations to require their own output as input.
2. CHECK FOR CONSTRAINT NODE INCOMPATIBILITY (CNI): In this fatal analytical condition, one node in a constraint relationship (PN or SN) is not identified as a node record.
3. CHECK FOR DUPLICATE NODE IDENTIFIERS: Each node identifier code must be unique. We must ensure that we have not inadvertently duplicated any.
4. CHECK FOR INVALID DUPLICATE CONSTRAINT RECORDS: The same pair of node identifiers in more than one constraint record usually indicates an error, with some exceptions. A start-to-start (SS) and a finish-to-finish (FF) constraint between the same two nodes could model a valid relationship. Two or more FF, two or more SS, or two or more FS constraints between the same two nodes is an error. An FS with either an SS or FF constraint between the same two nodes is also an error. These erroneous situations do not violate the CPM processes, but the resultant calculated data may be erroneous.
5. CHECK FOR CALENDAR CONFLICTS: Any date entered as a model parameter (PES, PEF, AES, AEF, FNL, etc.) must correspond to the appropriate calendar. Nonworkdays or holidays entered as model parameters should be corrected. This is not a fatal error, but it creates inconsistency in the model.
6. CHECK FOR OUT-OF-SEQUENCE UPDATING: Dates entered into the status parameters of activities (BES or BEF, PES or PEF, AES or AEF dates) should correspond to the constraint logic of the model. This is not a fatal error, but it does create inconsistency in the model.
7. CHECK FOR CALENDAR INCONSISTENCIES ON CONSTRAINTS: The calendar applied to lag values on constraints must correlate to the relationship type. In start-to-start (SS) constraints, the calendar should match that of the predecessor task because the lag represents a portion of this task. In finish-to-finish (FF) constraints, the calendar should match that of the successor task because the lag represents a portion of this task. The calendar of a finish-to-

start constraint with a lag value is independent of either the predecessor or successor.

II. SCHEDULE TRACKING AND CONTROL

A. APPROPRIATE BASELINE DATES FOR ACTIVITIES

1. Each activity node should have both a baseline start (BES) date and a baseline finish (BEF) date.
2. The baseline start date should precede the baseline finish date.

B. APPROPRIATE BASELINE DATES FOR EVENTS

True events nodes (DU = 0) should have either a baseline start date (BES) or a baseline finish date (BEF), depending on the node's bias. (An SEF value indicates a start event; an FEF value indicates a finish event.) Both dates are not necessary.

C. APPROPRIATE STATUS

1. Any event (node end) with baseline dates prior to the status date (SD) should have either an actual date (AES or AEF) or a projected date (PES or PEF) that is later than the status date.
2. Completed events nodes need only contain the appropriate actual date (AES or AEF depending upon the event's bias).
3. Any task that has an actual finish (AEF) date should also have an actual start (AES) date that precedes the AEF date.

APPENDIX E

Hammock Structures as Independent Project Models

Chapters 13 (Summarization) and 14 (Horizontal Integration) discussed hammock mechanisms at length. Both showed this method's significant flaws. This appendix will further explore its inadequacy, which stems from its fundamental premise.

The upper portion of Figure E-1 illustrates a very simple project model of three activities interfaced in sequence (A to B to C). These tasks summarize detailed effort, and we must decompose them to define the significant discrete work. The resulting detail model appears in the lower portion of Figure E-1.

We could consider the model in the upper portion of the figure as a hammock structure that models the detail tasks at a higher level. Does this hammock structure adequately model the detail tasks it represents? In order for this to be true, the interface of the detail work (i.e., that between A activities and B activities) must occur at the finish of the last detail predecessor to the start of the first detail successor. This is the fundamental premise of the hammock mechanism.

Once we break each of these higher-level tasks into its components, we find that the actual interface joins discrete tasks which may or may not be the last or first of their group. There are also many detail relationships and a single higher-level constraint cannot properly represent the various combinations at the detail level. Even in our simple example, with only three detail constraints between the activities of A and B, no single constraint can represent them collectively. If Task A slips 10 days in the hammock model, the effort of Task B will slip by 10 days, as well. In the detail model, however, the effects of a 10-day slip depend upon which specific task slips.

If activity A2 slips by 10 days, it has no impact on any of the B activities. B1, its immediate successor, will not suffer until A2 slips more than 20 days. In the second case, where A4 slips by 10 days, B4 also slips 10 days, pushing B2 and B3 back by 5 days, as well. Any slip to A4 will more directly affect A2's successors than will slips to A2 itself. If A5 slips at all, then B6, B5, and B7 will all slip day-for-day.

The slip had entirely different effects in each instance. In the first case, none of the finish nodes of B (B3 or B7) moved due to the 10-day slip; in the second case, a B task finish node slipped by 5 days; in the third case, one slipped by 10 days. Slips in A activities have no simple impact on B activities.

Figure E-1

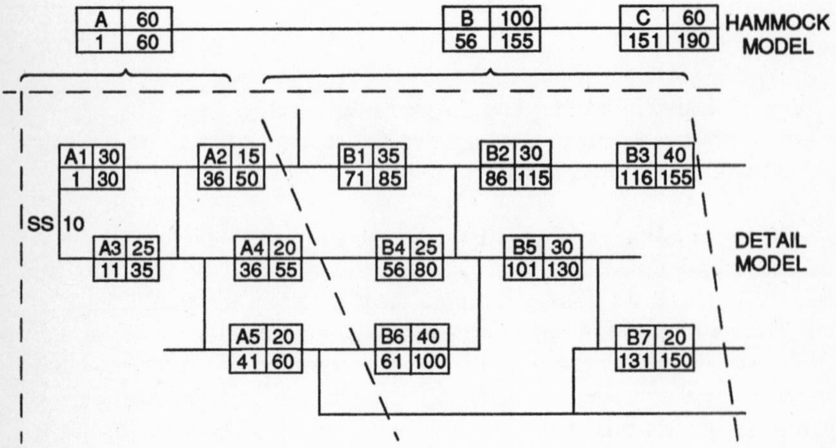

Notice further in Figure E-1 that the earliest start date of Task B (workday 56) creates an inconsistency with the latest finish date of A (workday 60). The relationship between the two summary tasks cannot be a finish-to-start relationship—the effort is overlapped. The correct relationship would require a pair of constraints (one SS and one FF), although even this would not accurately represent the interface.

These problems arise in a relatively simple model. In many cases, the final A activities might not interface at all to the first B activities. In fact, the detail interfaces may join any number of nodes, so only rarely could a single constraint or even a pair of constraints (SS and FF) between two hammock tasks adequately represent the detail. We can conclude that hammocks, or any summary mechanism, can adequately represent groups of activities and project milestones, but summary level constraints cannot perfectly replicate several detail-level constraints. The basic premise of hammock structures is false.

This creates a problem since the hammock model will yield very different results than the detailed model. Doing both would only create confusion. We should pursue the accuracy of the detail level and perform all analysis there. As long as we build the detail model at the proper level of significant interface, our analysis remains valid.

In conclusion, hammock structures fail in their promise to represent detail models accurately. They serve no valid analytical purpose. We could still use them to portray summarized sections of our project, but we cannot use them as standalone models. This strictly limits their utility in the project modeling processes. We have much more effective means for summarization (vertical integration) and horizontal integration (see Chapters 13 and 14).

APPENDIX F

Glossary of Acronyms and Abbreviations

Act	Activity—One of the three basic elements of a model that defines a discrete element of work on a project. It consumes time and/or resources.
AD	Activity Description—A textual description of each node in a project. [FIELD NAME]
ADM	Arrow Diagramming Method—A method of project modeling that defines nodes as the events of a project and activities as a relationship between a start node and a finish node. It also defines constraints as node-to-node relationships or nodes (finish of one activity—start of the next).
AEF	Actual finish date—The date when a task actually completes or finishes. [FIELD NAME]
AES	Actual Start date—The date when a task actually starts. [FIELD NAME]
a.k.a.	Also known as—Alias or alternative name.
AOA	Activity on arrow—See ADM.
AON	Activity on node—See PDM.
AOS	Amount of stretching—The number of work units an EF date is stretched by an external predecessor in the forward pass (for an EPSA) or an LS date is stretched by an external successor in the backward pass (for an ESSA). $\lvert AOS \rvert$ = SPTR – FPTR [FIELD NAME]

ASF — Activity stretching flag—A node flag field that indicates whether or not a node experienced external constraint stretching. The value *FPS* indicates stretching in the forward pass (SPTR > FPTR), whereas *BPS* indicates stretching in the backward pass (SPTR < FPTR). [FIELD NAME]

BEF — Baseline finish date—The date on which a project elects to finish an activity. Progress or status is measured against this date. [FIELD NAME]

BES — Baseline start date—The date on which a project elects to start an activity. Progress or status is measured against this date. [FIELD NAME]

BIA — Event bias—A field in the Complementary Scheduling Control and Tracking datafile that defines the start or finish bias of a task-dependent event. [FIELD NAME]

BP — Backward pass—The second of the CPM/PERT processes that determines the schedule of each node relative to its successors.

BPS — Backward pass stretching—A value in the activity stretching flag (ASF) field that indicates external constraint stretching in the backward pass: SPTR < FPTR.

CAL/cal — Calendar—A designator on each activity and constraint that defines the specific work pattern (e.g., nonwork days, holidays, work units, etc.) relative to the activity or constraint. [FIELD NAME]

CI — Calendar incompatibility—a nonfatal modelfile error flag indicating a conflict between a date defined in a record field by the user and the appropriate calendar (i.e., the date falls on a specified nonwork day or holiday).

CNI — Constraint node incompatibility—A fatal modelfile error flag indicating that a node identified in a constraint record either as the predecessor (PNI) or successor (SNI) corresponds to no node record.

CPF — Critical Path Flag—A flag field on both nodes and constraints that marks the path to a specified terminal node in the project with the least PTR. [FIELD NAME]

CPM — Critical Path Method—A project modeling process that calculates time reserve at each node end in the project. It consists of a forward pass that calculates each node's early dates, a backward pass that calculates each node's late dates, and time reserve calculations. It is a deterministic method, as it takes a single value for the duration of each task.

CR Calendar reserve—The amount of time between two dates that is excluded from the working calendar, including holidays, normal nonwork days (Saturdays, Sundays, etc.), and any off-shift time.

CSP Cumulative sequence of a path—The sum of the durations and lags from a start node to a finish node through a continuous sequence of constraints. Delays on any nodes on the path caused by other predecessors or date targets are also part of the cumulative total path length.

DACS Date altered compensation for stretching—A process that adjusts the Early Start date of an EPSA (a node with forward pass stretching) to maintain consistency with the task's duration (ES = EF – DU + 1). In the backward pass, it also adjusts the Late Finish date of an ESSA (a node with backward pass stretching) to maintain duration consistency (LF = LS + DU – 1).

DU Duration—The estimated length of time necessary to accomplish a task. [FIELD NAME]

DUWC Duration of work completed—The length of time a task has been on-going. It is the difference in work units between the actual start date of the task and the status date (DUWC = SD – AES + 1). [FIELD NAME]

EAD Event actual date—A field in the Complementary Scheduling Control and Tracking datafile that contains the actual date of a task-dependent event when it occurs. [FIELD NAME]

EB Event bias—Indicates which event in a node (the start or finish) is reported.

EBD Event baseline date—A field in the Complementary Scheduling Control and Tracking datafile that stores the baseline schedule date of a task-dependent event. [FIELD NAME]

ED Event description—A field in the Complementary Scheduling Control and Tracking datafile that stores a textual description of a task-dependent event. [FIELD NAME]

EF Early Finish date of a node—One of the two dates calculated in the forward pass, it defines the earliest a node can finish as determined by its predecessors. [FIELD NAME]

EPD Event projected date—A field in the Complementary Scheduling Control and Tracking datafile that stores a projected schedule date of a task-dependent event. [FIELD NAME]

EPSA External Predecessor Stretched Activity—An activity with an Early Finish date driven by a predecessor (FF or SF) other than its own duration. The difference between the ES of the task and its EF exceeds its duration.

ES Early Start date of a node—One of the two dates calculated in the forward pass, it defines the earliest a node can start as determined by its predecessors. [FIELD NAME]

ESP Early Start of a path—The Early Start date of the first node in a path. This forms the left boundary of the time available to the path.

ESSA External Successor Stretched Activity—An activity with a Late Start date driven by a successor (SS or SF) other than its own duration. The difference between the LS of the task and its LF exceeds its duration.

EV Event—One of the three basic elements of a model that marks the initiation or completion of a specific effort, consuming neither time nor resource.

FED Finish event description—A node text field that describes the event at the finish end of the node. [FIELD NAME]

FEF Finish event flag—A node alphanumeric field that defines the schedule level and the graphic shape of the event marker at the finish end of a node. It is primarily used in the generation of Gantt barcharts to depict necessary events that appear at the finish end of the node. [FIELD NAME]

FF Finish-to-finish constraint type—Defines a relationship between two nodes where the successor can finish once the predecessor finishes and the lag passes.

FFR Finish-end free reserve—The amount of time a node's finish (EF) date can slip and not affect any of the task's successors. [FIELD NAME]

FN/Fin Finish—The completion of a task or effort. It also appears as a flag value in the SNF field to indicate a properly targeted finish node (one with an FNL date present).

FNE Finish-no-earlier-than date—A date imposed on the model that sets the left-most boundary for a node's Early Finish date. [FIELD NAME]

FNF Finish node flag—A two-digit alphanumeric field on nodes that specifies properly date targeted logical finish nodes (FN) or improperly date targeted logical finish nodes (HF). [FIELD NAME]

FNL Finish-no-later-than date—A date imposed on the model that sets the right-most boundary for a node's Late Finish date. It simulates or models an external requirement of the task's completion. [FIELD NAME]

FON Finish-on date—A date target that sets both the Early Finish and the Late Finish dates of a node, ignoring all predecessor and/or successor relationships. [FIELD NAME]

FP Forward pass—The first of the CPM/PERT processes that determines the schedule of each node relative to its predecessors in the form of Early Start (ES) and the Early Finish (EF) dates.

FPS Forward pass stretching—A value in the activity stretching flag (ASF) that indicates external constraint stretching in the forward pass (SPTR > FPTR). [FIELD NAME]

FPTR Finish-end Path Time Reserve—The Path Time Reserve (PTR) measured at the finish end of each node (FPTR = LF − EF). [FIELD NAME]

FS Finish-to-start constraint type—Defines a relationship between two nodes where the successor can start once the predecessor finishes and the lag passes.

FSE Full screen edit—A method of interfacing with a computer datafile that portrays the data in columns under appropriate field headings, giving access to fields through cursor or tab and backspace keys. This user interface is fast but requires greater skill on the part of the user.

FSTR Finish specified time reserve—An alternate method of defining the Late Finish date of a finish node in which a numeric value in this field added to the node's Early Finish date determines a Late Finish date with a specified reserve. (LF = EF + FSTR) [FIELD NAME]

GPW Gantt plot writer—A software feature by which the user controls and formats Gantt barcharts.

IFS Interface source—A textual node record field that identifies the sources of cross-model interfaces. During horizontal integration, the software enters the modelfile name of the predecessor into this field on the successor node and the modelfile name of the successor into this field on the predecessor node. [FIELD NAME]

IJ — See ADM.

INI — Interface node identifier—A single-character alphanumeric node record field entered by the user to identify the interface nodes in cross-model constraints. The value *I* signifies such an interface. [FIELD NAME]

INT — Interface node type—A single-character alphanumeric node record field set by the software during horizontal integration to identify interface nodes as either predecessors or successors of a cross-model constraint. The value *P* signifies a predecessor, the value *S* signifies a successor. [FIELD NAME]

LAG — Lag—A measure of the delay on a constraint. It is an integer value added to the predecessor value to satisfy the successor. See each constraint type (FS, FF, SS, and SF). [FIELD NAME]

LF — Late Finish date of a node—One of the two dates calculated in the backward pass, it defines the latest a node can finish as determined by its successors. [FIELD NAME]

LFP — Late Finish of a path—The Late Finish date of the last node in a path. This forms the right boundary of the time available to the path.

LTR — Least time reserve—The least or smallest value of Path Time Reserve (PTR) in a project.

LS — Late Start date of a node—One of the two dates calculated in the backward pass, it defines the latest a node can start as determined by its successors. [FIELD NAME]

MPR — Multiple predecessor relationship—A condition in a model where one end of a node (start or finish) has more than one predecessor. We calculate dates for each independently and choose the latest early date.

MPW — Model plot writer—A software feature by which the user controls and formats model plots.

M/S — Milestone—A significant event in a project.

MSR — Multiple successor relationship—A condition in a model where one end of a node (start or finish) has more than one successor. We calculate dates for each independently and choose the earliest late date.

NI	Node identification code—A unique series of alphanumeric characters that identifies each node in a project, primarily to specify constraint relationships. [FIELD NAME]
NIS	Stretched node identifier—An alphanumeric field on constraints that specifies the node identification code (NI) of any external constraint stretched node. [FIELD NAME]
NVA	No value added—Any effort that returns no value to the project.
ODU	Original duration—The baseline estimate of the length of time necessary to accomplish a task with its current work scope. It is the difference between the task's baseline start (BES) date and its baseline finish (BEF) date. (ODU = BEF – BE + 1) [FIELD NAME]
OFFSET	Offset—A field in the SRP resource subrecord that indicates the point in time (measured from the task's start) at which the defined resource is first applied. [FIELD NAME]
O/T	Overtime—The quantity of time normally considered nonwork time. It is a project time reserve.
PC	Percentage complete—A parameter that estimates accomplishment of an on-going task as a percentage. [FIELD NAME]
PDM	Precedence Diagramming Method—A method of project modeling that defines fundamental nodes as the activities of a project. Events appear as node ends (each node has a start and a finish), and interface constraints take the form of node-to-node relationships.
PEF	Projected Finish date—A projected date that updates the schedule finish of on-going or near-term tasks. It is a reforecast of a task's expected finish date. [FIELD NAME]
PER	Percentage of task—A field in the Complementary Scheduling Control and Tracking datafile in which the user can specify the point of occurrence of a task-dependent event as a percentage of the task's duration. [FIELD NAME]
PERT	Project Evaluation and Review Technique—A project modeling process that calculates time reserve at each node end in the project. It consists of a forward pass that determines each node's early dates, a backward pass that determines each node's late dates, and time reserve calculations. It is a relativistic method, as a single value is calculated for the duration of each task as a weighted average from three time estimates.

PER# Period number (#)—A series of fields in the PRP resource subrecord (PER1, PER2, PER3, etc.) that indicate the quantity of a resource needed for each period, expressed in units like person-hours, person-months, etc. [FIELD NAME]

PES Projected start date—A projected date that updates the scheduled start of near-term tasks. It is a reforecast of a task's expected start date. [FIELD NAME]

PN Predecessor node—The preceding or first node in a constraint relationship; the activity that must occur first.

PNI Predecessor node identifier—The first or left-hand node identifier in a constraint. It identifies the earliest or first node. [FIELD NAME]

PRP Periodic resource profile—A method of linking resource requirements to individual tasks by first defining a period (weeks, months, etc.) and then specifying the task's resource needs for each period. This method works best for defining labor resource requirements.

PSR Padded span reserve—A time reserve that measures the difference between the estimated span of a specific task and its actual span.

PSRR Periodic summation of resource requirements—A process that sums the requirements of specific resources for a specified period to determine the total project requirements of all resources as a function of time.

PTR Path Time Reserve—The amount of time determined by the difference between the time available (TA) to a path and the cumulative sequence of that path (CSP) (PTR = TA – CSP).

PWR Percentage of Work Remaining—A parameter that measures how much work remains on an on-going task as a percentage estimate of how much of the task is left to complete. It is the difference between 100 percent and the estimated percentage complete (PWR = 100 – PC). [FIELD NAME]

RESID Resource identification—A field in both the SRP and PRP resource subrecords that names a specific resource. [FIELD NAME]

RESPAN Resource span or duration—A field in the SRP resource subrecord that defines the length of time (starting at the offset) for which an activity needs a resource. [FIELD NAME]

RESQTY Resource quantity— A field in the SRP resource subrecord that defines the amount or quantity of a resource needed for the period specified by OFFSET and RESPAN. [FIELD NAME]

RDU	Remaining duration—An estimate of how many work units it will take to finish an on-going task. [FIELD NAME]
ROE	Rate of erosion—The rate at which that the PTR of a path is diminishing expressed as a percentage (e.g., 25 percent of the time reserve is eroded) equal to the decrease in time reserve on a path divided by the original amount of time reserve.
ROP	Rate of progress—The rate at which a path is progressing expressed as a percentage (e.g., 25 percent of the path is complete) equal to the completed length of the path divided by its current total length.
S	Start—The point in time at which work on a path begins.
SD	Status date—A project date on which status is assessed. It is measured at the end of the work period on the finish date.
SED	Start event description—Node text field that describes the event at the start end of the node. [FIELD NAME]
SEF	Start event flag—Node alphanumeric field that defines the schedule level and the graphic shape of the event marker at the start end of a node. It is primarily used in the generation of Gantt barcharts to depict necessary events that appear at the start end of the node. [FIELD NAME]
SF	Start-to-finish constraint type—Defines a relationship between two nodes in which the successor can finish once the predecessor starts and the lag passes. This is an impractical constraint type.
SFR	Start-end free reserve—The amount of time a node's start date (ES) can slip and not affect any successor. [FIELD NAME]
SFS	Span from start—A field in the Complementary Scheduling Control and Tracking datafile in which the user specifies the point of occurrence of a task-dependent event relative to an offset from the start of the task. [FIELD NAME]
SN	Successor node—The succeeding or second node in a constraint relationship; the activity that occurs last in the interface.
SNE	Start-no-earlier-than date—A date imposed on the model that sets the left-most boundary of a node's Early Start date. It simulates or models an external delay on a task's initiation. [FIELD NAME]
SNF	Start node flag—A two-digit alphanumeric field on nodes that specifies properly date targeted logical starts (ST) and improperly date targeted logical starts (HS). [FIELD NAME]

SNI Successor node identifier—The second or right-hand node in a constraint. It is the latest node. [FIELD NAME]

SNID Stretched node identifier—A field on constraint records that identifies any node stretched as a result of an external constraint. [FIELD NAME]

SNL Start-no-later-than date—A date imposed on the model that sets the right-most boundary of a node's Late Start date. [FIELD NAME]

SON Start-on date— A date target that sets both the Early Start and the Late Start dates of a node, ignoring all predecessor and successor relationships. [FIELD NAME]

SOP Span of the pull-in—The number of work units a task is currently scheduled to be pulled in or advanced relative to the baseline schedule (SOP = BES - AES/PES or SOP = BEF – AEF/PEF).

SOS Span of the slip—The number of work units a task is currently scheduled to be slipped or delayed relative to the baseline schedule (SOS = BES – AES/PES or SOS = BEF – AEF/PEF).

SPTR Start-end Path Time Reserve—The Path Time Reserve (PTR) measured at the start end of each node (LS – ES). [FIELD NAME]

SRN Subrecord identification number—A field in the Complementary Scheduling Control and Tracking datafile set by the application to define the subrecord sequence of a task-related event. This value in combination with the task identifier (TI) uniquely identifies each event. [FIELD NAME]

SRP Step resource profile—A method of linking resource requirements to individual tasks by defining a series of step functions each containing the resource identification (RESID), the resource quantity (RESQTY), the delay in the resource need (OFFSET), and the duration of the resource need (RESPAN).

SS Start-to-start constraint type—Defines a relationship between two nodes in which the successor can start once the predecessor starts and the lag passes.

SSTR Start specified time reserve—An alternative method of defining the Early Start date of a start node that subtracts a numeric value in this field from the node's Late Start date to determine an Early Start date with a specified reserve (ES = LS – SSTR) [FIELD NAME]

ST/St Start—The point in time at which a task or effort begins. As an SNF flag value, it indicates a properly targeted start node.

SYM Symbol designation—A field in the Complementary Scheduling Control and Tracking datafile set by the user to define the geometric shape of a task-related event marker. [FIELD NAME]

TA Time available—The number of work units available to a path (TA = LFP − ESP + 1).

TI Task identifier—A field in the Complementary Scheduling Control and Tracking datafile set by the user to uniquely designate each task. [FIELD NAME]

TRW Tabular report writer—A software feature by which the user controls and formats a tabular report.

TYP Type—A two-character alphanumeric constraint field that defines the constraint type: FS, SS, FF or SF. [FIELD NAME]

UF User friendly—An interface with a computer datafile that portrays the data on a screen with each field highlighted and fully identified. The interface is very simple and requires only minimal training.

APPENDIX G

Modelfile Composition

Activity Records

Field Name	*Field Type*	*Field Description*
Critical Parameters		
NI	Alphanumeric	Node identification code
DU	Integer	Activity duration or span
AD	Text	Activity description
CAL	Alphanumeric-2	Calendar designation
ODU	Integer	Original activity duration
Calculated Dates		
ES	Date	Early Start date
LS	Date	Late Start date
EF	Date	Early Finish date
LF	Date	Late Finish date
Calculated Time Reserves		
SPTR	Integer	Start-end Path Time Reserve
FPTR	Integer	Finish-end Path Time Reserve
SFR*	Integer	Start-end free reserve
FFR*	Integer	Finish-end free reserve

*Not needed in standard application.
#Optional in standard application.

Field Name	*Field Type*	*Field Description*
On-Going Task Status Parameters		
DUWC#	Integer	Duration of work completed
PC#	Integer	Percentage complete
PWR#	Integer	Percentage of work remaining (100 – PC)
RDU#	Integer	Remaining duration
Special Flags		
SNF	Alphanumeric	Start node flag (ST, HS, null)
FNF	Alphanumeric	Finish node flag (FN, HF, null)
ASF	Alphanumeric	Activity stretching flag (FPS, BPS, null)
AOS	Integer	Amount of stretching
CPF	Alphanumeric	Critical path flag (S, F, SF)
Event Parameters		
SEF	Alphanumeric-2	Start-end event flag (shape and schedule level)
SED	Text	Start-end event description
FEF	Alphanumeric-2	Finish-end event flag (shape and schedule level)
FED	Text	Finish-end event description
Status and Control Dates		
BES	Date	Baseline start date
PES	Date	Projected start date
AES	Date	Actual start date
BEF	Date	Baseline finish date
PEF	Date	Projected finish date
AEF	Date	Actual finish date

*Not needed in standard application.

#Optional in standard application.

Field Name	*Field Type*	*Field Description*
Date Targets		
SNE*	Date	Start-no-earlier-than date predecessor (ES)
FNE*	Date	Finish-no-earlier-than date predecessor (EF)
FNL	Date	Finish-no-later-than date successor (LF)
SNL*	Date	Start-no-later-than date successor (LS)
FON*	Date	Finish-on date (EF and LF)
SON*	Date	Start-on date (ES and LS)
Structure Parameters		
S1	Alphanumeric	Structure field 1
S2	Alphanumeric	Structure field 2
S3	Alphanumeric	Structure field 3
S4	Alphanumeric	Structure field 4
S5	Text	Structure fields
S6	Text	Structure fields
Horizontal Integration Parameters Cross-Model Interface Nodes		
INI	Alphanumeric-1	Interface node identifier
INT	Alphanumeric-1	Identifies predecessor/ successor nodes
IFS	Text	Identifies source project
Variant Parameters		
SSTR	Numeric	Start node specified time reserve
FSTR	Numeric	Finish node specified time reserve
LAGPC	Numeric	Lag expressed as percentage of PN (SS constraint) or SN (FF constraints)

*Not needed in standard application.
#Optional in standard application.

Resource Requirements Subrecord

Field Name	*Field Type*	*Field Description*
Step Resource Profile (SRP)		
RESID	Alphanumeric	Resource identification code
OFFSET	Numeric	Delay in resource need
RESPAN	Numeric	Duration of resource need
RESQTY	Numeric	Amount of resource need
Periodic Resource Profile (PRP)		
RESID	Alphanumeric	Resource identification code
PER#	Numeric	Amount of resource need in period

Constraint Records

Field Name	*Field Type*	*Field Description*
Critical Parameters		
PNI	Alphanumeric	Predecessor node identifier
SNI	Alphanumeric	Successor node identifier
TYP	Alphanumeric	Constraint type (FS, SS, or FF)
LAG	Numeric	Constraint lag value
CAL	Alphanumeric	Calendar designation
Data Flags		
NIS	Alphanumeric	Stretched node identifier
CPF	Alphanumeric	Critical path flag (X)
Structure Parameters		
S1	Alphanumeric	Structure field 1
S2	Alphanumeric	Structure field 2
S3	Alphanumeric	Structure field 3
S4	Alphanumeric	Structure field 4
S5	Text	Structure field 5
S6	Text	Structure field 6

Complementary Schedule Tracking and Control Process

Field Name	*Field Type*	*Field Description*
Detail Task Record (Standalone Event Record)		
TI	Alphanumeric	Unique task identifier
NI	Alphanumeric	Parent activity node identifier
AD	Text	Task description
BES	Date	Baseline start date
PES	Date	Projected start date
AES	Date	Actual start date
BEF	Date	Baseline finish date
PEF	Date	Projected finish date
AEF	Date	Actual finish date
S1	Alphanumeric	Structure field 1
S2	Alphanumeric	Structure field 2
S3	Alphanumeric	Structure field 3
S4	Alphanumeric	Structure field 4
S5	Text	Structure field 5
S6	Text	Structure field 6
Associated Event Subrecord		
SRN	Integer	Subrecord sequence number
ED	Text	Event description
SYM	Alphanumeric-2	Symbol designation
BIA	Alphanumeric-1	Start/finish bias
EBD	Date	Event baseline date
EPD	Date	Event projected date
EAD	Date	Event actual date
PER	Integer (+)	Percentage of task
SFS	Integer (+/–)	Span from start

APPENDIX H

Conventions

1. TIME FLOWS FROM LEFT TO RIGHT

All graphic depictions of model or schedule information follow a common convention for date chronology. Time flows from the left to the right. Examine the nodes in Figure H-1, which illustrates this convention. Node UCLA starts before node UTEP, so its start lies to the left of UTEP's. However UTEP finishes before UCLA, yet its finish end lies to the right, creating a contradiction in our convention. This occurs because the entire duration of UTEP (20 days) takes place while UCLA is on-going. As both node boxes are the same size, both ends of this node cannot follow the convention.

As one of its primary benefits, this convention allows us to tell predecessors from successors in constraint relationships, so whenever we are confronted with a contradiction, we must first satisfy the constraints. For this same reason, the finish of DUKE must lie to the right of the finish of UNLV to properly establish the predecessor and successor of the finish-to-finish constraint. In this way, we can meet the intentions of the convention, even though the data may sometimes contradict it.

Figure H-1

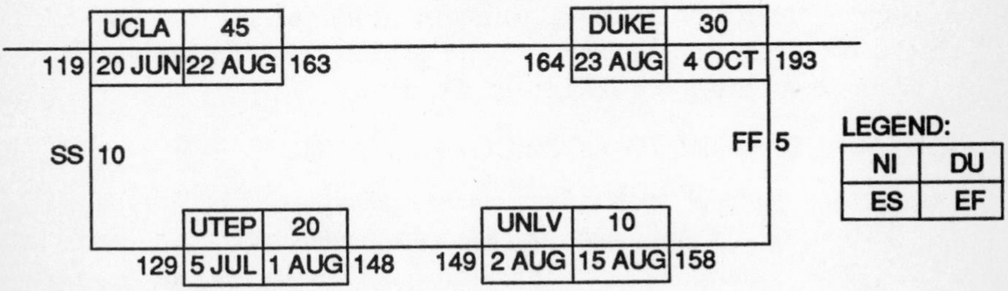

2. START-FINISH DATE CONVENTION

This convention provides a frame of reference for each date. Start dates (ES, LS, BES, PES, AES, etc.) reflect the beginning of the work period. In the case of workday units, this would be the start of the business day. Finish dates (EF, LF, BEF, PEF, AEF, etc.) reflect the end of the work period. In the case of workday units, this would be the finish of the business day. This corresponds to the normal perception of dates by most people. Therefore all date math and CPM equations must take this into account.

3. PRECEDENCE—NODE (ACTIVITY) CONVENTION

In Precedence, nodes represent activities. Graphically, nodes appear as geometric shapes, typically rectangles as they are best suited for holding many fields of data. This is accomplished by subdividing the node box into compartments, each of which holds a specific field of data (see Figure H-2). The left and right ends of the graphic node pertain to the start and finish, respectively, of the activity, which is also used for accurate portrayal of constraints.

Figure H-2

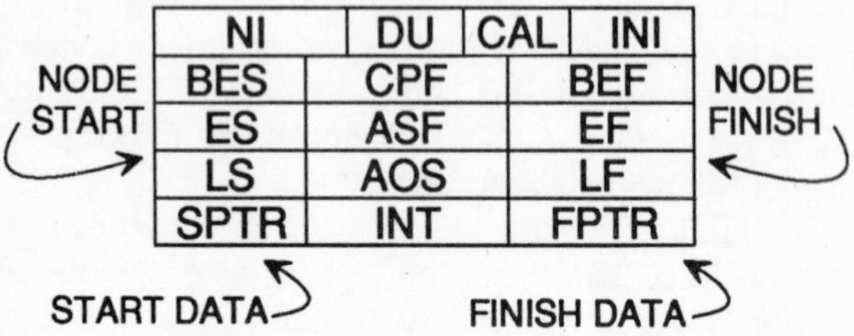

4. PRECEDENCE—NODE END (EVENT) CONVENTION

In Precedence, we model events as node ends. Each node has two—a start and a finish. We add additional fields to each node record to contain essential event parameters. A special durationless node (DU = 0) called an events node marks the initiation or completion of parallel paths. We treat it just like any other node (i.e., it has two events, one measuring starts and one measuring finishes—see Figure H-3), except that it has no associated activity.

5. PRECEDENCE—NODE-TO-NODE CONSTRAINT CONVENTION

In Precedence, we model relationships between activities as node-to-node records. The first, or left-most, node is called the predecessor. The second, or right-most, node is called the successor. Precedence constraints join the events (start or finish) of nodes. As each node has two events, four possible combinations of constraint types exist.

Figure H-3

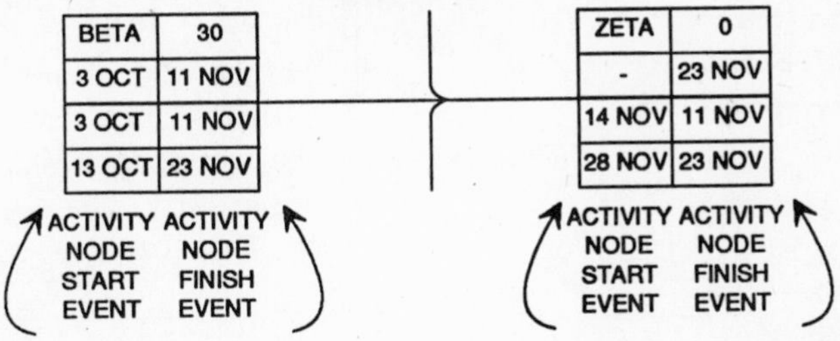

A. Finish-to-Start (FS) Constraint. An FS constraint defines a relationship between two serial tasks where the predecessor must finish before the successor can start. Any lag value on the constraint delays the start of the successor additionally (see Figure H-4).

Processing Equations:

Forward Pass: ES of SN = EF of PN + LAG + 1
Backward Pass: LF of PN = LS of SN – LAG – 1

Figure H-4

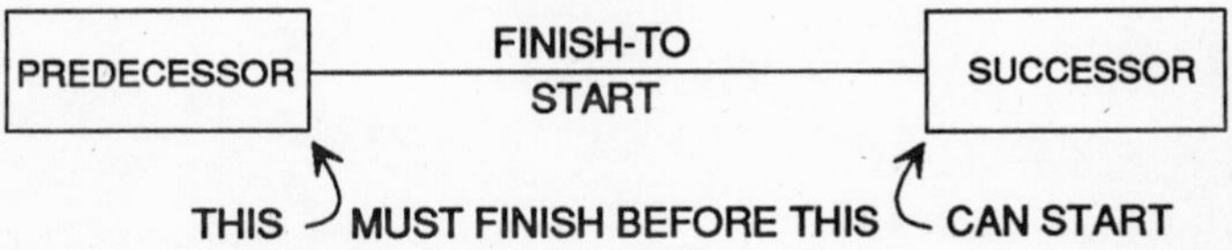

B. Finish-to-Finish (FF) Constraint. An FF constraint defines a relationship between two tasks where the predecessor must finish before the successor can finish. It establishes a relationship of parallel effort where the successor can start independently of the predecessor, but it cannot finish until the predecessor finishes. A lag value is necessary on a FF constraint to define how much work remains in the successor once the predecessor finishes—the true measure of the relationship (see Figure H-5).

Processing Equations:

Forward Pass: EF of SN = EF of PN + LAG
Backward Pass: LF of PN = LF of SN – LAG

Figure H-5

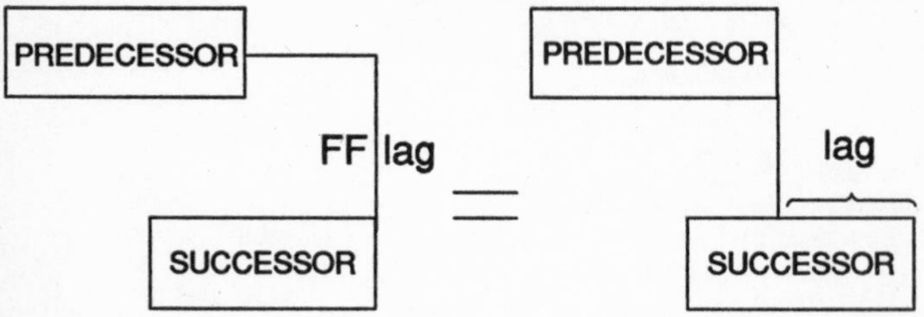

C. Start-to-Start (SS) Constraint. An SS constraint defines a relationship between two tasks where the predecessor must start before the successor can start. It establishes a relationship of parallel effort in which the successor can start independently of the finish of the predecessor, but it depends upon a portion of the predecessor being finished. A lag value on an SS constraint is necessary to define how much of the predecessor must be accomplished before the successor can start—the true measure of the relationship (see Figure H-6).

Processing Equations:

Forward Pass: ES of SN = ES of PN + LAG
Backward Pass: LS of PN = LS of SN – LAG

Figure H-6

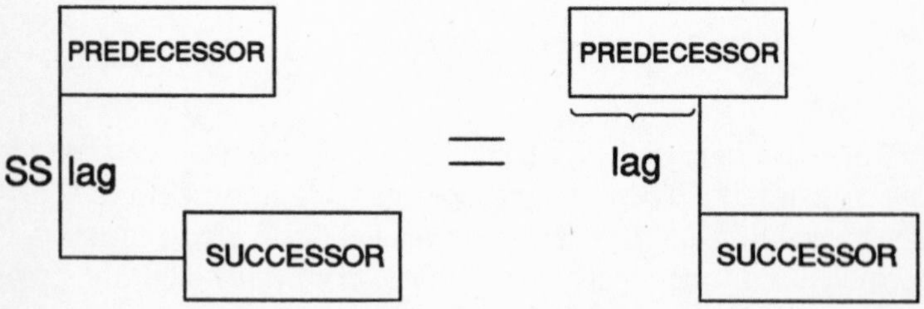

D. Start-to-Finish (SF) Constraint. An SF constraint defines an impractical relationship between two tasks where the predecessor must start before the successor can finish. It establishes a relationship of parallel effort in which a portion of the predecessor must be accomplished before the remaining portion of the successor can finish. The lag is therefore a complex number, part predecessor and part successor, which makes it difficult to control and manage. Therefore, it is recommended to avoid using this type of constraint relationship (see Figure H-7).

Processing Equations

Forward Pass: EF of SN = ES of PN + LAG – 1
Backward Pass: LS of PN = LF of SN – LAG + 1

Figure H-7

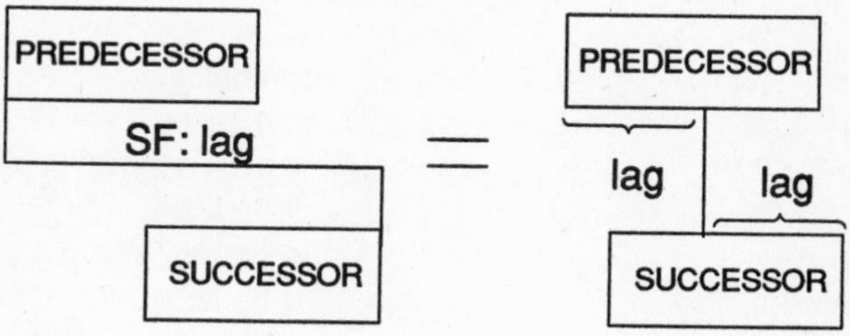

INDEX

A
Activities, 25-26, 299
 completed, 170-76
Actual finish date, 171-75
Actual start date, 170-71
Alphanumeric fields, 197, 302
Anomaly status conditions, 270-75
Attributes, 25

B
Backward pass, 79-103, 112
 duration and finish-to-start equations, 80-83
 events nodes, 101-2
 finish nodes and date successors, 93-101
 finish-to-finish equation, 83-84
 multiple successor relationships, 90-92
 start-to-finish equation, 88-90
 start-to-start equation, 85-88
Base node attributes, 26

C
Calendar, 43-49, 122, 126, 132, 196
 incompatibility, 213, 230-31, 410
 5 day week, 47
 node and, 209
 6 day week, 48
 utilities, 209-11
Calendar counting, 43-49
Calendar datafile, 207-11, 281
Calendar reserve, 105, 131-36, 144-45
 cost consideration of, 134
Cause-and-effect evaluation, 405
Change, 168-69
Code fields, 239-40, 282, 309-10, 410-11
Code field summarization, 282-91, 309-10
Common finish dates, 98
Common start dates, 70
Completed activities, 170-76
 actual finish date, 171-75
 actual start date, 170-71
Computers, role of, 195-242
 advantage of in CPM, 195-96, 240-41, 393
 basic requirements of, 407-17
 database function and size, 203-7, 279
 data sort and selection, 231-40
 modelfile flags, 215-23
 output, 331-62, 412-17
 project model database structure, 196-203
 project modelfile errors, 223-31
 system requirements, for CPM, 417
Constraint record parameters, 409
Constraints, 29-37, 39
 control of, 343-45
 and critical path, 117-20
 cross-model, 319
 external constraint stretched dates, 111-13
 finish-to-finish, 449
 finish-to-start, 149, 449
 lag values, 30, 40, 149-52, 157, 164
 multiple between two tasks, 36-37
 start-to-finish, 450

start-to-start, 118, 450
stretching, 111
summarization of, 292-300
types of, 30-37
Conventions, 447-51
Cost, 134, 136
baseline cost, 135
Cost reserve, 9
Cost schedule integration, 182
CPM in Arrow, 24
CPM vs. PERT (Project Evaluation and Review Technique), 403-5
CPM in Precedence
Critical path, 120, 128, 165
defined, 113-14
flag, 216-18
and longest path compared, 120-21
Critical Path Method (CPM), 17-19, 393-97
computer and the CPM, 240-42, 393, 407-17
CPM project modeling, 255
defined, 19
and PERT compared, 403-5
purpose of, 19, 105
Cumulative sequence of the path, 7

D

Data
entry, 203, 212, 237, 311
record change process, 205-6, 295
sharing of, 312
sorting and selection, 231-40, 283, 299, 334-37, 348-49
Database, 196, 301-2, 310
flag example, 220-23
functions, 203-7
record, 197
size of, 279
structure, 196-203
summarization of, 279-300
Date fields, 197-99
Date math, 41-43
Dates, 198
date altered to compensate for stretching, 154
date math, 41-43
date successors, 93
imposed dates, 159-61
precedence-node conventions, 448
projected start and finish, 176, 185-86, 261-63
project status, 259
start-finish date convention, 448
Dates altered to compensate for stretching, 153-55, 164
Date targets determined by time reserves, 147-49, 164
Duration, 26
Duration equation, 52-53
Durations of tasks, 7, 26, 175
calculating, 189, 191-92, 211, 278, 363
percentage complete method, 176-80
remaining duration, 180-82

E

Earned value, 389-91
Effective rate of time reserve erosion, 13
Estimated finish date, 182-84
Event attributes, 25
Events, 26-28, 299, 421
event summarization, 285-89
Events nodes, 27-28, 74-75, 101-2, 161
Excessive target dates, 162-63165
External predecessor stretched activity, 60, 77
External successor stretched activity, 88

F

Facilities, 363-65, 375
modeling of, 367
Far term tasks, 251
Fields, 197, 302,
alphanumeric, 197, 302
code fields, 239-40, 282, 309-10, 410-11
coding structure, 234
date fields, 197

field coding for selection and sorting, 233
field format in reports, 333, 340-43
numeric, 199
summary of coding structure, 290
text field, 199, 302
File passing, 312-16
Finish dates, 96-101, 182-83, 186, 262, 271
Finish date successor, 96
Finish Event Flag, 26
Finish-no-later-than dates, 96-98
Finish node flags, 220
Finish nodes, 93-96
Finish specified time reserve, 147
Finish-to-finish constraint, 33-35
Finish-to-finish forward pass equation, 57-60
Finish-to-start constraint, 30-31, 149-51
Finish-to-start equation, 53-55
Flags, 215
modelfile flags, 215
start and finish flags, 219
stretching flags, 218
Float, 6, 9
span altered float, 143-44
Forward pass, 51-77, 112
equations, 52-62
events nodes, 74-75
multiple predecessor relationships, 63-65
set forward scheduler, 76
start nodes and date predecessors, 65-74
Free reserve, 136-43

G

Gantt barcharts, 283, 293, 331, 350-58, 373, 415
basic format, 350-55
from-until dates, 357
legends, 358
paging, zoning, order, and selecting data, 356-57
record attributes window, 355-56
summarized Gantt barcharts, 357-58
Glossary, 429-39

H

Hammock structures, 279-82, 299, 309, 425-27
Hanging events nodes, 161-62, 165
Hanging finish, 147
Hanging start, 148
Histograms, x-y plots, and pie charts, 359-60
Holidays, 208-9
Horizontal integration, 311-30
file passing, 312-16
multimodel integration, 318-28
nested decompositional integration, 316-18

I-L

Imposed dates, 159-61, 165
Interface nodes, 321
Labor, 363, 365, 367, 382
pool, 371
requirement summation and leveling, 379-87
Lag of the constraint, 30
Lags as a percentage of effort, 151-53, 164
Lag values on finish-to-start constraints, 149-51, 164
Late dates, 79
Longest path and critical path compared, 120-21

M

Management, 331, 395-97
See also Project management
horizontal integration of, 311-30, 411
simplifying, 311
types of, 3, 4, 279
vertical integration of, 279
Management graphics, 363-65, 391
Margin, 6
Material, defined, 365. *See also* Resources
Milestone, 28
symbols, 296
Model attributes, 25

Modelfile composition, 441-45
Modelfile errors, 223-31
Modelfile flags, 215-23
critical path, 216-18
database example, 220-23
start and finish node, 219-20
stretching, 218-19
Modelfiles
composition, 441-45
defined, 197
Modeling
See also Project modeling *and* Modeling varients
computer modeling, 195-242
earned value, 389
layout and format, 339
in practical project environment, 243-78
successful modeling methods, 339
Modeling process, change and status in, 167-94
completed activities, 170-76
near-term task status, 185-91
on-going tasks, 176-85
Modeling techniques, in practical project environment, 243-78
anomaly status conditions, 270-75
baseline schedule, 252-55
level of model detail, 249-52
preproject schedule-model relationship, 244-45
projected start and finish date rules, 261-63
schedule-model, 246-49, 263-70
status in the model, 259-61
what-if analysis, 275-76
Modeling varients, 147-65
dates altered to compensate for stretching, 153-55
date targets determined by time reserves, 147-49
excessive target dates, 162-63
hanging events nodes, 161-62
imposed dates, 159-61
lag values on finish-to-start constraints, 149-51
lags as a percentage of effort, 151-53
negative lag values, 157-58
time reserve nodes, 155-56
zeroing time reserve paths, 158-59
Model parameter summary, 38-39
Model plots, 339-50
basic layout, 339
constraint control, 343-44
display controls, 348-50
horizontal position control, 347-48
node format and control, 340-43
vertical position control, 344-47
Multimodel integration, 318-28
multiple predecessor/successor interface nodes, 325-28
submodel integration using cross-model constraints, 319-21
submodel integration using common nodes, 321-25
Multi-to-one cross-model relationship, 325
Multiple predecessor relationships, 63-65
Multiple successor relationship, 85, 90-92

N

Near term tasks, 185-91, 251
projected finish dates, 186
projected start dates, 185-86
Negative lag values, 157-58, 165
Negative PTR or time reserve, 107
Nested decompositional integration, 316-18
Node ends, 26
Node identification code, 26
Nodes, 25, 39
events nodes, 74-75, 101, 161
finish nodes, 93, 95
node identification code, 26
node record parameters, 408
predecessor nodes, 29, 39, 325
start nodes, 65, 67
successor nodes, 29, 39, 325
time reserve nodes, 109-10, 155-56
Numeric fields, 199

O
One-to-multi cross-model relationship, 325
On-going tasks, 176-85
earned value, 182
estimated or projected finish date, 182-84
percentage complete method, 176-80
remaining duration, 180-82
Output products, 331-62
Gantt barcharts, 350-58
histograms, x-y plots, and pie charts, 359-60
model plots, 339-50
tabular reports, 331-39
Overloads, 381
Overtime, 15-17, 105, 133

P-Q
Padded span reserve, 144
Parent-child relationship tasks, 244, 290, 304
Paths, 110-11, 114,
path length, 121, 126, 129, 277
zero time reserve path, 158-59
Path of activities, 7
Path length determination, 121-22
Path time reserves, (PTR), 12-14, 105, 122-23, 134, 155
calculation of, 127
defined, 8
erosion of, 257-58
time reserve decisions, 21
Percentage complete method, 176-80
Periodic resource profile, 368, 371
Periodic summation of resource requirements, 372-75
PERT in Arrow, 24
PERT in Precedence, 24
PERT vs. CPM, 403-5
Pie charts, 359
Precedence model diagrams, 25
Predecessor of the constraint, 29
Predecessor equation summaries, 62
Predecessors of the activity's start and finish, 30
Proactive management, and reactive compared, 3-4
Problem solving, 114
Project
assessment of, 389-90
cost, 16, 363, 387-88
defined, 1, 110
management, 2, 38, 134
model conventions, 447-51
model evaluation, 419-23
modelfile, 197, 298
modeling. *See* Project modeling
planning, 1-21, 244, 246
record modelfiles, 199-200
reports, 331
schedule, 243, 378
sharing modelfiles, 312
summarization of, 279-300
variables, 5, 6, 10, 11, 127, 168, 364,
Projected finish date, 176, 186
Projected start date, 185-86
Project management
defined, 3
managing project reserves, 124-27
and reactive management compared, 3-4
Project modelfile errors, 223-31
calendar incompatibility, 228, 230-31
constraint node incompatibility, 226-27
duplicate node identifiers and constraint pairs, 227-28
loops, 223-24
out-of-sequence updating, 228-30
unloops, 224-26
Project modeling, 1-21
activities, 25-26
Arrow vs. Precedence techniques, 399-402
choice of techniques and date convention, 23-25
constraints, 29-37
CPM process, 17-19, 21
database structure, 196-203

elements of successful, 394-97
events, 26-28
misconceptions, 393-94
PERT vs. CPM techniques, 403-5
planning and management, 1-3
price of successful, 393
project model, 4-6, 20
project planning and management, 1-3, 19-20
project time reserves, 6-9, 14-17, 21, 105-29
purpose of, 249-50
qualitative evaluation of models, 419-23
reactive and proactive management compared, 3-4
by resources, 363-92
seventh planning step, 9-14, 2021
time reserve decisions, 14-17, 21
Project reserves, managing, 124-27
Project status, 169
Project status dates, 259
Project time reserves, 6-9, 105-29
critical path, 113-21
external constraint stretched dates, 111-13
as management parameter, 122-24
path length determination, 121-22
path vs. node time reserve, 109-10
paths, 110-11
project reserves, managing, 124-27
Quantitative risk analysis, 405

R

Reactive management, 3-4
Reconciliation, examples of, 267-70
Record modelfiles, 199-203
Remaining duration, 180-82
Resource files, 370
Resource leveling or allocation, 375-79
Resource modeling, 363-92
cost of resources, 387-88
earned value, 389-91
labor requirement summation and leveling, 379-87
periodic summation of resource requirements, 372-75
resource leveling, 375-79
resource loading of tasks, 364-69
resource pools, 370-72
Resources, 363-65, 391
allocation of, 375
cost of, 387-88, 392
periodic resource profile, 368, 371
requirements, 372
resource pools, 370-72
Resource subrecords, 365
Risk analysis, 123, 405
Rolling wave task decomposition, 6

S

Scheduling, 246, 248, 266, 300-309
baseline schedule, 252-54, 303, 305
below model level, 300-309
control of, 279, 291, 409-10, 423
Schedule-model reconciliation, 246-49
Schedule-model relationship, 246, 263-64, 267, 276
Set-back scheduler, 103
Set forward scheduler, 76
Side path, 149
Slack, 6
Software, 231, 237, 241-42, 311, 331
Span altered float, 143-44
Spans of tasks. *See* Durations of tasks
Specified time reserves, 147
Start dates, 70, 170
Start Event Flag, 26
Start-finish convention, 24
Start node flag, 220
Start nodes and date predecessors, 65-74
common start dates, 70-72
deriving early starts from successors, 72-73
finish-no-earlier-than dates, 70
start nodes, 65-68
start no-earlier-than dates, 68-70
Start no later than dates, 68, 98
Start-to-finish constraint, 35-36

Start-to-finish forward pass equation, 60-62
Start-to-start constraint, 32-33
Start-to-start forward pass equation, 56-57
Status, 169, 259-63
Step resource profile, 366
Stretched activity, 60, 77, 112
  amount of stretching, 218
  dates altered to compensate for stretching, 153-55
Stretched node identifier, 219
Stretching flags, 218-19
Submodels, 311-316, 318-25, 328
Successor of the constraint, 29
Successor interface nodes, 325-28
Summarization, 279-300
  code field summarization, 282-91, 309-10
  constraint summarization, 292-300
  hammock structures, 279-82, 309

T

Tabular reports, 331-39
  field or parameter control, 332-34
  record selection, order, and zoning, 334-37
  special reports, 337-39
Tasks, 363-65
  breaking down of, 317
  far term, 251
  near term, 251
  on-going, 176, 184-85
  parent-child relationship, 244, 290, 304
  resource loading of, 364-69
  status, 185-86, 190-91, 298, 391
Text field, 199, 302
Time convention, 447
Time reserve, 8, 12-14
  See also Project time reserves *and* Critical Path Method
  date targets determined by, 147-49
  decisions, 14-17
  as management parameter, 122-24
  nodes, 155-56, 164-65
  paths, zeroing, 158-59
Time units, 207
Tools, defined, 365

U-Z

Underload, 381
What-if analysis, 275-76, 411
Work, 15, 16, 247, 363, 381, 391
  See also Labor
  calculating available workdays, 133
  modeling of, 367
  overtime, 15, 105, 133
  sequence of, 6, 17, 126
  workdays, 15, 131-32, 207-8
  workload, 381
Workday pattern, 207-8
X-Y plot, 359-60
Zeroing time reserve paths, 158-59, 165